Viral Safety and Evaluation of Viral Clearance from Biopharmaceutical Products

Developments in Biological Standardization

Vol. 88

This series Dev Biol Stand begins with Vol. 23 and is the continuation of both «PROGRESS in Immunobiological Standardization, Vols 1-5» and «SYMPOSIA SERIES in Immunobiological Standardization, Vols 1-22».

KARGER

Basel · Freiburg · Paris · London · New York ·
New Delhi · Bangkok · Singapore · Tokyo · Sydney

Viral Safety and Evaluation of Viral Clearance from Biopharmaceutical Products

Natcher Conference Center, Bethesda, Maryland, USA
June 14-16, 1995

Volume Editors

Fred Brown
Plum Island Disease Center
Greenport NY, USA

A. Lubiniecki
SmithKline Beecham Pharmaceuticals
King of Prussia PA, USA

Proceedings of a Symposium organized and sponsored by the
Food and Drug Administration (FDA)
International Association of Biological Standardization (IABS)
National Institute of Allergy and Infectious Diseases (NIAID)
U.S. Department of Agriculture (USDA)
National Vaccine Program Office (NVPO)

33 figures and 73 tables, 1996

KARGER

Basel · Freiburg · Paris · London · New York ·
New Delhi · Bangkok · Singapore · Tokyo · Sydney

Scientific Committee

Chairpersons:

Kenneth B. Seamon, FDA, USA
Ruth Wolff, FDA, USA
Anthony Lubiniecki, PhRMA, USA

Members of the Scientific Committee:

W. Berthold, EFPIA
P. Burd, FDA
K. Clouse, FDA
M. Dinowitz, PhRMA
T. Hayakawa, MHW
K. Hellman, FDA
L. Henchal, FDA
F. Horaud, EU
A. Khan, FDA
J. Löwer, PEI
R. Lundquist, FDA
P. Minor, NIBSC
B. Murphy, NIAID
M. Nobuhara, JPMA
A. Ridgway, BB (Canada)
D. Tankerslay, FDA
C. Vincent, FDA

Organization of the Symposium

Introduction

Session I: *Cell Characterization*
Chairpersons: T. Hayakawa and P. Noguchi

Session II: *Assays for Screening Contamination*
Chairpersons: I. Hewlett and G. Vicari

Session III: *Evaluation of Viral Clearance*
Chairpersons: T. Kohama and J.S. Robertson

Session IV: *Manufacturing Considerations*
Chairpersons: R. Devine and A.S. Lubiniecki

Session V: *Regulatory Expectation for Data on Viral Testing and Process Evaluation for Phase I Studies and Product Licensing in the United States, Europe and Japan*
Chairpersons: E. Griffiths and R. Wolff

Breakout Session A: *Sensitivity and Specificity*
Chairpersons: A.S. Khan and B. Murphy

Breakout Session B: *Adventitious Contaminants Observed during Characterization and QC Testing: Findings and Follow-up*
Chairpersons: P. Minor and N. Finter

Breakout Session C: *Adenoviral Gene Therapy Issues*
Chairpersons: A. Ridgway and P. Noguchi

Breakout Session D: *Lessons Learned from Related Product Categories*
Chairperson: J. Finlayson

Breakout Session E: *Adventitious Agent Screening Assays*
Chairpersons: A. Lewis and A.C. Lee

Breakout Session F: *Adventitious Agents from Animal-Derived Raw Materials and Production Systems*
Chairpersons: K. Hellman, C. Vincent, J. Honstead and R. Rohwer

Breakout Session G: *Retrovirus Clearance for Early Clinical Trials*
Chairpersons: K. Stein and P. Martin

Breakout Session H: *Process Evaluation*
Chairpersons: J. Fogh and A. Moreira

Summary and Conclusion to the Meeting

Developments in Biological Standardization

Bibliographic Indices. This publication is listed in bibliographic services, including Current Contents® and Index Medicus.

Drug dosage. The authors and the publisher have exerted every effort to ensure that drug selection and dosage set forth in this text are in accord with current recommendations and practice at the time of publication. However, in view of ongoing research, changes in government regulations, and the constant flow of information relating to drug therapy and drug reactions, the reader is urged to check the package insert for each drug for any change in indications and dosage and for added warnings and precautions. This is particularly important when the recommended agent is a new and/or infrequently employed drug.

Printed in Switzerland by Médecine et Hygiène, Genève
ISBN 3-8055-6391-4

Contents

Brown F, Lubiniecki AS: Viral Safety and Evaluation of Viral Clearance from Biopharmaceutical Products. Dev Biol Stand. Basel, Karger, 1996, vol 88, p 1

Introduction

Brown F, Lubiniecki AS: Viral Safety and Evaluation of Viral Clearance from Biopharmaceutical Products. Dev Biol Stand. Basel, Karger, 1996, vol 88, pp 3-4

Viral Safety and Evaluation of Viral Clearance from Biopharmaceutical Products: Goals of the Meeting

K. Seamon

CBER/FDA, Rockville MD, USA

This meeting was developed about one year ago as part of discussions associated with the development of the ICH Viral Safety Document. It was suggested that there were many issues associated with the ICH document on Viral Safety that would benefit from an open discussion of the current science relating to viral risk and assays to ensure viral safety of biopharmaceuticals. As Dr. Zoon pointed out, it is critical that the best science and data are used in developing any regulatory guidance including regional and international guidelines. We therefore embarked upon organizing this conference with the following goals in mind:

1. To bring together scientists to provide updates on the current state of knowledge about assays for viral contamination and procedures for inactivation and/or removal.
2. To have discussions that identify the risks associated with the processing of biopharmaceuticals.
3. To present data on the actual experiences of the testing and industrial laboratories.
4. To discuss these data in the context of the clinical setting and how this risk relates to the individual patient.

The FDA uses workshops for a variety of reasons which include: (i) the provision of scientific updates on novel or rapidly developing fields, (ii) the presentation and discussion of current regulatory guidance; (iii) the discussion of issues, and (iv) help in promoting policy development, which has to be grounded in good fundamental science and data. It is critical that these data be discussed in an open and transparent forum so that everyone involved can appreciate the concerns and issues that are taken into account during development of regulatory policy and development.

Many different issues will be discussed over the next three days. One major issue that we all recognize is the need to ensure the safety of all biological products. That is the primary goal of the meeting. However, we recognize that this is an evolving field. Novel cell substrates and production systems are being developed and it is important that we discuss new risks associated with these systems. Production systems are becoming more sophisticated and assay systems are more sensitive and specific. Even with this increased sophistication, there are still examples of contamination. However, as you will hear today and tomorrow the appropriate selection of tests and validation can and does ensure the safety of the regulated product.

Many new therapies are being developed for specific clinical settings that may require a different level of risk assesment. Therapies are being investigated which may bring promise for treatment to patients with life-threatening diseases which have no alternative therapies. Discussions will focus on the appropriate level of characterization for these settings which will ultimately help promote the development of novel therapies for public health.

Another goal of this meeting is to discuss specifically information related to the ICH document on viral safety. The experts working on this document met yesterday and will meet on Saturday to discuss the results of the meeting. As you will hear from Dr. Hayakawa this document has been developed to apply specifically to products derived from characterized cell banks. This clearly represents a large number of biopharmaceutical products but does not include many of the blood products, viral or bacterial vaccines, or newer cellular or gene therapies. It is important that, as we listen to discussions on viral safety during the next three days, we distinguish the levels of risk that are associated with different biopharmaceutical products and develop our guidance accordingly.

Previous workshops have resulted in regulatory guidance documents. The FDA-sponsored workshops in 1991 and 1992 resulted in the development of new PTC documents. The IABS-sponsored meeting in 1993 helped facilitate the development of the ICH document on genetic stability.

We expect that this workshop will facilitate reaching consensus on the ICH Viral Safety document which will then be adopted and implemented by FDA and the other regions. We also expect that information discussed here will have imput in other areas of guidance for other products not included on the scope of the ICH document.

In closing, let me urge this group to focus on the science and data during your discussions. The meeting will be data rich and I particularly want to thank our colleagues from industry who are willing to share their data with us about some real life experiences. This meeting was designed to bring together scientists from different sectors of industry, including other regulatory agencies, government laboratories, testing laboratories, and industry. As we move forward in our discussions it will be important to focus on areas of consensus, identify areas where data are lacking, and provide recommendations. Over the next few years we will have an unprecedented opportunity to bring new and novel therapies to the public. It is incumbent upon all of us that we develop guidance that can help facilitate the development of these exciting products for the public health. However, it is critical that we keep the welfare and health of the patients who will be receiving these products uppermost in our deliberations.

Dr. K. Seamon, CBER, FDA, 1401 Rockville Pike, HFM-20, Rockville, MD 20852-1448, USA.

Brown F, Lubiniecki AS: Viral Safety and Evaluation of Viral Clearance from Biopharmaceutical Products. Dev Biol Stand. Basel, Karger, 1996, vol 88, pp 5-7

Safety of Biopharmaceuticals: a Current Perspective

P.D. Parkman

Parkman Associates, Kensington MD, USA

Keywords: Biologics, biotechnology products, safety, viruses, regulatory

Abstract: The impetus for this conference comes in large measure from the continuing development of the ICH Viral Safety document concerned with testing and evaluation of the viral safety of biotechnology products derived from characterized cell lines of human or animal origin [1]. My opening remarks for this conference were, I think, suggested by the organizers in the belief that George Santayana's quote, «Those who cannot remember the past are condemned to repeat it» has validity for this effort. In this brief review I have tried to recall some past unfortunate episodes and outline the lessons that they might have for us. I have also attempted to outline a number of issues which I believe are critical to the success of efforts to guarantee, as much as humanly possible, the safety that we seek balanced against our ability to develop and use important new products.

Evaluation of the safety of cells and cell lines with respect to viruses has been made mamy times [1, 2, 3] and has been discussed in international conferences. It has probably been discussed in connection with regulations and guidance documents everywhere that biologics are produced. The focus for this conference relates to the detection, removal and inactivation of indigenous or extraneous viruses that might be present, particularly in human and animal cells and cell banks used for the production of recombinant DNA products and monoclonal antibodies. The issue is one of concern, not only to our hosts here at the Center for Biologics Evaluation and Research and the other sponsoring organizations, but also to the producers of biologics. Both regulatory agencies and the manufacturers share the responsibility for assuring that safe and effective products reach the marketplace.

The sessions that we embark upon over the next three days will dissect in great detail the issues raised by the real presence of unwanted viruses. They will also address the theoretical possibility that currently unknown viruses might exist in biopharmaceutical products.

This is an important task. Past efforts to ensure product safety have been quite successful. Historically, there are relatively few examples of significant product-related problems resulting from the use of cell-culture-produced material. Through caution, the application of good science and sometimes sheer luck, the biologics enterprise has avoided serious adverse events. Problems more often have arisen because of a product's active principle than from the substrates used in their preparation. One is reminded of examples such as the occurrence of living virus in improperly inactivated Salk poliovirus vaccine [4]. Another example, although not a cell-culture product, is the increased neurovirulence that accompanied the unrestricted serial passage of the yellow fever virus [5].

Nonetheless, there have been striking instances of contamination introduced though both cultured cells and added materials. The yellow fever vaccine, used during World War II and stabilized with pooled human serum, produced many thousands of cases of hepatitis in recipients [6]. This led to a greater realization of the hazards in using human materials in products.

Other episodes of the presence of indigenous unwanted viruses occurred in experimental and licensed vaccines produced from embryonated eggs containing avian leucosis virus[7]. Yellow fever vaccine produced in chick embryos and a variety of experimental products prepared from chick embryo cell cultures must have contained such viruses. A classic, and a particularly frightening example, was the episode in which monkey kidney cells were used containing the simian virus SV-40 [8]. Such cells were used to produce licensed inactivated Salk poliovirus vaccines and experimental live Sabin vaccines. These viruses were undetected when these preparations were first made in the mid-1950s and early 1960s and administered to many thousands of people [9]. Fortunately, despite the oncogenic potential of each of these viruses in some hosts, careful epidemiologic follow-up studies have not shown oncogenic risk for human vaccinees. It is from such experiences that we know the issue of unwanted viruses is not entirely theoretical.

Today the biologics enterprise is exceptionally vigorous and characterized by many new products in development. New and perhaps less experienced producers have entered the field. Novel and complex substrates are being investigated. These factors could result in an enhanced opportunity for the inadvertent introduction of unwanted viruses into products.

The avian leukosis and SV-40 episodes that I cited above alert us to two additional points. First, the vaccine producers and testers who made these products that very probably regularly contained these viruses, were following all the applicable cautions and rules of their time. It was not disregard of risks or carelessness on their part, but a lack of scientific knowledge that led to these agents being administered to thousands of subjects. These scientists did not know everything.

I am sure that we, today, probably do not know everything that may be important to this issue either. Even with all our marvellous science, so much more advanced than that of the 1950s and 1960s, we cannot ensure absolute safety.

A question that you should consider in your deliberations is how close to knowing everything and dealing with every theoretical possibility should one be? Risks and benefits must be calculated. Caution comes naturally to regulators. On the other hand, it is important not to allow excess conservatism in dealing with theoretical risks to hinder the development of important new products. Also important in your deliberations is the question of what testing is appropriate for experimental vaccines as opposed to those produced for commercial distribution.

I do not claim to have all the answers to these questions. I think no one person does. I believe that the best way to pick one's path through these problems, as we have done in the past, is to muster the best scientific knowledge and use the best scientific minds available and to define the problems and their potential solutions. This is best done by organising a gathering such as we will have over the next few days.

References

1 Meyer HM: The control of viral diseases: Problems and prospects viewed in perspective. J Peds 1968;73:653-675.

2 Marcus-Sekura CM: Introduction to the issues: Detection of retroviruses and associated risks, in Brown F, Esber E, Williams M (eds): Continuous Cell Lines – An International Workshop on Current Issues. Dev Biol Stand. Basel, Karger, 1992, vol 76, pp 137-140.

3 WHO Study Group: Petricciani JC, Hennessen W (eds): Cells – Products – Safety: Background Papers from the WHO Study Group on Biologicals. Dev Biol Stand. Basel, Karger, 1987 vol 68, pp 1-81.

4 Nathanson N, Langmuir AD: The Cutter incident. Poliomyelitis following formaldehyde-inactivated poliovirus vaccination in the United States during the Spring of 1955. I. Background. II. Relationship of poliomyelitis to Cutter vaccine. III. Comparison of the clinical character of vaccinated and contact cases occurring after use of high rate lots of Cutter vaccine. Am J Hyg 1963;78:16-28.

5 Fox JP, Lennette EH, Manso C, Aquiar JR: Encephalitis in man following vaccination with 17D yellow fever virus. Am J Hyg 1942;36:117-142.

6 Fox JP, Manso C, Penna HA, Pare M: Observation on the occurrence of icterus in Brazil following vaccination against yellow fever. Am J Hyg 1942; 36:68-116.

7 Harris RJC, Dougherty RM, Biggs PM, Payne LN, Goffe AP, Churchill AE: Contaminant viruses in two live virus vaccines produced in chick cells. J Hyg (Cambridge) 1966;64:1-6.

8 Sweet BH, Hilleman MR: The vacuolating virus, SV-40. Proc Soc Exp Biol Med 1960;105:420-427.

9 Murray R: Contemporary problems in regulating the potency and safety of viral vaccines, in First International Conference on Vaccines against Viral and Rickettsial Diseases of Man. Scient. Publ. N° 147, Washington, DC, May, 1967, Pan American Health Organization, WHO, pp 577-580.

Dr. P.D. Parkman, Parkman Associates, 4101 Glenridge Street, Kensington, MD 20895, USA.

Brown F, Lubiniecki AS: Viral Safety and Evaluation of Viral Clearance from Biopharmaceutical Products. Dev Biol Stand. Basel, Karger, 1996, vol 88, pp 9-11

Current Issues in Viral Assays and Viral Clearance Evaluation

A.S. Lubiniecki

SmithKline Beecham Pharmaceuticals, King of Prussia PA, USA

Abstract: Viral contaminants are less amenable than chemical contaminants to analytical detection due to the lower sensitivity, specificity and precision of existing assays. In combination with the potential for biological amplification of viral contaminants in cell substrates or product recipients these analytical limitations lead directly to potential risk of virus infection of product recipients. While less than perfect, a broad spectrum of assays is available to characterize cell banks for the presence of adventitious and endogenous agents. Each product lot can be tested by an appropriate subset of assays for selected viruses of concern. Engineering and procedural controls over facilities, equipment and operations are necessary to minimize the possibility of viral contamination. Processes can be designed to include unit operations to remove or inactivate specific viruses of potential concern; the efficacy of these measures can be evaluated. While each of these measures has value, each also has limitations. None is solely adequate by itself to the task of preventing potential exposure to viral contaminants in product recipients.

Biopharmaceuticals (purified characterizable macromolecules) are prepared by using a combination of such preventive measures. Specific measures will vary from product to product, depending on the nature and origin of the cell substrate, cell bank characterization results, raw materials used, physicochemical properties of the biopharmaceutical, effectiveness of assays, available facilities, and the agents of concern. Thus, flexibility in approach is necessary, resulting in different combinations of measures for each set of circumstances. So far, this approach has been successful in preventing viral infection in large populations treated to date with cell culture-derived biopharmaceuticals.

Current issues for discussion include appropriate assays for adventitious and endogenous agents, appropriateness of current control systems and the scientific basis for standards.

I would like to introduce three general approaches to the control of virus contamination in biopharmaceutical products: analytical, preventive, and viral clearance evaluation. Each approach has its own strenghs and limitations. It is important to understand each in order to design a control system properly for each product. None of the approaches is self-sufficient to ensure the safety of the product from the perspective of viral contamination. In general, a combination of approaches is employed to ensure product safety.

This conference is being held partly to assess the current scientific knowledge concerning viruses with respect to the preparation of biotechnological/biological products on the eve of harmonization of regulations for this area. It is therefore important that the participants of this conference attempt to provide answers to five key questions:

What is known?

What is believed to be related to biopharmaceutical product safety?

What is reasonable and important to control?

Are current controls effective?

Are additional controls necessary?

Analytical Approaches

Virus detection assays have been available for well over half a century. In general, positive results are meaningful, but negative results are ambiguous. It is rarely clear whether a negative result reflects inadequate sensitivity for the desired virus, selection of a host with too narrow a specificity, poor assay precision, limited sample size, sample instability, or simply no virus present. For endogenous retroviruses, these possibilities are further compounded by the effects of variable gene expression for three or more critical genes. In short, it is not possible to test virological quality in biopharmaceutical products due to these limitations.

Participants at the conference should consider whether there exists a sound scientific basis and rationale for performing current tests. It should also be considered whether existing assays are generally adequate for detection of viral contaminants for biopharmaceutical products. Finally, are there areas for further research and development of viral assays?

Prevention Approaches

These fall into several broad categories, including cell bank characterization, raw material certification, and engineering and procedural controls. The goal is to minimize the portals of entry for viruses. Usually they are very effective, but not failsafe. Evidence of this statement will be presented.

Participants should ask whether existing paradigms for cell-bank characterization, raw material certification, and the design/operation of facilities and equipment as they exist today are adequate to ensure biopharmaceutical safety. Are there needs in these areas which are not being met?

Viral Clearance Evaluation

These measures have been widely used and are believed to be effective. It is generally felt that the applicability of virus clearance data are limited to members of the same virus family and occasionally to a subset of the same virus family. Many inactivation steps are effective against some but not all virus families. Virus removal steps are effective for a selected range of viruses with specific physicochemical properties. Multiple orthogonal approaches to inactivation and removal are generally considered more effective than single steps.

Participants at the conference should consider whether existing paradigms for evaluation of virus clearance are adequate. Are the approaches and standards used for pathogenic viruses of human source materials (e.g. plasma) appropriate for detective retrovirus-like particles of biopharmaceuticals? In the absence of evidence of viral contamination, is characterization of process clearance using model viruses useful? Are the results of such model studies «for information only», or are standards needed? If so, what is the basis for such standards?

General Issues

In addition to these specific questions, I encourage the participants to consider three additional questions:

- do data exist which indicate that biopharmaceuticals have virus safety problems found with cenventional biologicals?
- is the combination of analytical, preventive, and clearance approaches the most appropriate method to ensure virus safety for biopharmaceuticals?
- should the approach taken for a given product reflect the risks observed and the needs of the product recipients?

Dr. A.S. Lubiniecki, SmithKline Beecham Pharmaceuticals, 709 Swedeland Rd, P.O.B. 1539, UE-0428, King of Prussia, PA 19406-0939, USA.

Brown F, Lubiniecki AS: Viral Safety and Evaluation of Viral Clearance from Biopharmaceutical Products. Dev Biol Stand. Basel, Karger, 1996, vol 88, p 13

SESSION I

Cell Characterization

Chairpersons: T. Hayakama, NIH, Japan
P. Noguchi, FDA/CBER, USA

Brown F, Lubiniecki AS: Viral Safety and Evaluation of Viral Clearance from Biopharmaceutical Products. Dev Biol Stand. Basel, Karger, 1996, vol 88, pp 15-18

Issues Realted to Harmonization of Testing Requirements for Viral Safety

T. Hayakawa

Division of Biological Chemistry and Biologicals, National Institute of Health Sciences, MHW, Tokyo 158, Japan

Key words: ICH document, harmonization, viral safety, biotechnology drugs.

Abstract: The current situation concerning the development of an internationally harmonized document on the technical requirements for the viral safety of biotechnology drugs (ICH viral safety document) is described. Major issues relating to harmonization of testing requirements for viral safety include: (i) clarification of scientific rationale, (ii) the way and extent of virus tests for cell line qualification and cell cultivation process evaluation; (iii) discussion of the scope, extent and frequency of virus testing on the unprocessed bulk harvest; (iv) discussion of the choice of viruses for virus clearance elvaluation or process characterization study.

INTRODUCTION

Recently, three regulatory authorities and three pharmaceutical manufacturers associations from the EU, USA and Japan, and the Canadian regulatory authority have been engaged in developing an internationally harmonized document on technical requirements for the viral safety of biotechnology drugs (ICH viral safety document). The first draft document, which incorporates the major points of the current EU, USA and Japan guidelines, was prepared by a biotechnology drug study group of the Ministry of Health and Welfare (MHW) of Japan in June, 1993. There has been considerable progress in developing the harmonized document. A copy of the current draft 4 document is being circulated at this meeting. In this paper, the current situation in the ICH viral safety document will be described.

Results and Discussion

Scope and general considerations

The ICH document deals with the testing and evaluation of the viral safety of biotechnology products derived from characterized cell lines of human or animal origin propagated in vitro. Three principal complementary approaches have been considered to control the potential viral contamination of biotechnology products:

- selecting and testing cell lines and other raw materials for the absence of undesirable viruses;
- testing the capacity of the production processes to remove viruses;
- testing the product at appropriate stage of manufacture for the absence of contaminating viruses.

Cell line qualification: testing for viruses

An important issue which we are currently discussing is the clarification of scientific rationale as well as the extent of virus tests to be performed once only at various cell levels, including the master cell band (MCB), the working cell band (WCB) and the production cells cultured up to or beyond the maximum level. It is important to clarify what sampling points, testing items and testing articles are necessary and appropriate for the characterization of cell lines as well as for the evaluation of cell cultivation process(es) with respect to viral contamination.

All interested parties have naturally agreed that extensive screening for both adventitious and endogeneous viral contamination should be performed on the MCB, taking into consideration the source and nature of the individual cell line and possible contaminating viruses. There is also consensus with respect to performing appropriate endogenous virus tests on the production cells which are obtained at the end of the maximum defined limit of in vitro cell age under conditions that reflect those used for full scale production. However, we will need further discussions on a suitable sampling point for such a purpose as well as on the definition of such a cell level.

There are still many disagreements as to the nature and extent of testing for the working cell bank and for the cells at the end of production runs with regard to tests for adventitious virus to be performed once on cell banks. One of the issues to be clarified is whether certain virus tests should be performed on the WCB even when appropriate adventitious virus tests are designed for the cells obtained at the end of production runs.

Some interested parties believe that testing for adventitious agents should be performed on the WCB. They have claimed that this would provide some assurance to all parties that, were a contaminant to be found in the cells at the end of production runs, the source of that contaminant would probably not be the cell bank. However, others have argued that when reviewing data supplied for new drug approval, the absence of any adventitious viruses can usually be expected for cells at the end of production runs, and such results can be automatically accepted to ensure the absence of any adventitious viruses at the WCB level. The general requirement for adventitious virus tests on the WCB will result in redundant tests in many cases. If any adventitious viruses are detected in the cells at the end of production runs during the course of developmental stages to establish a suitable cell cultivation process for drug production, the process including the WCB should be checked carefully to determine the cause of the contamination and completely redesigned if necessary. Therefore, even if any undesirable events occurred during developmental stages, when data are submitted for new drug approval, it will always be anticipated that the data demonstrate that any undesirable contamination is not detected in the cells at the end of production runs and thus the established production process is confirmed to be free of adventitious viruses.

In relation to this, it is of interest to know about manufacturers' experience regarding the frequency of occurrence of adventitious viral contamination during drug development or routine production runs.

Testing for viruses in unprocessed bulk harvest

The results and extent of virus tests performed on the MCB, WCB and especially on the cells at the end of production runs will certainly affect the scope, extent and frequency of virus testing on the unprocessed bulk harvest or cell lysate as a starting material for further processing to obtain a purified desired product. However, it should be noted that the single virus tests at various cell levels deal mostly with cell qualification issues and the absence of adventitious virus in the cells does not always provide constant assurance of the virus-free status of every subsequent lot of harvest material. The other most probable route of introduction of adventitious viruses into harvest materials would be the use of naturally contaminated biological reagents such as media components. To obtain the best reasonable assurance of freedom from viral contamination in production, without adding any irrational redundant virus tests, it would be important to discuss the extent and frequency of adventitious virus tests in relation to the extent of qualification and control of raw materials. It should be noted that some parts of this discussion are closely related to GMP matters.

Process evaluation of viral clearance

Two separate approaches will be considered for process evaluation of viral clearance. The first is evaluation studies to determine the clearance of a virus known to be present in the cell lines. The other is process characterization studies which provide some level of assurance that adventitious agents which might gain access to the production process would be cleared.

A major issue in performing a virus clearance evaluation or process characterization study is to determine which viruses should be used. Such viruses tentatively fall into three categories: (i) «relevant viruses», (ii) «specific model viruses», (iii) «non-specific model viruses».

The purification and/or inactivation process should be shown to be capable of removing and/or inactivating possible contaminating viruses. Such a demonstration can be made using relevant viruses or specific «model» viruses.

When the purpose is to evaluate the capacity of the manufcturing process to remove and/or inactivate viruses in general, i.e. to evaluate the robustness of the clearance process for a wide range of viruses rather than for one specific virus of interest, and when the «relevant» or specific «model» viruses used do not cover a wide range of physicochemical properties, process characterization studies should be performed with a panel of non-specific model viruses, representative of the different possible virus types (RNA or DNA, enveloped or non-enveloped, small or large). The rationale of such a choice of viruses for the evaluation of virus clearance should be the subject of extensive discussions.

Dr. T. Hayakawa, Division of Biological Chemistry and Biologicals,
National Institute of Health Sciences, 1-18-1 Kamiyoga, Setagaya-Ku, Tokyo 158, Japan.

Brown F, Lubiniecki AS: Viral Safety and Evaluation of Viral Clearance from Biopharmaceutical Products. Dev Biol Stand. Basel, Karger, 1996, vol 88, pp 19-24

Viral Safety of Biological Products and the Concept of Validation

F. Horaud

Institut Pasteur, Paris, France

INTRODUCTION

I presume that young colleagues currently working in the field of manufacturing and control of new biotechnological products are probably wondering why viral safety of these category of biologicals is such an important concern? Why do products that are now highly purified and characterised need such sophisticated experimental studies to convince regulatory authorities that they are not capable of transmitting viruses to patients? Moreover, biologicals derived from cell culture, such as rDNA molecules or monoclonal antibodies (MAb), are reputed as safe since so far they have not been involved in clinical accidents, in contrast to biologicals derived from human material that have transmitted severe diseases. The answer to the above questions is not simple and we will try to explain briefly why the cautious and vigilant attitude of regulatory organisations and public health authorities is justified. Since this meeting is devoted only to products derived from cell culture, the particular safety aspects of biologicals having animal or human tissues or fluids as source material will be not discussed.

HISTORICAL BACKGROUND

The first category of products derived from cell culture was viral vaccines. Since viruses need cells for their growth, the development of viral vaccines was dependent on the introduction of a new technology for cell substrate preparation [1]. It is therefore easy to understand that the discovery by Enders, Weller and Robbins that poliovirus is able to replicate efficiently in non- neural tissue was a milestone in the history of biologicals [2]. Between 1953, when the first batches of inactivated polio vaccine became available, and the late 1980s, an important number of human and veterinary viral vaccines were developed by using cell culture as substrate [1]. Since scientific knowledge on cell biology and viral contaminants of cells in culture was limited at this time, the regulations established by WHO for the control of cell substrates used for the production of vaccines mainly concerned the detection of adventitious lytic viruses in primary cell culture. The real debate concerning the criteria of acceptability of cell substrates has been generated by the discovery in 1960 of simian virus 40 (SV40), a DNA oncogenic virus, as a common contaminant of polio and other vaccines. SV40 was present, as an adventitious agent, in the primary cell culture of rhesus monkey kidney used for the preparation of polio vaccine [3]. During this time, at the beginning of the 1960s, the mechanisms of oncogenic cell transformation began to be understood and the fear of transmitting cancer by the cells used in the preparation of vaccines was a major cause for concern.

Another event that addressed safety questions in connection with the cell substrate used in the preparation of vaccines was the development by Hayflick and Moorhead of human diploid cell lines having a finite life span [4]. This new and very important achievement was accepted after considerable discussion in which the assessment of cancer risk was the focus of the debate [5] .

During the decades 1960 and 1970, the dominant ideology concerning cell substrates was the acceptability of «normal cells», that is, primary culture or diploid cell lines having a finite life span. This view was clearly stated in the U.S. Regulations for the Manufacture of Biological Products prohibiting the use of heteroploïd continuous cell lines (CCL) derived from in vitro cell transformation or from neoplastic tissue as substrate for the preparation of biologicals [6].

The Major Shift of the 1980s

Since the assessment of safety requirements for cell substrates is a complex problem devolving on public health authorities, the cautious attitude of these institutions in accepting major changes in policy is justified. Decisions of the control authorities are based mainly on scientific knowledge and risk-benefit analysis. However, other considerations can influence the policy of regulatory bodies, such as the experience gained in the clinical use of various products and their safety records. Modifications of the safety regulations are justified when major scientific and technological progress, opening avenues for obtaining a new generation of biologicals products, is achieved.

In the late 1970s, the advantages of using CCL in the production of biologicals became evident [7]. However, the acceptability of CCL as substrate in the production of viral vaccines was controversial in the late 1960s and early 1970s, since the oncogenic potential of heteroploïd cells having an infinite life span was already well known. However, at this time the molecular biology of oncogenic transformation was little understood and the scientific community was not prepared to accept easily the use of transformed cells in the production of biologicals destined for clinical use in humans. The hesitation in the approval of CCL for the production of biologicals came mainly from the idea that the final preparations might contain enough residual cellular nucleic acid capable of «inducing cancer» in humans receiving the product.

At this point it is important to underline that the acceptability of novel technology of cell culture used in the manufacturing of biologicals evolved slowly and was progressively influenced by various factors such as (i) new achievements which gave impetus to the development of new medicinal products such as r-DNA and hybridoma technology; (ii) improvement of the technique of mass cell cultivation; (iii) characterisation of new viruses, as was particularly the case with rodent retroviruses present in the hybridoma cells and, last but not least, (iv) advancement in the understanding of the molecular basis of cell oncogenic transformation. This new direction became an industrial reality only when the whole field of molecular separation and protein purification underwent a radical renewal as has been witnessed during the past 20 years. The significant scientific and educational effort to make the new biotechnological products acceptable is illustrated by the number of scientific meetings devoted to this topic and presented in Table 1.

There is no doubt that the most important achievement in the late 1970s was the preparation by Wellcome of interferon prepared on the continuous cell line Namalva, derived from Burkitt's lymphoma and harbouring the genome of EB virus [8]. It was the first time a cell culture-derived product destined for human use was prepared on cells obtained from a human cancer which also contained a large segment of an oncogenic virus (EBV). The advantages of preparing interferon on Namalva cells were weighed against the risk represented by the neoplastic nature of the cells containing genetic information of an oncogenic virus in their chromosome. This assessment was done on the basis of the results obtained by validation studies of the capacity of the purification method to eliminate various viruses such as canine hepatitis, mouse CMV, polyoma and infectious polyoma DNA, polio, Sendai, etc. The results generated by these experiments justified the clinical use of the product and allowed the regulatory institutions to release a marketing authorization.

Table 1: Principal international meetings on the safety of products derived from CCI.

1.	Cell substrates, their use in the production of vaccines and other biologicals, in Petricciani JC, Hopps HE, Chaple PJ (eds): Adv Exper Med Biol. New York and London, Plenum Press, vol 118.
2.	Joint ESACT-IABS Meeting on the use of heteroploïde and other cell substrates for the production of biologicals, in Horaud F, Hennessen W (eds): Dev Biol Stand. Basel, Karger, 1981, vol 50.
3.	Production and explosion of existing and new animal cell substrates, in Spier R, Griffiths B, Hennessen W (eds): Dev Biol Stand. Basel, Karger, 1984, vol 60.
4.	Abnormal cells, new products and risk, in vitro, cellular and developmental biology, in Hopps HE, Petricciani JC (eds): Gaithersburg, MA, Publ; Tissue Culture Association, 1985.
5.	Cells, products and safety: background papers from the WHO Study Group on Biologicals (this volume includes the final document of WHO meetings on cells for the preparation of biologicals), in Petricciani JC, Hennessen W (eds): Dev Biol Stand. Basel, Karger, 1986, vol 68.
6.	Hayflick L, Hennessen W (eds): Continuous cell lines as substrates for biologicals. Dev Biol Stand. Basel, Karger, 1989, vol 70.
7.	Virological aspects of the safety of biological products, in Horaud F, Brown F (eds): Dev Biol Stand. Basel, Karger, 1990, vol 75.
8.	Viral Safety and Evaluation of Viral Clearance from Biopharmaceutical Products, in Lubiniecki A, Brown F (eds): Dev Biol Stand. Basel, Karger, 1996, vol 88.

What was also new in the Wellcome approach of making interferon on neoplastic cells was the use of reconstruction experiments in which the worst case scenario was simulated by spiking the unprocessed bulk with large amounts of different viruses and evaluating their elimination during the purification procedure. This type of experiment is now considered a basic technology in the evaluation of the viral safety of biotechnological products.

Since the use of CCL in the preparation of viral vaccines was particularly important, during the late 1970s we studied in our laboratory at the Institut Pasteur the fate of HeLa cell DNA during the preparation of purified inactivated polio vaccine [9]. By performing spiking experiments with large amounts of DNA, we were able to quantify its elimination during the purification process. The results of this study showed clearly that the amounts of DNA in purified viral vaccines can be reduced to very small quantities, insignificant in terms of biological activity.

During the development of MAb and rDNA products destined for clinical use, the technology of viral safety validation by spiking experiments was further developed. It is now a common technology in the validation of viral safety of products derived from cell culture, or tissues and fluids having a human or animal origin.

Introduction of Guidelines

The complexity of the problem raised by the safety of the new products derived from cell culture generated major concern for the government regulatory organizations. The iatrogenic transmission of viruses also contributed to this reaction, such as the contamination with HIV of haemophiliacs treated with coagulation concentrates or the appearance of Creutzfeld-Jakob disease among young patients treated with human growth hormone extracted from human pituitary gland [10]. The emergence of new infectious diseases [11] and the isolation of new viruses was also a source of questions that were raised concerning the viral safety of new biologicals.

The first time viral safety was debated by regulatory organizations was in the European Union. In the late 1980s, the European regulatory system for pharmaceutical products was created. The European Union was in a special situation since it was in the position of creating a new central regulatory system that would not be in conflict with the national regulatory procedures of the member states. Special attention was given to the new biotechnological products obtained by cell engineering or by processing human plasma, since these categories of products raised special consideration in connection with viral safety. The objective was to market products having the same quality and safety in all member states. The Biotechnology/Pharmacy Working Group (BPWG) of CPMP gave priority to the guidelines concerning the viral safety of biotechnological products.

The approach was complex since it was the first time a philosophical and practical guide for ensuring the viral safety of biologicals had been elaborated. The BPWG decided to draft a general document covering all biological products, those derived from cell culture as well as products extracted from human or animal tissues or fluids [12]. The technology of validation studies ensuring viral safety was described and the theoretical considerations necessary to evaluate the results of such studies were developed. However, particular aspects of viral safety raised by different products were approached in other guidelines, such as those concerning plasma derivatives or monoclonal antibodies. The experience gained in the field of validation studies clearly showed that revision was necessary and in 1996 the viral safety guideline was updated and released by CPMP.

It is now interesting to comment on the guidelines developed at the International Conference on Harmonization (ICH). It is well known that the aim of the ICH is to obtain a regulatory consensus between Japan, USA and Europe to accelerate and facilitate the licensing and marketing of pharmaceutical products between these three partners. In this light, it has been decided to give priority to some categories of new products, such as rDNA pharmaceutical molecules, monoclonal antibodies and other new biotechnological products obtained by cell cultivation. This is the reason that ICH guidelines on viral safety of products derived from the cultivation of animal cells are now beiong revised.

It is perhaps important to explain the difference between the basic concept of European viral safety guidelines and those being drafted by the ICH expert group. As mentioned above, the CPMP guidelines comprise a basic general document covering all the products in which concern over viral safety is present, while the ICH document pertains only to biological products derived from animal-cell cultivation. For this reason the ICH project of guidelines is more prescriptive than the European document. To implement the ICH objectives, it is more appropriate to have guidelines precisely indicating the different tests and the technology that should be used to assess the viral safety of the product candidate for licensing in the geographical areas of the partners of ICH.

In conclusion, it is my personal feeling that the viral safety of biological products will remain in the next decade as a major topic in the strategy of developing new medicinal molecules. The explanation of this situation resides in the inherent viral risks included in the development and manufacture of new biologicals. In addition, consumer pressure concerning viral safety has been influenced by the transmission of viral pathogens via medicinal products and the change in the epidemiological patterns of some diseases, notably the outbreak of BSE in the U.K. A vigilant attitude by the regulatory and public health authorities, scientists and producers is of paramount importance in the manufacture of safe products and in the search for new and useful pharmaceutical products.

References

1 Horaud FN: Viral vaccines and cell substrate: a «historical debate», in Plotkin A, Fantini B (eds): Vaccinia, Vaccination, Vaccinology. Jenner, Pasteur and their successors.. Paris, Elsevier, 1996, pp 217-220.

2 Enders JF, Weller TH, Robbins FC: Cultivation of Lansing strain of poliomyelitis virus in culture in various human embryonic tissues, Science 1949;109:85-87.

3 Keereti S, Nathanson N: Human exposure to SV40: review and comment. Amer J Epidemiol 1976;103: 1-12.

4 Hayflick L, Moorhead PS: The serial cultivation of human diploid cells strains. Exper Cell Res 1961;25:585-621.

5 Hayflick L: History of the cell substrates used for human biologicals., in Hayflick I, Hennessen W (eds): Continuous cell lines as substrates for biologicals. Dev biol Stand. Basel, Karger, 1989, vol 70, pp 11-26.

6 U.S. Departement of Health, Education and Welfare, Public Heath Service: Regulation for the Manufacture of Biological Products, title 42, part 73. DHEW pub. N° (NIH) 71-161, formerly PHS pub. N° 437, revised 1971-1976.

7 Cell substrates, their use in the production of vaccines and other biologicals, in Petricciani JC, Hopps HE, Chaple PJ (eds): Adv Exper Med Biol. New-York and London, Plenum Press, vol 118.

8 Finter NB, Fantes KH, Lockier MJ, Lewis GD and Ball GD: The DNA content of crude and purified human interferon prepared by Wellcome Biotechnology Limited, in Hopps HE, Petricciani JC (eds): In vitro Cellular and Development Biology. Monograph N° 8, USA, Tissue Culture Association , 1985, pp 125-128.

9 Horaud F, Crainic R: Characterisation and quantitation of cell DNA contaminating inactivated poliovaccine, in Hopps H, Petricciani JC (eds): In vitro Cellular and Development Biology, Monograph N° 8, USA, Tissue Culture Association , 1985, pp 110-114.

10 Brown P, Preece MA, Will RG: «Friendly fire» in medicine: hormones, homografts, and Creutzfeld-Jakob disease. The Lancet 1992;340:24-27.

11 Morse SS: Factors in the emergence of infectious disease: Emerg Inf Dis 1995;1:7-15.

12 EEC Regulatory Document. Note for Guidance. Validation of Virus Removal and Inactivation Procedures. Biologicals 1991 (revised 1996) 19:247-251.

Prof. F. Horaud, Institut Pasteur, 28 rue du Docteur-Roux, F-75724 Paris Cedex 15, France.

Brown F, Lubiniecki AS: Viral Safety and Evaluation of Viral Clearance from Biopharmaceutical Products. Dev Biol Stand. Basel, Karger, 1996, vol 88, pp 25-29

Mammalian Cells and their Contaminants

P.D. Minor

National Institute for Biological Standards and Control, Potters Bar, Herts, UK

Key words: Mammalian cell (contaminants).

Introduction

Virus contamination of biologically derived medicinal products has been a major issue from the earliest production of viral vaccines in tissue culture systems, and the manufacture of products derived from human blood or plasma, where potentially serious infections are known to have been transmitted. The difficulties associated with viral vaccine production in the 1950s, where as many as 75% of all cultures might be discarded because of obvious cytopathic adventitious viral infections, led to the development of strategies for avoiding or detecting contamination of the product which formed the starting point for methods used in the modern biotechnology industry.

Contamination in the Production of Viral Vaccines

Live and killed poliovaccines were originally produced on primary cultures of cells derived from the kidneys of wild-caught monkeys where the donor animals were frequently infected with viruses, some of which were highly pathogenic for humans. Herpes simiae or B virus is a herpes virus of macaque monkeys [1]. Between 1932 and 1987 there were 28 cases of infection reported in individuals who had close contacts with primates, of which 20 were fatal and seven had severe sequelae. It is possible that asymptomatic infection occurs, but exposure can clearly have major consequences, so that strenuous efforts are made to prevent the virus contaminating the cultures used for production. The precautions include (i) the use of African Green Monkeys for production, as they are not susceptible to infection by B virus, (ii) veterinary examination and quarantine of animals intended for use, and (iii) in vitro or in vivo culture systems for detection of the virus. The in vitro culture system used involves rabbit kidney cells and illustrates the fact that the detection system must be tailored to the virus or viruses sought. Similar considerations apply to Marburg virus [2], a filovirus related to Ebola which was responsible for an outbreak in monkey handlers and their contacts in 1967, in which seven of 31 cases were fatal. Effective quarantine of animals is now imposed to guard against further incidents of this type, and in Germany, where the outbreak occurred, manufacturers are required to test for presence of the virus in guinea pigs. The need to use rodents or rodent cells to detect a virus of primates having a major pathogenic effect in humans demonstrates one of the potential limitations of general virus screening tests, in that rodent cells are not the obvious choice for viruses with this host range.

A further relevant example from the production of poliovaccines is provided by SV40, a papovavirus which is an extremely common infection of macaques [3]. The virus is slow-growing relative to polio and infection may therefore be inapparent; in the appropriate host such as the hamster it is oncogenic. Other members of this virus family are also oncogenic, and therefore of concern. It is estimated that essentially all individuals given poliovaccine between 1954 and 1961 also received live SV40. This applies even where the vaccine was formalin inactivated, as SV40 is relatively resistant to the treatments used in preparing inactivated poliovaccine. Several studies were carried out to investigate possible adverse consequences and most involved follow-up for about two years, although one carried on for 19 years. The long-term study, which is the most appropriate to detect any oncogenic effects, involved 1,073 individuals, 153 of whom had received inactivated vaccine by injection. The remainder received live vaccine by mouth, which is a poor route for infection with SV40. While no adverse clinical consequences were identified, it could be argued that the relevant study group was small. However, in view of the large number of individuals who are believed to have received infectious SV40 and the absence of any corresponding increase in related cancer, it is widely believed that the contamination was without clinical consequences.

It has recently been reported [4] that SV40-like sequences can be detected at significant frequency in patients suffering from the asbestos-related cancer, mesothelioma. The sequences clearly resemble SV40 more than the equivalent human papovaviruses BK or JC, and if they are indeed from SV40 the source from which the patients acquired the infection is not clear. The possibility exists that both asbestos exposure and SV40 infection are required for the development of mesothelioma, and synergistic effects of this kind would make it more difficult to identify clinical consequences of infection.

Contamination of Cells used in the Production of Products of Biotechnology

Murine monoclonal antibodies were among the first products of modern biotechnology manufactured in mammalian cells to be considered for human use and the virological aspects of their safety were of specific concern [5]. A list of viruses which were known to grow in mice was devised and divided into two categories (Table 1). Group 1 comprised viruses either known to cause human disease or able to infect human cells, while group 2 contained other murine viruses. Some group 2 viruses such as polyoma were potent oncogenic viruses in mice, while others such as MVM or Sendai were very common infections of laboratory mice. Table 2

Table 1: Viruses potentially infecting rodent cells.

	Virus	Virus family
Group 1	Hantavirus Lymphocytic choriomeningitis virus (LCMV) Rat rotavirus Reovirus type 3 Sendai virus	Bunya Arena Rota Reo Paramyxo
Group 2	Ectromelia virus K virus Kilham rat virus Lactate dehydrogenase virus (LDH) Minute virus of mice (MVM) Mouse adenovirus Theiler's virus Mouse hepatitis virus Mouse rotavirus (EDIM) Polyomavirus Rat coronavirus Retrovirus Sialoacryoadenitis virus Thymic virus HI (Toolan) virus	Pox Papova Parvo Toga Parvo Adeno Picorna Corona Rota Papova Corona Retro Corona Herpes Parvo

shows the results of some virological assessment of mice received at NIBSC from a breeder of laboratory animals working to high standards, and illustrates that evidence of infection with both MV and Sendai virus was found in all groups. The view developed that hybridoma lines should be tested for contamination with all the viruses shown in Table 1 by appropriate techniques, such as antibody production tests in mice or rats or other suitable methods. The cells should be free of any detectable virus other than endogenous retroviruses which, as they are reactivated from the host genome, are regarded as intrinsic to the cell. Contamination of cells for making products of biotechnology has been reported [6], as summarised in Table 3, and includes some hybridoma cell lines. While the frequency is not necessarily high, it is clearly not zero.

Table 2: Virological assessment of commercially supplied laboratory mice.

	Mouse strain		
	C57B1/6	**C51B2/10Sc**	**SnC3H/He**
Sendai virus	+	+	+
Mouse hepatitis virus	–	–	–
Pneumonia virus of mice	–	–	–
Ectromelia	–	–	–
Minute virus of mice	+	+	+

Table 3: Viruses detected in cells for the manufacture of products of biotechnology[1].

Virus detected	Assay system
LMCV	Antibody production in mice
MVM	Antibody production in mice
Sendai virus	Antibody production in mice
EDIM	Antibody production in mice
LDH	Antibody production in mice
Bovine viral diarrhoea virus	Bovine viral screen
Porcine parvovirus	Porcine parvo test

[1] Data from Moore [6].

Strategies for the Prevention of Viral Contamination of Products of Biotechnology

The sources of viral contamination of cells used for production include (i) the species of origin, which may be infected with latent or cytopathic virus, (ii) materials used in production which may include biological materials such as serum, trypsin or antibodies used in purification, but may also include chemicals contaminated by imperfect rodent control, and (iii) breakdown of GMP. The strategies which have developed to deal with viral contamination of products of biotechnology are the same as those used for other biologicals, and include three approaches. Firstly the source material is tested for freedom from viruses. This includes the reagents used in production as well as the production cell banks themselves. Thus bovine serum supplements would be tested for freedom from common bovine viruses and ideally subjected to virus inactivation procedures, while cell banks would be tested for the presence of adventitious viruses. Secondly, the production process would be evaluated for its ability to remove or inactivate viruses. Thirdly, the final product or appropriate intermediates would be tested for viral contamination. Depending on the product and the source, emphasis might be placed more strongly on one of these approaches than the others, but a single approach alone is insufficient. Screening of source materials is essential but the systems available are not perfect; a cell culture system can detect only those viruses able to grow in the cells concerned and other methods are in general even more restricted. Moreover, statistical considerations mean that testing a sample is insufficient to prove the absence of contamination in bulk, and testing of raw materials for contamination by viruses as a result of imperfect rodent control may be very difficult. Evaluation of the process for the removal or inactivation of viruses gives a measure of the capability of the process which is more or less approximate depending on the design and interpretation of the studies, and for some, although not all, products of biotechnology the process may be insufficiently rigorous to remove viruses. Testing of intermediates or final products is subject to some of the limitation of testing of starting materials. Nonetheless, for viral vaccines, testing of starting material and final product has given materials of high safety. Products of biotechnology, so far as is known, have not transmitted viral infections.

Conclusions

The virological contamination of biological products has been a major concern in the production of both viral vaccines and blood products. Existing data demonstrate that contamination of cells and harvests by viruses can occur for products of biotechnology, and while the frequency may be low it is not zero. The concerns and the strategies elaborated to deal with viral contamination, including screening, sourcing and process evaluation have been applied to products of biotechnology, with a high degree of success.

References

1 MMWR: B virus infection in humans. Pensacola, Florida. MMWR 1987;36:289-296.

2 Gordon Smith CE, Simpson DIH, Bowen ETW, Zlonik I: Fatal disease from vervet monkeys. Lancet 1967;ii:1119-1121.

3 Shah K, Nathanson N: Human exposure to SV40: Review and Comment. Am J Epidemiol 1976;103:1-12.

4 Carbone M, Pass HI, Rizzo P, Marinetti M, DiMuzio M, Mew DJW, Levine AS, Procopio A: Simian virus 40-like DNA sequences in human pleural mesothelioma. Oncogene 1994;9:1781-1790.

5 CPMP. Ad Hoc Working Party on Biotechnology and Pharmacy: Guidelines on the production and quality control of monoclonal antibodies of murine origin intended for use in man. TIBTECH 1988;6:65-68.

6 Moore WA: Experience in cell line testing. Dev Biol Stand 1992;76:51-56.

Dr. P.D. Minor, N.I.B.S.C., Blanche Lane, South Mimms, Potters Bar, Herts. EN6 3QG, U.K.

Brown F, Lubiniecki AS: Viral Safety and Evaluation of Viral Clearance from Biopharmaceutical Products. Dev Biol Stand. Basel, Karger, 1996, vol 88, p 31

SESSION II

Assays for Screening Contamination

Chairpersons: I. Hewlett, CBER/FDA, USA
G. Vicari, Istituto Superiore di Sanita, Italy

Brown F, Lubiniecki AS: Viral Safety and Evaluation of Viral Clearance from Biopharmaceutical Products. Dev Biol Stand. Basel, Karger, 1996, vol 88, pp 33-35

Introduction to the Issues: Assays for Screening Contamination

I.K. Hewlett

CBER/FDA, Bethesda MD, USA

Key words: Adventitious viruses, contamination, screening assays.

Since the first introduction of biological products, there has been concern regarding transmission of infectious disease by products derived from biological substrates, due to sporadic incidents involving vaccines, hormones and blood products. This has resulted in the implementation of assays for screening substrates and products for contamination by viruses or other adventitious agents [1, 2]. Viral contamination may arise from (i) the original source of the cell line (ii) naturally contaminate reagents such as serum components or (iii) from adventitious introduction during manufacturing.

A common strategy towards assessment of cell substrates and products for freedom from adventitious viruses and viral contaminants has been testing for viruses that may be known or suspected to contaminate the starting culture. When a relatively high likelihood of the presence of a virus is suspected, based on the origin of the cell or the production process, it may be necessary to perform tests for specific agents. For example, it would be appropriate to screen hybridoma cell lines likely to harbour EBV with assays specific for this agent. Additionally, biological reagents such as commercial lots of foetal bovine serum may be known or suspected to be contaminated with viral contaminants. They would include bovine viral diarrhoea virus (BVDV), parainfluenza virus-3 and infectious bovine rhinotracheitis virus [3]. The use of screening assays for these agents may be relevant in this situation.

Viral contaminants from cells of rodent origin that deserve special attention because of their ability to cause acute illness in humans are members of the Hantagenus and Arenavirus families. These viruses are capable of establishing persistent infection in cell lines and can remain undetected for a long time.

Parvoviruses can also establish infections of continuous and some primary cultures. Parvovirus contamination, observed more in cells of porcine and rodent origin, would be difficult to eliminate due to the small size and high stability of these agents. Precautions should be taken to exclude them from reagents and starting cultures. An example of adventitious contamination with a murine parvovirus will be discussed later in this session.

Other viruses that deserve mention are the Filoviruses i.e. the Marburg and Ebola viruses associated with monkeys from Africa and the Philippines. These viruses grow well in mammalian culture [4] and the absence of overt cytopathic effects can result in occult contamination. Current detection methods include antibody staining and electron microscopy, although nucleic acid probes have also been developed. Both viruses are known to be highly pathogenic to man. However, based on the historical safety of products from members of this virus family, there may be no perceivable hazard for biopharmaceutical products at the present time.

The major concern, however, in designing screening assays and strategies for detection of adventitious viruses continues to be the potential contamination of cell substrates by unknown or unanticipated agents for which there are no sensitive assays. This has also been a challenge with regard to designing suitable removal/inactivation methods and screening strategies. An evaluation of these assays and their utility in screening is especially important since validation of methods for virus removal and inactivation are based on appropriate screening assays for the various agents of concern. Therefore, a periodic re-evaluation of the issues and strategies for screening adventitious viruses is instrumental in meeting the needs of this critical and evolving area in biotechnology product development.

In this session, we have brought together experts who will discuss both the general and specific issues in virus screening assays as well as providing case study presentations involving both rDNA derived products and viral vaccines. We will hear an overview from Dr. Robertson on the current issues in developing such strategies. Questions he will address include (i) when should virus testing be performed? (ii) what are the viruses that should be tested? and (iii) what are the assays that should be performed? I think we all agree that these are some of the key questions that we should bear in mind during this session and the course of the meeting.

Dr. Lee will provide some insights from the standpoint of viral vaccines where testing would be performed for the seed stocks, cell substrates and the bulk/final container product. Since in vivo and in vitro assays are generally complementary, he will discuss the need for both types of tests for detection of adventitious viruses.

Dr. Garnick will present a case study involving parvovirus contamination of a rodent cell substrate, in this instance the minute virus of mice (MVM). This incident illustrates the effective control of a potentially large scale contamination, through timely implementation of appropriate general and specific screening assays, i.e. PCR and culture.

These talks will be in preparation for a break-out session later in the meeting where open discussion of these issues should generate some recommendations and conclusions that can be reported back to this group.

Finally, I should like to add that with the advances that have been made in the development of a variety of rapid and sensitive assay methods, it is timely and appropriate for both the biotechnology industry and regulatory authorities to re-examine the use of current screening strategies and their application in the manufacturing process.

References

1 Shah K, Nathanson N: Am J Epidemiol 1976;103:1-2.

2 Ward JW: Transfusion-Associated (T-A)-AIDS in the United States, in Brown F (ed): Virological Safety Aspects of Plasma Derivatives. Dev Biol Stand, 1993, vol 81, pp 41-43.

3 Hyclone Laboratories: Art to Science in Tissue Culture. 1988;7 (2):3-7.
4 Siegert R: Marburg virus. Virology Monographs 1972;11:97-153.

Dr. I.K. Hewlett, Chief, Laboratory of Molecular Virology, Division of Transfusion Transmitted Diseases, CBER, FDA, 1401 Rockville Pike, Rockville, MD 20852-1448, USA.

Brown F, Lubiniecki AS: Viral Safety and Evaluation of Viral Clearance from Biopharmaceutical Products. Dev Biol Stand. Basel, Karger, 1996, vol 88, pp 37-40

Strategy for Adventitious Agent Assays

J.S. Robertson

National Institute for Biological Standards and Control, Potters Bar, Herts, UK

Key words: Adventitious viruses, virus assays, MAP tests, antibody production assays.

Abstract: There are three important questions to be addressed when considering a strategy for adventitious virus screening.
(i) What viruses should be tested for? This will depend on the species of origin of the production cell although careful consideration has to be given also to the entire manufacturing process for the possible introduction of a contaminating virus from other biological materials used in production, e.g. bovine serum.
(ii) What tests should be performed? Tests for viruses are highly varied. General tests act as «catch-all» types of assays. These are complemented by a variety of tests for individual viruses or types of viruses.
(iii) At what stage(s) during the manufacturing process should virus testing be performed? Manufacturing stages which should be tested for virus contamination include extensive testing of the master cell bank and limited testing of the working cell bank and end of production cells.

There are three important questions to be addressed when considering a strategy for adventitious virus screening: (i) what viruses should be tested for? (ii) what tests should be performed? and (iii) at what stage(s) during the manufacturing process should virus testing be performed?

What Viruses?

The nature of the viruses to test for will depend on the species of origin of the production cell. Thus, tests for murine viruses will be important when the cell line is of murine origin, e.g. a murine hybridoma secreting a monoclonal antibody, and similar considerations apply if human, insect or a cell line from any other source is used. The variety of murine viruses which should be tested for received considerable attention during the 1980s when guidelines were first drafted for biopharmaceuticals derived from cell lines. The consensus opinion among regulators is that all known murine viruses should be tested for when a murine cell is used. The guidelines of the European CPMP [1] placed the 23 known murine viruses into two categories: those known to cause infection and disease in man, i.e. known zoonotic viruses, and those with no evidence for disease in man. The former group contains the Hantaan, lymphocytic choriomeningitis (LCMV), rat rotavirus, reovirus type 3 and Sendai viruses. While the latter group are not known to cause disease in man, many of them can be propagated in tissue culture in cells of human or of non-human primate origin. While the list of murine viruses is extensive, it is recognised that some of them are quite rare and the probability of them arising in cell cultures is remote.

If a cell of human origin is involved in production, then there should be tests for human viruses. In the case of human-murine heterohybridomas, both human and murine viruses must be tested for. The human viruses in the above example would be the lymphotropic human viruses HIV, HTLV, EBV, CMV, HHV6 and HHV7. Indeed, EBV is likely to have been used in the initial immortalization of the human lymphocyte.

For production based on insect cell lines, consideration must be given to arthropod-borne viruses, e.g. *Flaviviridae,* some of which, St Louis encephalitis, yellow fever and dengue, have been shown to be capable of persisting in the insect cell line Sf9 used for baculovirus vector expression [2].

However, it is not only the origin of the production cells that has to be considered. The possible introduction of a contaminating virus via the use of a biological material in the manufacturing process must also be appraised. For example, bovine serum is a common component of cell culture media and although serum-free medium is often used during production, it is probable that the cells will have been exposed to bovine serum at some stage. Thus, tests for the viruses frequently contaminating bovine sera, i.e. bovine viral diarrhoea virus (BVDV), infectious bovine rhinotracheitis virus (IBRV) and parainfluenza 3 virus, may be required depending on the extent of validation of the serum itself. Tests for bovine polyoma virus may also be warranted since there is evidence of it contaminating bovine sera and also for it being a zoonotic agent [3]. In addition, a further reason for testing for murine viruses would be if a murine monoclonal affinity column is used during processing. In summary, the origins of all cells and biological materials used at any stage of production must be carefully considered for their potential to contaminate the production process with viruses of specific species of origin. Appropriate tests should be performed to identify any such contaminants.

What Tests?

Tests for viruses are highly varied. The approaches used involve general tests which may indicate the presence of one or more of a large variety of different viruses and specific tests which are used to assess the presence of individual viruses.

The general tests can be considered to be 'catch-all' tests, designed to reveal the presence of a virus without necessarily identifying it. These involve in vitro and in vivo testing. The in vitro tests involve the application of the cells, disrupted cells or culture supernate, on to an indicator cell line and examination of the indicator cells for any signs of viral infection, e.g. cell death, cytopathic effect, abnormal morphology, ability to adsorb red blood cells, etc. Typically, at least three distinct virus-susceptible indicator cell lines would be used; a cell line derived from the same species of origin as the test cell, a human cell line and another virus susceptible cell line chosen to reveal the presence of any virus which may be a potential contaminant, e.g. a bovine virus from bovine serum.

In vivo testing involves the inoculation of the material into test animals, typically rodents, which are then monitored for weight gain and general health. The test material should also be inoculated on to the CAM and into the amniotic and allantoic cavities and the yolk sac of embryonated hens' eggs which are amenable to the growth of a variety of viruses. After incubation of the inoculated eggs, the fluids are tested generally by haemagglutination assay.

One of the most widely used specific test currently in use in the biopharmaceutical arena is the antibody production test. When applied to mice, it is known as the MAP (mouse antibody production) test, or the RAP test in rats and HAP test in hamsters. In this test, the sample is inoculated into suitable numbers of the animals which are then tested for serological conversion to specified viruses. Reagents are now available for assaying for the presence of specific antibodies to most rodent viruses. Additional specific tests may be applied based on a special feature of the virus, e.g. lactate dehydrogenase (LDH) virus can be assayed by virtue of its LDH enzymic activity. Enzymic activity can also be used as a marker for the presence of a retrovirus by assays for reverse transcriptase.

Electron microscopy is an extremely useful adjunct in virus testing and is used especially for the detection of (primarily endogenous) retrovirus particles. Molecular techniques are also increasingly being applied in viral diagnostics and although DNA hybridisation may be useful, the increased sensitivity of genome amplification methods means that they are or will be the approach of choice. Detection of viral genomes is especially useful when no infectivity assay is available although there are few data yet to correlate a positive gene amplification result with infectivity.

When to Test?

The manufacturing stages at which tests for virus contamination should be performed include all or combinations of the master cell bank (MCB), the working cell bank (WCB), end of production cells and possibly cells cultured beyond the end of production. Most emphasis is placed on establishing a master cell bank free from adventitious agent contamination. It is on these cells that a wide battery of tests for all relevant viruses, involving both general 'catch-all' types of tests and specific tests, e.g. a MAP test, will be applied. Thorough characterization of the MCB in this way is extremely important. There is generally little requirement to perform extensive testing of a WCB if the MCB has been thoroughly characterised although some testing, e.g. in vitro and in vivo tests, should be considered. Testing of production material after fermentation should be performed at least on a limited basis to test for viruses which may become apparent after cell growth in a bioreactor or which may have derived from the use of contaminated media. Testing of cells passaged beyond the normal limits of production is generally performed to test for endogenous or latent viruses.

There is good consensus among manufacturers and regulators for the approach to virus testing described above. This broad approach to viral testing is due to the diverse nature of viruses and their detectability. Viruses are not easy to detect and most tests have their limitations. There is no single simple assay which can be guaranteed to detect any or all viruses. Indeed, the absence of any guarantee that a contaminant virus within a production cell line will be detected, if present, means that direct testing forms only a part of viral safety assurance and, as described in other papers in this volume, safety is reinforced by demonstrating that the purification process is capable of eliminating any virus which may have gone undetected. In this respect, it is vital to remember that past incidents of viral contamination have derived from viruses not known to be present within the production system. That is the importance of an approach involving a variety of both general and specific tests applied at more than one stage of manufacture in combination with viral elimination steps during subsequent processing to assure the safety of the product.

References

1. Ad Hoc Working Party on Biotechnology/Pharmacy, Committee for Proprietary Medicinal Products, European Commission. Production and quality control of monoclonal antibodies (revision). Note for guidance. 1995;III/5271/94.
2. Zhang PZ, Klutch M, Muller J, Marcus-Sekura CJ: Susceptibility of the Sf9 insect cell line to infection with adventitious viruses. Biologicals 1994;22:205-213.
3. Schuurman R, van Steenis B, Sol C: Bovine polyomavirus, a frequent contaminant of calf serum. Biologicals 1991;19:265-270.

Dr. J.S. Robertson, N.I.B.S.C., Blanche Lane, South Mimms, Potters Bar, Herts EN6 3QG, England.

Brown F, Lubiniecki AS: Viral Safety and Evaluation of Viral Clearance from Biopharmaceutical Products. Dev Biol Stand. Basel, Karger, 1996, vol 88, pp 41-47

Issues of Biological Assays for Viral Vaccines

C.K. Lee

The Salk Institute, Government Services Division, Swiftwater, PA, USA

Key words: Viral vaccine assays.

Abstract: The major test materials in viral vaccine production are seed virus, cell substrate and vaccine. The seed virus should be clearly identified as to source strain and its freedom from adventitious agents. The cell substrate to be used for the production of vaccine is generally tested following the recommendation that appeared in «Points to Consider in the Characterization of Cell Line Used to Produce Biologicals». It should be noted that a shift in emphasis has taken place from the earlier versions of the «Points to Consider». The tests for vaccine include bulk and final container tests. These tests, in principle, are performed following 21 CFR 600-680.
In vitro and in vivo tests for detecting adventitious viruses are complementary; hence, both types of tests are desirable. Some of the practical points to consider in designing and performing the tests for adventitious viruses are discussed.

Major Test Materials in the Production of Viral Vaccines

1. Seed virus

Generally, a minimum of two types of tests are needed for seed virus. These are tests for identity and adventitious agents. The seed virus for the production of vaccine should be clearly identified as to source strain to verify the integrity of the strain characteristics and also to verify its freedom from adventitious agents such as viruses, bacteria, fungi, mycoplasma, etc. [1, 2].

2. Cell substrate

The shift of emphasis in assuring the safety of the cell substrate has been made during the last decade according to « Points To Consider » (PTC) issued by the FDA [1]. Until 1987, the normalcy of the cell substrate was emphasized. Consequently, a few diploid cell lines such as WI-38 and MRC-5 human cells, FRhL-2 non-human cell, and some primary cell cultures were used for the production of human vaccines [3]. However, from 1987, emphasis was placed on the identification and characterization of the cell substrate, and testing of the bulk and final product. This allows use of a cell line that would previously have been rejected as a substrate simply because it showed abnormal karyology, infinite life or abnormal growth potential or malignancy. But the products derived from such a cell line should undergo rigorous testing to demonstrate freedom from detectable contaminants such as residual cellular DNA [4, 5, 6]. In addition, 1993 PTC focuses on the validation of the manufacturing process for removal and/or inactivation of adventitious agents, and revised or eliminated some of the tests previously recommended [7].

Table 1 shows the summary of the tests recommended for the cell substrate by PTCs issued by the FDA since 1984. «Yes» indicates that the particular test is recommended whereas «No» indicates that the particular test is not recommended or needed. «Yes/No» means that the particular test is still recommended but in some cases is not necessary.

Table 1: The tests recommended in «Points to Consider in the Characterization of Cell Lines Used to Produce Biologicals» issued by the FDA since 1984.

Test	1984	1987	1993
Tumorigenicity	Yes	Yes	Yes/No
Karyology	Yes	Yes	Yes/No
Mycoplasma	Yes	Yes	Yes/Spiroplasma
Bacteria & fungi	Yes	Yes	Yes
Test for viruses			
1) Test in culture	Yes	Yes	Yes
2) Test in animal			
Suckling mouse	Yes	Yes	Yes
Adult mouse	Yes	Yes	Yes
Guinea pig	Yes	Yes/No	Yes/No
Rabbit	No	Yes/No	Yes/No
Monkey	No	No	Yes/No
Embryonated chicken egg	Yes	Yes	Yes
3) Retrovirus	Yes	Yes	Yes
Oncogene expression	No	Yes	No
Test for DNA (product)	No	Yes (10 pg/dose)	Yes (100 pg/dose)
Validation of viral elimination	No	No	Yes

Tumorigenicity is no longer recommended for continuous cell lines of rodent origin because virtually all of these lines are tumorigenic. Karyology still applies to diploid cell lines but it is not recommended for continuous cell lines. A spiroplasma test in addition to the mycoplasma test is now recommended for the biological products made in insect cell lines. With regard to oncogene expression, 1987 PTC recommended tests such as DNA hybridization studies to evaluate the presence of activated oncogenes to provide additional assurance of safety. However, 1993 PTC dropped this recommendation because oncogenes may be involved in normal cell growth. Traditionally, cell lines free from contamination of adventitious viruses were mostly used for the production of viral vaccines. However, as continuous cell lines have been introduced for production, it has become necessary to qualify cell lines that produce virus-like particles or even infectious viruses. This has resulted in studies to validate the effectiveness of the manufacturing process in eliminating those viruses. This validation of viral elimination is now recommended and may be accomplished by the downstream processing steps such as purification processes [7].

3. Vaccine

The third category of the major test materials is the vaccine itself, which consists of Bulk and Final Container [1].

a) Bulk

Sterility to check bacteria and fungi, mycoplasma, tissue culture safety to check adventitious viruses, animal safety to check adventitious viruses and bacteria, infectivity to check the titre and neurovirulence to check the neurotropic property of the vaccine virus are frequently used for testing the bulk material of the vaccine [2].

b) Final container

Tests that apply to the final container include general safety to check extraneous toxic contaminants, sterility, potency, pyrogenic substance/LAL endotoxin, stability, residual moisture when the product is freeze-dried, identity and nucleic acid contamination test, especially when the cell substrate is other than a diploid cell line [2].

Tissue Culture Safety Test as In Vitro Assay to Detect Adventitious Viruses

1. Number and type of cell lines needed

In general three cell lines are recommended as a minimum [3, 4, 7]. At least one human cell line and one monkey cell line should be included in addition to the cell being characterized or used for the production of a vaccine. However, depending upon the source and the passage history of test materials, additional cell lines are very often needed. For example, if an insect or insect cell line is involved as a source or in the passage history of test materials, one additional cell line which supports growth of the insect viruses should be included [8, 9].

2. Neutralization of vaccine viruses by antiserum

When the vaccine virus causes a cytopathic effect (CPE) of the indicator cells, a difference in the type of CPE or the time of CPE occurrence may be helpful in distinguishing CPE caused by vaccine virus from that caused by adventitious virus. However, it is not easy practically and may be impossible to distinguish CPEs. Therefore, it is desirable to neutralize vaccine virus with high titred antiserum before testing in cell systems susceptible to vaccine virus CPE. Neutralization may also be used to prevent interference caused by the vaccine virus. Two principles should be followed in the preparation of antiserum [10]:

a) The immunizing antigens used for the preparation of antiserum shall be grown in a cell line other than that used for testing the vaccine. For example, to produce antiserum to neutralize polio vaccine virus to be tested in a monkey kidney cell system, the immunizing polio virus should be grown in a cell other than a monkey kidney cell. By the same token, when a human cell system is used for testing, the immunizing polio virus should be grown in a cell line other than a human cell.

b) The origin of antiserum should be from a species other than the one in which the vaccine virus was passaged. In the case of polio vaccine, since the vaccine virus was passaged in monkey kidney cells, the antiserum should be produced in a system other than primate.

3. Inoculation and observation

A dilution of the test material in medium not to exceed 1:4 should be used. The area of surface growth of the indicator cells should be at least 3 sq cm/ml of test inoculum [10].

The cultures are observed for at least 14 days by checking for CPE. At the end of the observation period at least one subculture of fluid from one of the indicator cell lines (usually a monkey cell line) is made in the same tissue culture system and the subculture is observed for at least 14 days. At the end of the observation period the culture should be tested for haemadsorption. This test, originally developed by Vogel & Shelokov [11], is performed using erythrocytes from guinea pigs, chickens and from monkeys or humans. The erythrocytes derived from each of these species may be used individually or an equal mixture of erythrocytes from the three species may be used. Tests should be read after the cultures have been incubated for 30 min. at 3-4°C and again after 30 min. at 34-37°C. Additional readings may be made after the cultures have been incubated at room temperature [3, 4, 7, 8].

Animal Safety Test to Detect Adventitious Viruses

The presence of latent or inapparent viruses in a test article may not be detected by an in vitro (tissue culture safety) test. The presence of these viruses may be detected, however, by an in vivo test [8, 9].

1. Number and type of animals needed

Some of the test systems frequently recommended for detecting adventitious agents are the suckling mouse, adult mouse, guinea pig, rabbit and embryonated chicken egg [8, 10].

a) Suckling mouse safety test

Suckling mice are used to detect possible contamination of the test material with Coxsackie viruses. At least 40 suckling mice less than 24 hours old are needed per test. This includes 20 additional suckling mice for the subpass of emulsified skeletal tissues of all mice surviving the original 14-day test.

b) Adult mouse safety test

Adult mice are used to detect lymphocytic choriomeningitis virus. At least 20 adult mice, each weighing 15 to 20 grams, are needed per test.

c) Guinea pig safety test

Guinea pigs are used to detect lymphocytic choriomeningitis virus. At least five guinea pigs, each weighing 350 to 450 grams, are needed per test. In addition this test detects myobacterium tuberculosis with additional 21-day observation.

d) Rabbit safety test

Rabbits are used to detect B virus. At least 10 rabbits, each weighing 1500 to 2500 grams, are needed per test.

e) Embryonated chicken egg safety test

Embryonated chicken eggs are used to detect myxoviruses and arboviruses. Forty embryonated eggs, 10 to 11 days old, are needed per test by the allantoic route. This includes 20 additional embryonated eggs for the subpass of the allantoic fluids from the original three-day test. Forty embryonated eggs, six to seven days old, are needed per test by the yolk sac route. This includes 20 additional embryonated eggs for the subpass of the yolk sac material harvested from the original nine-day test.

2. Neutralization of vaccine viruses by antiserum

When the vaccine virus causes clinical symptoms or death of the test animals, the vaccine virus should be neutralized with high titred antiserum prepared following the same principles discussed under Tissue Culture Safety Test.

3. Inoculation and observation

Table 2 is a summary of in vivo safety tests which shows observation period, observation parameter, route of inoculation and inoculum per animal.

Table 2: In vivo safety test.

Test animal	Observation period (days)	Observation	Route of inoculation	Inoculum per animal (ml)
Suckling mouse	14 14 (subpass)	Clinical signs and death	i.c./i.p.	0.01/0.1
Adult mouse	21	Clinical signs and death	i.c./i.p.	0.03/0.5
Guinea pig	21	Clinical signs and death	i.c./i.p.	0.1/5.0
Rabbit	21	Clinical signs and death	i.d./s.c.	1.0/9.0
Embryonated chicken egg	3 3 (subpass) 9 9 (subpass)	Viability and haemagglutinin viability	allantoic yolk sac	0.5 0.5

The animals are observed daily during a specified period for clinical signs (including death) and the embryonated eggs are observed for viability and/or haemagglutinin. For both suckling mouse and adult mouse safety tests each mouse that dies after the first 24 hours of the test, or is sacrificed because of illness, is necropsied and tissues are subinoculated into at least five additional mice which will be observed for the specified days (14 days for the suckling mouse test and 21 days for the adult mouse test).

For the guinea pig safety test each animal that dies after the first 24 hours of the test, or is sacrificed because of illness, should be necropsied and its tissues subinoculated into at least three additional guinea pigs using the same routes of inoculation. If clinical signs suggest infection with lymphocytic choriomeningitis virus, serological tests should be performed on blood samples of the test guinea pigs to confirm the clinical observations. For the rabbit safety test, each rabbit that dies after the first 24 hours of the test, or is sacrificed because of illness, should be necropsied and the brain and organs removed and examined [10].

Conclusion

Seed virus, cell substrate and vaccine are the major test materials in the production of viral vaccines. Identity and freedom from adventitious agents tests are recommended for seed virus. Continuous cell lines may be used for vaccine production as long as procedures establish adequate characterization of the cells, rigorous testing of the vaccine product, and validation of the process to remove or inactivate adventitious agents. Testing of cell substrate and vaccine is performed following «Points To Consider for Cell Substrate» and 21 CFR 600-680. In vitro and in vivo tests are both recommended for detecting adventitious viruses since they are complementary to each other. Tissue culture safety testing requires a minimum of three cell lines but additional cell lines are frequently needed. The cultures are observed for CPE and haemadsorption. High titred antiserum is needed when a vaccine virus causes CPE. For in vivo tests the animals are observed for clinical signs, including death, whereas the embryonated eggs are observed for viability and/or haemagglutinin.

Acknowledgements

I wish to thank Dr. Harold Lupton, Mrs Roxane Martin and Miss Grace Lee for the critical reading of the manuscript and Mrs. Noreen Colonese for typing the manuscript.

References

1 Lee CK: Human Vaccine Development and Approval Process in the USA. J Biomed Res 1991;1(1):46-51.

2 Code of Federal Regulations, Title 21, Parts 610.10-610.30, 1995. US Government Printing Office, Washington, DC 20402, USA.

3 Points to Consider in the Characterization of Cell Lines Used to Produce Biologicals, June 1, 1984, US Food and Drug Administration, Rockville, Maryland 20857, USA.

4 Points to Consider in the Characterization of Cell Lines Used to Produce Biologicals, November 18, 1987, US Food and Drug Administration, Rockville, Maryland 20857, USA.

5 Petricciani JC: Safety issues relating to the use of mammalian cells as hosts, in Perkins FT, Hennessen W (eds): Standardization and Control of Biologicals Produced by Recombinant DNA Technology. Dev Biol Stand. Basel, Karger, 1985, vol 59, pp 149-153.

6 Magrath DI: Safety of vaccines produced in continuous cell lines, in Brown F, Horaud F (eds): Virological Aspects of the Safety of Biological Products. Dev Biol Stand. Basel, Karger, 1991, vol 75, pp 17-20.

7 Points to Consider in the Characterization of Cell Lines Used to Produce Biologicals, May 17, 1993, US Food and Drug Administration, Rockville, Maryland 20857, USA.

8 Facklam TJ, Geyer S: The preparation and validation of stock cultures of mammalian cells, in Chiu YH, Gueriguian JH (eds): Drug Biotechnology Regulation. Marcel Dekker, 1991, pp 73-75.

9 Grist NR, Ross CAC, Bell EJ, Stott EJ: Virus isolation, in Davis FA (ed): Diagnostic Methods in Clinical Virology, 1966, pp 17-25, 104-111.

10 Code of Federal Regulations, Title 21, Part 630.18 and 630.35, 1995. Tests for Safety, US Government Printing Office, Washington, DC 20402, USA.

11 Vogel J, Shelokov A: Adsorption-hemagglutination test for influenza virus in monkey tissue culture. Science 1957;126:358-359.

Dr. C.K. Lee, The Salk Institute, Government Services Division, Swiftwater, Pennsylvania 18370, USA.

Brown F, Lubiniecki AS: Viral Safety and Evaluation of Viral Clearance from Biopharmaceutical Products. Dev Biol Stand. Basel, Karger, 1996, vol 88, pp 49-56

Experience with Viral Contamination in Cell Culture

R.L. Garnick

Genentech, Inc., South San Francisco, CA, USA

Key words: Cell culture, investigation, Minute Virus of Mice, MVM, PCR, viral contamination.

Abstract: Genentech has had direct experience with two contaminations of large-scale cell cultures by Minute Virus of Mice (MVM). No definitive source of either contamination has been identified, although the conclusions of the investigation were consistent with a raw material used in cell culture as the source. Effective analytical barriers were developed following the first contamination using polymerase chain reaction (PCR) and cell culture-based methodology for MVM, and these were employed on a routine basis. This approach was effectively used to detect and contain the second contamination, thus dramatically minimizing its impact. Close interactions with regulatory authorities were critical in the positive management of these situations.

Historical Perspective on Recombinant DNA Products

Some historical perspective is needed on the issue of contamination of recombinant DNA (rDNA) products with adventitious agents when considering the impact of such episodes. The earliest rDNA products were produced in bacteria (*Escherichia coli*) and included insulin, growth hormones and interferons. In developing the fermentation systems for these products, it was clearly recognized that contamination by foreign bacteria, bacteriophage and yeast/fungi had to be precluded, and significant advances in fermenter design and operation practices and procedures were rapidly developed to minimize these occurrences. Nevertheless, contamination did occur on a sporadic basis and this was often traced to equipment failures and/or personnel. The use of contained fermenter systems was demonstrated to limit the spread of these contaminations effectively. Significantly, the contaminations themselves were usually readily apparent and easily detected by classical microbiological methods.

The success of the biotechnology industry in producing relatively simple non-glycosylated proteins was enhanced in the early 1980s by the industry's ability to produce large glycosylated proteins such as tissue plasminogen activator (t-PA), Factor VIII, and DNase. These molecules were produced in rodent-derived cell lines (CHO-K1, BHK-21) in 100-10,000 litre cell culture facilities. Since the early

1980s, such facilities have used literally millions of litres of complex media, air and other gases as well as immeasurable quantities of raw materials in their production. This single factor weighs heavily in the prediction of contamination by adventitious agents. It was apparent from the onset of this technology that viral contamination of cell cultures from these sources was a serious threat. Thus, extensive measures were put in place to prevent the potential for contamination. Exhaustive characterization of the master cell banks for endogenous viruses (retroviruses) as well as for other adventitious agents was performed as well as routine lot-by-lot testing of the cell cultures for mycoplasma, viruses and bacteria/fungi. The result of these programmes established a very favourable track record with relatively few contaminations, and these were usually bacterial/fungal. Although reverse transcriptase activity was detected, no retroviral infectivity was observed. Notably, there were only sporadic, often unconfirmed, reports of adventitious viral contaminations, although MVM was previously reported by others [1].

The concerns with respect to viral contamination listed in Figure 1 revolve primarily around the difficulty inherent in detecting low levels of these contaminations and, as a result, the potential to unknowingly contaminate the facility in which rDNA materials are produced.

Genentech's Experience with MVM

Genentech has had two experiences with the contamination of large-scale cell cultures by MVM. These two experiences were completely different in their outcomes and serve as a model for dealing effectively with adventitious viral contaminations. The characteristics of this virus are presented in Figure 2.

The first episode was detected at a late stage by the general viral screening assay outlined in Figure 3. In this assay, a slight positive reaction in the haemadsorption (HAd) test was detected after five days in a cell culture and ultimately the source of the HAd was identified as MVM by a fluorescent antibody test. Unfortunately, the late detection resulted in the apparent spread of the virus to multiple fermenters operating simultaneously, which required an extensive facility decontamination and led to significant loss of product as well as time-consuming investigations and regulatory involvement. During this episode, a tremendous amount of analytical effort went into developing sensitive and selective methods for detecting MVM so that

Viral contamination

- *Historical problem in cell culture*
- *Viruses are not easily detected at low titer in cell culture*
- *Small particles - not easily removed by filtration*
- *Difficult to inactivate*
- *Potentially pathogenic*
- *Regulatory authorities' reactions*

Fig. 1: Concerns regarding viral contamination of cell cultures.

Minute Virus of Mice

- *Autonomous parvovirus*
- *Isolated in 1966*
- *Mice are the natural host. Infection is common (20%) and clinically inapparent*
- *Virus excreted in urine and feces*
- *Two strains of MVM having different specificities:*
  - *MVM(p) - prototype strain grows in fibroblasts*
  - *MVM(i) - immunosuppressive variant grows in T-lymphocytes*
- *Non-enveloped, icosahedral virus, 20 nm diameter*
- *Replication is dependent on factors transiently expressed during the S phase of growing host cells*
- *Genome is linear single-stranded DNA molecule that is ~5,100 bases - entire sequence is known*

Fig. 2: Characteristics of MVM.

these tools could be used to isolate the source of the virus and perform as an early detection system to prevent the spread of putative future contaminations. Literally thousands of hours of effort went into the development of both a PCR (DNA)-based method as well as a 324K cell culture-based infectivity assay. The

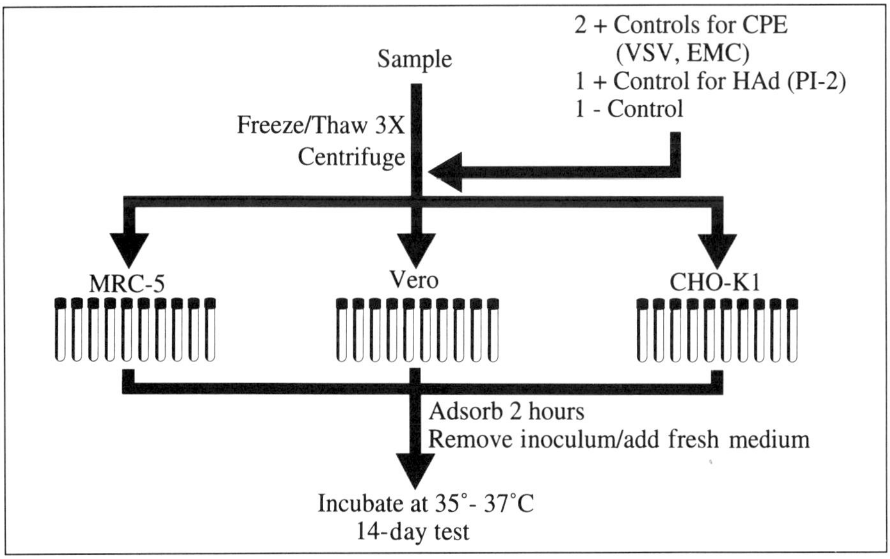

Fig. 3: Schematic of the viral screening assay. CPE = cytopathic effect; VSV = vesicular stomatitis virus; EMC = encephalomyocarditis virus; HAd = haemadsorption; PI-2 = parainfluenza virus type 2.

results of this work were that validated PCR and 324K assays were put into routine operation in the Quality Control Department immediately following the first contamination episode.

A long period of time elapsed before MVM was again detected. However, this time the situation was completely different. In the second episode, MVM was detected by PCR before breaking containment in a single fermenter while other fermenters were in simultaneous operation. The single contaminated fermenter was thoroughly decontaminated while the other fermenters continued in operation. Exhaustive studies demonstrated no spread of the virus and resulted in minimal loss of material. Based on the barrier approach and the early detection results, regulatory interactions with the Food and Drug Administration (FDA) Center for Biologics Evaluation and Research (CBER) were highly productive and co-operative. Nevertheless, exhaustive investigations were conducted once again to attempt to identify the source of the MVM. The scope of this investigation is shown in Figure 4, and the results are summarized in Figure 5. The conclusion was that all available evidence was consistent with the MVM being introduced through a non-homogeneous vector delivered through a raw material.

Assay Methodology

Three different methods for the routine detection of viruses are currently used at Genentech on each production cell culture. These include the general viral screening method described in Figure 3, and the PCR and 324K cell culture-based assays. A comparison of these methods, presented in Figure 6, demonstrates the complementary nature of these methods.

The use of a PCR-based method to evaluate routine cell cultures is fraught with a number of serious issues. Because this method serves as an early warning system, concern over both false positives and false negatives must be addressed in a careful manner. These considerations are presented in Figure 7 and were important factors

Scope of MVM investigation

- *Quality assurance internal GMP compliance audits*
- *Compressed gases, water*
- *Equipment - process flow*
- *Raw materials quarantined/survey vendors*
- *Personnel (access, training and survey)*
- *Working Cell Bank*
- *Pest control*
- *MVM isolate characterization*
- *Environmental controls*
- *Campaign documentation*

Fig. 4: The investigation of the MVM contamination involved multiple aspects, processes, and documentation.

Investigational results

- *Early detection of MVM coupled with effective containment/decontamination procedures prevent spread of the virus*
- *Facilities can continue to operate effectively*
- *Two different strains of MVM were isolated*
- *No identifiable source was ever found - raw materials/air/environment/personnel*
- *All available evidence is consistent with a nonhomogeneous contaminant delivered through a raw material*

Fig. 5: Conclusions from the investigation of the MVM contamination.

during the development and validation of this assay. The DNA primers used for the detection of MVM were chosen from the published sequence of MVM(p) and are located in the DNA coding for non-structural proteins of the virus that are highly conserved among the known parvoviruses. This homology is shown in Figure 8, which indicates the alignment of the 21 nucleotide (21 mer) primer in the forward direction. The location of the 774 base pair (bp) amplicon produced by the forward and reverse primers is illustrated in Figure 9, and a representative PCR agarose gel from the analysis of production cultures is shown in Figure 10. This latter figure demonstrates the use of MVM DNA-spiked controls and shows the clear detection of the second MVM contamination after 168 hours into the culture. A major advantage of the PCR method is its speed. Reliable results can be obtained within eight

Methodology for the detection of viruses

	PCR	*324K*	*Viral Screening*
Specificity	MVM(i) MVM(p) MVM(Genentech)	MVM(i) MVM(p) MVM(Genentech) KRV, H1 (Rat Parvoviruses)	Very broad array of viruses
Sensitivity for MVM	0.1 $TCID_{50}$/mL 2000 genome eq/mL	10 $TCID_{50}$/mL	1 $TCID_{50}$/mL (PI-2) 0.1 $TCID_{50}$/mL (Mumps) 100 $TCID_{50}$/mL (Reo-3) Unvalidated MVM
Test Time	8 Hours	21 Days	14 Days
Implementation Schedule	Each 12K Fermenter tested 2 times prior to harvest	Each 12K Fermenter sampled preharvest	Each 12K Fermenter sampled preharvest
Operations	Routine Assay	Routine Assay	Routine Assay

Fig. 6: Comparison of the characteristics of the PCR, the 324K, and the viral screening assays used for the detection of viruses. $TCID_{50}$ = median tissue culture infectious dose; Reo-3 = reovirus type 3.

Considerations for PCR Assay

- *False positives - PCR can detect one molecule - may not be infectious*
  - *– Specific:* *carryover from previous PCR amplification*
  - *– Non-specific:* *primers may anneal with genomic DNA*
- *False negatives - Inhibition of PCR reaction (SDS, EDTA, phenol salts [Mg], pH)*
  - *– Spike MVM DNA into each sample, positive control*

Fig. 7: The PCR assay can generate false positives and false negatives, which must be considered in the interpretation of the assay results.

hours, thus providing ample time to adjudicate false positives by multiple analyses. However, it must always be recognized that a positive PCR result may not be indicative of an infectious viral particle; a suitable infectivity assay must also be used. As previously discussed, Genentech developed a 324K cell culture-based assay that is highly sensitive to MVM as well as to other rodent parvoviruses. The details of this assay are outlined in Figure 11.

Vulnerability Considerations and Conclusions

Based on the studies presented here and in the literature, it is apparent that CHO-K1 and BHK-21 cells are viable hosts for MVM, and that modern cell culture

```
1351                          1    ................. ACTGCCTGACACAAGAACCTG ............
ss.mvmpcg:1314-1363           1314 AAACTAACCAACTTTTC ACTGCCTGACACAAGAACCTG CAGAATTTTTGC
ss.mvmicg:1315-1364           1315 AAACTAACCAACTTTTC ACTGCCTGACACAAGAACCTG CAAGATTTTTGC
ss.gne17:1192-1241            1192 AAACTAACCAACTTTTC ACTGCCTGACACAAGAACCTG CAAAATTTTTGC
ss.gne52:1189-1238            1189 AAACTAACCAACTTTTC ACTGCCTGACACAAGAACCTG CAAGATTTTTAC
ss.mpv:1318-1367              1318 AAACTAACCAACTTTTC ACTGCCTGACACAAGAACCTG CAAGATCTTTGC
ss.parvo.h1:1317-1366         1317 AAACTAGCCAACTTTTC CATGGCTAGCACCAGAACCTG TAGAATCTTTGC
ss.parvo.lu3:1320-1369        1320 AAACTAACCAACTTTTT ACTGGCTGATACAAGAACCTG TAGAATCTTTGC
ss.parvo.fel1:1323-1372       1323 AAACTAACTAACTTTGA TCTTGCAAATTCTAGAACATG TCAAATTTTTAG
ss.parvo.can:1328-1377        1328 AAACTAACTAACTTTGA TCTTGCAAATTCTAGAACATG TCAAATTTTTAG
ss.parvo.por:1316-1365        1316 ATGCTACCAACATTTAA TATTAGCAATACAAGAACATG TAAAATATTCAG
ss.parvo.bov:1348-1397        1348 AACCTACAATGATTTGG TGGTACATTTGTCTGATTTGA TGTTAATGCTTG
```

Fig. 8: Comparison of the forward PCR primer alignment with the DNA sequences of several parvoviruses: mvmpcg [MVM(p)]; mvmicg [MVM(i)]; gne 17 and gne 52 (strains of MVM isolated at Genentech); mpv (Mouse parvovirus); parvo.h1 (H1 rodent parvovirus); parvo.lu3 (Lu3 rodent parvovirus); parvo.fell (Feline parvovirus); parvo.can (Canine parvovirus); parvo.por (Porcine parvovirus); and parvo.bov (Bovine parvovirus).

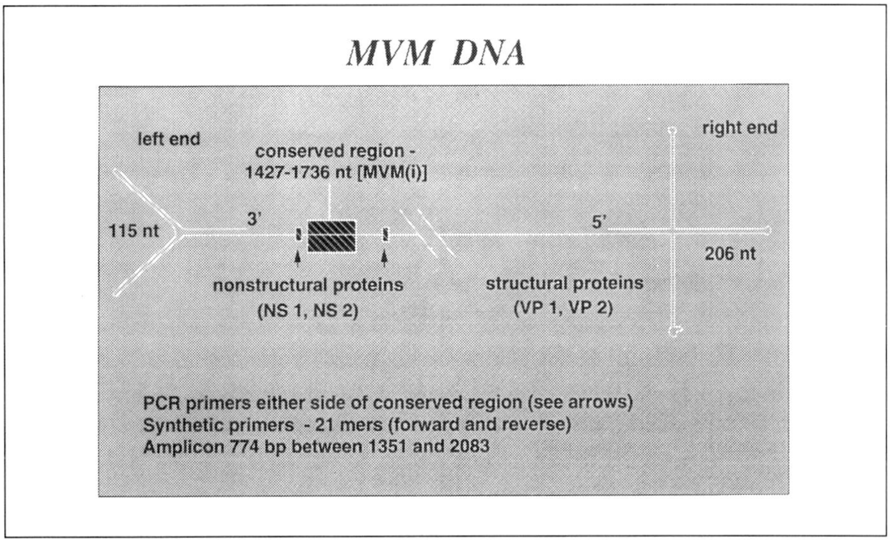

Fig. 9: MVM genomic DNA organization illustrating the more conserved left end of the genome coding for non-structural proteins and the more variable right end of the genome coding for structural proteins. PCR primers were designed near the conserved region in the left end of the genome. They produce a 774 bp fragment when MVM DNA is present during PCR amplification. The MVM DNA contains approximately 5000 nucleotides (nt).

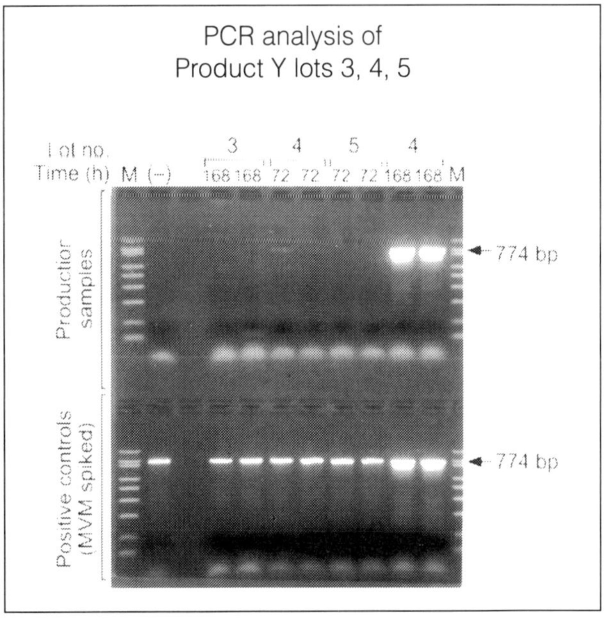

Fig. 10: Typical agarose gel using PCR analysis, which detected MVM definitively at 168 h.

324K assay

- *324K cells - clone of SV40 transformed human newborn kidney cells*
- *Inoculate ~25% confluent 324K cells and incubate*
- *At Days 7 and 14, subculture to fresh 324K cells*
- *Observe periodically for CPE*
- *Final observation for CPE at Day 21*

Fig. 11: Outline of the 324K cell-based infectivity assay.

processes are vulnerable to MVM contamination. Moreover, because of the scale of such processes in which literally millions of litres of media/air/raw materials are used per year, these processes are susceptible to MVM contamination. Based on Genentech's experiences, the most probable source of MVM introduction into such a process is through a non-homogeneous contaminant in one of the hundreds of raw materials used, and such a source is virtually impossible to detect through testing (particularly true of powdered media). Another consideration is that poor facilities or vendor rodent control systems could introduce such a contaminant at any point in the raw materials production process that would be impossible to detect. Thus, despite careful attention to current Good Manufacturing Practices (CGMPs), it is probable that more contaminations will occur in the future.

Based on these considerations and the experiences described in this paper, it is clear that the best approach currently is to construct an analytical barrier using PCR or other methods to provide early in-process detection to prevent the spread of contamination between fermenters and downstream processing.

Reference

1 Nettleton PF, Rweyemamu MM: The association of calf serum with the contamination of BHK21 clone 13 suspension cells by a parvovirus serologically related to the Minute Virus of Mice (MVM). Arch Virol 1980; 64: 359-374.

Dr. R.L. Garnick, Genentech, Inc., 460 Point San Bruno Blvd., South San Francisco, CA 94080, USA.

Brown F, Lubiniecki AS: Viral Safety and Evaluation of Viral Clearance from Biopharmaceutical Products. Dev Biol Stand. Basel, Karger, 1996, vol 88, p 57

SESSION III

Evaluation of Viral Clearance

Chairpersons: T. Kohama, NIH, Japan
J.S. Robertson, NIBSC, UK

Brown F, Lubiniecki AS: Viral Safety and Evaluation of Viral Clearance from Biopharmaceutical Products. Dev Biol Stand. Basel, Karger, 1996, vol 88, pp 59-71

Relationship between Nature and Source of Risk and Process Validation

W. Berthold, W. Werz, J.K. Walter

Process Development, Department Biotech Production, Dr. Karl Thomae GmbH, Biberach, Germany

Key words: Risk validation, process validation.

INTRODUCTION

Mammalian cells represent one of the most valuable sources of biological agents for treatment of human and animal diseases. They can produce large proteins or complex structures.

Since such complex biologicals are difficult to handle and to analyse, much of the confidence in their safety cannot rely on the analysis of the final preparation, but has to resort to the assessment of the process by which the biological preparation has been produced and the raw materials used. If the process has been shown to generate an acceptable product, the likelihood of unforeseen or unforeseeable hazards must be minimal, i.e. acceptably low. However, even if a well-established process of manufacturing is established, every individual production lot is unique because it is derived from a unique set of raw material batches, at a given time, produced in distinct facilities by operators known by name.

This presentation provides some thoughts regarding safety, especially freedom from viruses, that are considered as the most difficult risks to assess because viruses cannot be removed with absolute certainty from biopharmaceutical products.

Why is virus contamination a concern for biologicals ?

The assurance that biological products are free from any virus and therefore acceptable with regard to the hazards they cause is a difficult goal to reach because of the extensive diversity of viruses, their different nature and unforeseeable presence or absence. That viruses represent a real risk to patients has become public knowledge by recent difficulties encountered with products derived from human blood. A number of incidents have occurred since 1985 in preparations derived from human multidonor blood despite special methods that have been applied to the manufacture of these biologics to remove undetected virus. In these cases inactivation methods have apparently either not been used or have not been as effective as expected.

Why can virus be of concern?

For a number of reasons, the ideal biopharmaceutical preparation should be a pure efficacious protein devoid of any contamination by other biologicals such as micro-organisms. Thus, for proteins, sterility or freedom from microbiological contamination can be obtained by sterile filtration. However, complete freedom from viruses cannot be obtained in this way, because these are of the same order of size as proteins and integrity of filters is always difficult to prove.With small, stable chemical drugs, freedom from infections agents can be achieved comparatively easily by methods that destroy, i.e., denature proteins completely such as by extremes of pH, denaturant or heat.

Biopharmaceuticals consist of proteins, long chains of amino acids that are folded in a specific way to give a unique shape. The stability of comparatively weak bonds is responsible for the conformation or shape. Since proteins are the building blocks of all living organisms, proteinaceous biopharmaca have many properties in common with such organisms. The range of « harsh » chemical and physical conditions that can be employed to separate proteins from one another and from live micro-organisms is very limited. The dilemma for products so similar in physicochemical properties to whole viruses is that absolute destruction of all viruses (« virus sterility ») can only be by methods that also destroy the protein structure.

Terminal «virus sterilisation» of biological drugs is not meaningful because the beneficial effect of the protein drug is destroyed together with any virus that is present. Therefore, only prevention and partial inactivation techniques suitable for maintaining the biological properties of the product while disrupting viral structures can be applied.

Stragety of Viral Safety

Safety from potential viral contamination of biopharmaceutical products can be obtained. Instead of one drastic « sterilising » treatment, several methods contribute towards removal of virus. The three **major strategies** that in combination can provide the acceptable safety level for a biological are given in Table 1:

- prevention of contact with hazardous virus;
- establishment of a process that can eliminate potentially present virus;
- strategy of in-process control testing on production lots.

The three areas of activity recommended in Table 1 appear simple and logical. However, in detail they do represent some difficulties of compliance, often specific for the protein in question and the chosen cell culture system:

- Use clean raw materials and equipment for manufacturing. Give special attention to the production cells (in an MCB) and endogenous virus in these cells.

Table 1: Complementary approaches for viral safety of biologicals.

Prevention	• Use of raw material certified free of detectable virus – Master Cell Bank[1] – Biological additive in media[2] – Process equipment[3]
Pocess validation	• Design a process • Demonstrate[1] its efficiency to – avoid risk of contamination – eliminate or reduce potential virus load – inactivate potentially contaminating virus
Adventitious virus testing	• Survey individual production runs in appropriate tests (IPCs[2]) – raw material (incl. WCB[1]) – samples from chosen process steps[2]

1 Virus testing only once for each MCB and each manufacturing process.

2 Individual lots of raw material and every individual production run should be assessed or tested as part of the quality control release of individual batches of product.

3 Cross contamination within one run from early to late fractions (sequestering of steps), from run to run (cleaning validation) and from neighbouring runs in multipurpose facilities (contained systems) should be avoided.

- Use methods capable of removing and inactivating virus in the purification steps of the product. This capability has to be demonstrated in a validation study using a model virus to show the effect.
- Devise a strategy of virus testing (screening) in the manufacture of individual batches of product to safeguard against unpredictable adventitious virus.

Risk assessment of virus properties

In trying to grade viruses (Table 2) according to their relative importance to man, those viruses most relevant have two properties:

1. infectious for humans
2. replication competent in the production cells.

This type of virus could occur at very high viral titres in the starting material. Hence, the choice of a suitable host cell may consider animal species of which as little overlap or amphotropism of virus to the human species is known. This is true for micro-organisms such as E.coli, yeast, fungi and plants that do not replicate human viruses but also do not share with mammalian cells a number of post-translational processing steps for proteins.

For mammalian (and insect) cells a certain overlap of viruses can be expected. Thus, such cells can be subjected to a screening programme to identify replication-competent viruses that could be of concern. This leads to a list of viruses for which product batches can be tested specifically (positive control). This strategy is devised to reduce the effort of adventitious virus testing in batch mode while providing assurance of absence of a wide range of viruses of concern.

Table 2: Grading viruses according to risk.

Parameters	**Subcategories**
Animal species	Insect < vertebrates cold < warm blooded (Bird) < Mammal < Primate < Human
Properties	species specific-amphotrop human «banal» (endemic) or vaccinated < pathogenic
Symptoms	Infection: acute, latent, helper dependent, pseudotype

Viruses that may also be of concern are those that cannot replicate in the production cells but may be carried passively through the production process without being destroyed. Such viruses are either endogenous like retroviruses that may or may not have an intrinsic infectious character. They could also be adventitious viruses that could gain access to a preparation from some (biological) raw material or from human operators. These viruses may be of less concern because the titres are relatively low and are decreased by dilution.

A third category of viruses of possibly even less concern is an adventitious type that does not replicate in the production cells and is not infectious for man. The difficulty arises that for many viruses it is not known in detail how dangerous such viruses can be to man. Thus viruses of this kind could be considered as a breach of good manufacturing practice.

Risk assessment in the manufacturing process

If viruses cannot be eliminated categorically from a product a systematic risk assessment (Fig. 1) of the potential virus contamination (Table 3) starts with the available knowledge of the raw materials from which a biological will be produced. A manufacturing process consists of two major parts: (i) the propagation of cells and (ii) isolation of the desired natural product from the raw material produced by the cells. The final dosage form represents a stable pure form of the product which is then made available to patients. Biotechnological products are most commonly used as parenteral drugs and therefore have to be applied by passing through the immune barrier of the human patient. Therefore, safety considerations of potential infections rank high.

Application-associated risks

To assess potential risks to which a patient is exposed to such a biological product, it is important to control and assess materials that may inadvertently contact the preparation. This includes the application device and the accumulated biological history of each drug substance.

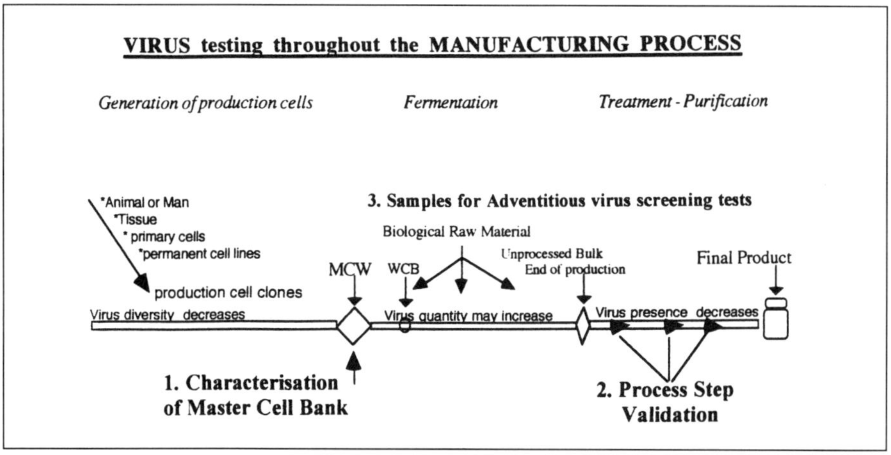

Fig. 1: Overview of virus sources and process test points.

Table 3: Potential sources of virus.

Virus sources	Production step
Biological raw materials media additives enzymes	*Cell propagation* « trypsinisation » treatment of product purification
Operator	*Cell propagation* purification
Environment air rinse water	*Cell propagation* purification
Application syringes	*Bedside application* multiple use container/patients

Prevention

The most dominant potential source of viruses is the living cell itself used for production. There are two complementary approaches to limit the hazard from this source, namely cell banking and raw material screening.

Cell banking concept

The first is the concept of cell banking the intended production cells to establish a homogeneous pool of samples of cells from which the product shall be derived in a reproducible process. In contrast to extracts from whole tissues, cadaver tissues or animal cell tissues, it is the great advantage of recombinant DNA-engineered cells that isolated cell populations can be chosen both for their synthetic properties and freedom from risks.

Components of recombinant cells: the gene: The history of the gene of interest from a recombinant cell may not be relevant because the history of potential carry-over of infection from the original organisms is terminated by the extraction of nucleic acids. Methods of extraction do not allow virus to persist. However, in some cases it may still be considered that the nucleic acids themselves, if isolated from infectious virus, could carry some infectivity as well.

Components of recombinant cells: the host cell: Of greater concern is the choice of the host cell which can be directly linked to its primary source, namely the animal or human tissue from which it has been isolated. Some animal species maintain endogenous (latent) virus components in their genome. Well known are retrovirus-like sequences and particles in mouse and hamster cells (hybridomas and CHO cells). Today they have become manageable so that such cell types are well accepted as a source for biopharmaceuticals.

Exposure to numerous lots of cell culture media that usually contain biological additives, again isolated from animal or human sources, provides ample opportunity for undetected viral infection. Thus any particular clone derived from a transfection experiment may well be regarded as a potential carrier of hazardous material. It is a major milestone in the development of any biological to establish a so-called master cell bank (MCB), that is a bank of many equal aliquots which are all representative. Thus a number of vials can be used for extensive testing of the suitability of the cell line for production of the wanted biological as well as for the freedom from hazardous infections or toxins.

Working cell bank: To expand the prospective useful lifespan of such a finite batch of cells, Working Cell Banks (WCB) can be laid down as a first step from MCB to a specific lot of product. The maximal time for which the cells can be kept in culture (the so-called limit of useful cell age i.e. the maximal number of generations), can be defined by experiments assessing cell-line stability (gene stability is a topic of the ICH process that has been dealt with successfully so far) and absence of latent virus emergence (a test related to the qualification of the MCB cells). Thus an MCB consisting of a few hundred vials can feed a reproducible production for decades providing cells within a window of cell age from MCB to maximal limit of acceptable cell age.

The establishment of such a cell bank system (MCB and WCB) is also a topic of the ICH process because it represents the very foundation for the reproducibility of a protein pharmaceutical regarding safety and identity. Having established a master cell bank every individual production run or whole campaigns to obtain a specific final preparation of reproducible quality can be related to a vial from this MCB.

Assay strategy

Second is the adoption of a tight standardized screening net for viruses that could potentially occur. The number of viruses for which specific tests have to be performed to demonstrate their absence in an MCB is defined by regulatory recommendations. Both these measures result in a conclusion of acceptablility for the cells to provide a safe product.

Biological raw materials

Other raw materials mainly used in the complex nutrient media that mammalian cells require for growth, may also potentially carry virus. Here, the most likely candidates are biological additives isolated from animal tissue or blood. For these raw materials it is also necessary for risk assessment to obtain information on the source, the way it has been purified – if at all – and whether such treatment has been shown to eliminate virus. Examples of such raw materials are whole animal sera, albumin, transferrin, and more or less crude fractions of blood or tissue such as fetuin, fibronectin, collagen, trypsin or lipid mixes. Quality control of such raw materials is not always easy and should be considered carefully because of the extensive travelling of biological components. Certificates may not always reflect the true history. It was observed that in times of shortage of acceptable foetal calf serum due to a suspicion of a contaminant in the product of certain countries the available amount of desirable serum e.g. from New Zealand greatly exceeded the annual production of New Zealand.

The remaining sources of potential viral contaminations are the endogenous viruses in the cell bank and adventitious virus that could find its way into the production media by handling of the non-sterile dry raw materials before their dissolution and filtration to obtain sterility.

Improvements towards prevention of access: The propagation of production cells in vitro can only be obtained if the cells receive the necessary ingredients in the medium. If they required an unknown growth factor, serum from foetal bovine sources is usually added to the nutrient mix. This serum is of undefined composition and usually untraceable history. It has been produced from multidonor animal herds and subjected to the vagaries of slaughter house hygiene in the countries of origin and has been stored, split, pooled and transported. Hence, from very early on manufacturers have made efforts to restrict the requirement of production cells for undefined biological additives such as serum. Today production cells have been obtained that can be grown at normal productivity in media that are entirely defined by chemicals. This is a major advance for the safety of biological products because one major source of concern has been eliminated.

Equipment cross-contamination: Possibilities of viral infection should also be prevented by re-use of equipment. Similar considerations should be given to an unnoticed or late detected viral contamination. The subsequent run should not be contaminated by running through the identical equipment without proper cleaning of those surfaces in contact with the previous production fluid. Hence, the re-use procedure of equipment even within the same product line has to be considered for cleaning validation. The cleaning procedures should be validated to be virocidal. In addition there may be a multi-use facility where the same consideration could apply to runs in parallel with different products. In this case the segregation and sequestering of operations have to be considered carefully.

Viral hazards downstream

Once the desired product is secreted in cell culture media it has to be removed from the cells and subjected to a purification and formulation process leading to a final dosage form. This operation per se may also have some potential for the introduction of virus to the product. There is firstly the possibility that operators carrying infections could introduce some virus. Another source may be the use of biological materials in extraction procedures or in affinity chromatography such as antibody or heparin. In the latter case it is very unlikely that anything infectious can be carried through to the affinity column because the manipulation of purification and coupling of the biological to the column are usually very adverse conditions regarding the viability of viruses. Other sources of adventitious viruses could be in liquid media because sterility filtration may not permit the access of bacteria but some viruses, being considerably smaller than bacteria, could still pass through sterility filtrations.

Process validation downstream

As only preventive measures can be taken in fermentation and product biosynthesis the downstream part of the manufacturing process itself has to be validated in order to assess its potential to eliminate or clear virus. To design a purification process in such a way that a minimum number of steps leads to both the purity of the protein, the desired product and efficient elimination of all biological impurities requires great skill. The general difficulty resides here in the physico-chemical properties of the product. The properties that maintain the beneficial effects of a product are very similar to those carried by all viruses. Therefore only a limited spectrum of techniques can be used for virus inactivation. To assess the capability of individual process steps to remove viruses they have to be tested with live model viruses for which clearing factors can then be calculated.

It is one of the major tasks of the ICH to provide guidelines for the choice of an appropriate virus for these validation studies. It is also important to obtain agreement on the technical performance of the validation in downscale experiments and the statistical significance of tests to show clearance factors. Global agreement is sought for acceptability of these experiments. It is hoped that this could avoid many unnecessary costs for validating the same production processes in different regions of the world caused by redundant testing in specific ways according to different national rules. Details of process validation will not be discussed here because a number of papers at the conference will deal with this subject.

On the prevention side it is important to consider a very specific type of virus contamination that is provided by the manufacturing environment. It is important to consider measures that can be taken to avoid contamination within one particular production run from an earlier phase of purification, e.g. before significant viral inactivation steps have been made, with samples from near the end of the purification process when no further virus clearing step separates the product from the patient.

Adventitious virus testing

Although the manufacturing process has been validated in many aspects to remove and reduce virus, this exercise cannot give sufficient assurance in individual runs where cells could not be exposed to viruses of concern or when they are exposed to fresh raw materials and media or when cells are handled in a specific facility and at a specific time. Hence, analogous to the testing of sterility of biological products, adventitious virus testing should be performed with every production lot. Despite all the care that can be taken to prevent access of hazardous virus material to products, adventitious virus testing is a useful and necessary screening method for increasing the level of safety for individual runs and individual products.

The sampling points for test samples subjected to adventitious virus testing along the progressing manufacturing process can be carefully chosen (Table 4). The master cell bank once established as free from virus is the starting point. Raw materials in particular from biological sources have to be screened for freedom of virus by the manufacturer of these additives or the manufacturer of the biological biopharmaceutical or sufficient assurance has to be provided that if it is a purified protein like insulin the material has been subjected to a validated process that can eliminate viruses.

In line with the previously mentioned category of viruses that could replicate in production cells, attention for adventitious virus testing is given to cell culture fluids

Table 4: Sampling points for virus testing.

Sampling point	Tests for virus
Master Cell Bank	List of viruses from guidelines **In vitro,** **In vivo** MAP, (RAP, HAP) **In vivo** mice, guinea pig, eggs **EM** **Retrovirus** RTase, S+L-, XC
Raw material[1]	**(Additive's) species of origin** specific or screening tests on cells of the species
Working Cell Banks	**Adventitious virus** on test cells: Production cell Human cell, Monkey cell (Additive's cell) **Retrovirus**
Unprocessed bulk harvest **Post production** cells[4]	**Adventitious virus** **Retrovirus**
«**old**» production cells[2] Late expanded **cells**	***Latent virus*** **(Adventitious virus, retrov.)**
Final product (?)[3]	**Adventitious virus**

1 Used for preparation of WCB or in production runs starting from MCB or WCB.

2 Old cells may be from production runs (as post production cells) or from a separate culture kept in continuous culture for a long period and prepared for this analysis of « limit of cell age » only (as late expanded cells). Extensive testing performed as part of the qualification of the MCB regarding absence of latent virus, inducible by cultivation or production conditions.

3 A very large sample volume for testing would be required for statistics of a sufficiently sensitive detection of low virus titres.

4 Cells at the end of a typical production run are tested to determine the virus load in case of retrovirus like partide bearing production cells. Only few harvests need to be tested (validation).

containing cells. In particular the last fraction immediately before the cell-free processing of the product (unprocessed bulk harvest, post-production cells) is important. In addition cells can be tested after long-term cultivation in medium in which cells are generated for manufacturing (seed strain) to ascertain that the medium and cultivation conditions are not stimulating any latent virus to emerge. For this purpose particular points have been defined as maximal end of production or late expanded cell bank cells. Such samples may be useful to test for cell-bank-associated viruses which can only be identified as such when these cells have to be tested free from adventitious virus due to handling inadequacies. Hence, such a result may disqualify this particular sample but not the MCB in general and the sample generation and test has to be repeated.

Virus categories which are replication-competent in the production cell and are infections in humans are the main concern. Therefore, the species and tissue type of assay cells have to be considered carefully. As a minimum, the production cells themselves or very close relatives have to be used together with human cells (primary or human cell lines) or a primate cell line for which a great number of viruses are known to be infective. Other cell lines may be added depending on what type of additives in the media the cells or the product have been exposed to. For example, if bovine serum has been used, a bovine cell line should be added to the infectivity tests in the adventitious virus testing.

Discussion

In addition to these operations of the manufacturers, consideration to the accumulated risk for any patient has to include that associated with the application (Table 5). Parenteral drugs are usually applied with syringes; inappropriate use of syringes may constitute a serious hazard. In particular it would be advantageous to use single-use syringes because multiple use of syringes is known to transmit viral diseases from patient to patient. This is one of the major sources of infection with HIV in some parts of the world.

Although this three-prong approach appears to be logical and straightforward (Fig. 1), some limitations have to be taken into consideration because of the diverse properties of viruses. The quantitative measurement of virus is a common theme used in all three approaches. However, only for expected (known) viruses can any definitive answer be obtained from such assays because of limits of detection. There are limits to methods of inactivation of virus without damaging the protein product. For the selection of model viruses and suitable elimination steps a weighting of risk regarding the (un)likelihood of virus contamination and the types of virus that might be involved may help the work with «relevant» steps and viruses.

As in most screening tests, an absolute certainty of absence of virus can never be obtained for theoretical reasons. However, the probability of the presence of a specific virus can be reduced to an acceptably low level. While the testing of raw material and final product will only avoid the use of contaminated substances, the purification steps may eliminate virus. The choice of viruses for this type of study is essential for risk assessment. Hence, the attempt is to concentrate on well characterised viruses for risk assessment of the process steps and to design strategies best suited to minimize the potential presence of those viruses that are either likely to occur or are especially pathogenic.

Table 5: Source of risk related to potential dose for patients.

Closeness to patient	**Source of (perceived) risk**	**Estimated virus dose**
Bedside application	– Skin of patient – Hygiene of staff – Single/multiple use of syringes	variable
Filling formulated bulk (aseptic cell-free handling)	– Container and closures – Biological excipients (HSA) – Operational hazards (lyophilization: long open)	small
Downstream: purification (cell-free handling)	– Biological raw materials: Antibody/Heparin etc. affichromatography – Used equipment – Neighbourhood of untested (contaminated) CCF	small
Upstream: fermentation (handling of life cells)	– Biological raw material in media – Filter passage of environmental virus – Operators or untested products	large

Potential sources of virus are limited by the species of origin and the country, the tissue and the kind of treatment of the biological material. The type of virus that could be of concern could be placed in an order of grades of concern (Table 3). Parameters for differentiation can be pathogenicity and the likelihood of occurrence.

Assessment of «pathogenicity» is a rather difficult term for biopharmaceuticals. In common usage it relates to a normal healthy adult. However, the state of the immune system of the individual patient is unknown and depends on the individual life history of exposure. Hence, «banal» infections may be very serious in some patients. Thus, pathogenicity may be an unreliable criterion. As every virus seems to make «illicit» use of the host cell's metabolism, all viruses may be considered pathogenic rather than «commensial» or even «symbiotic» – a term that may apply to therapeutic vectors. Nonetheless, in its common usage considering the extent of negative impact of a viral infection, it is still useful to classify viruses as «pathogenic» and make them specific targets for screening and validation purposes.

Conclusion: Acceptance of a Biological as Safe

While a major effort to obtain low risks of viral contamination rests on prevention, i.e. testing of raw materials and final product to detect compromising virus, it is the purification of the product that can remove viruses if a certain viral burden cannot be avoided (as in the case of murine cells) or even if no virus has been observed. In validation efforts, an intention to address all viruses in screening tests is a theoretically endless effort and is not a practical alternative. This limitation of practical testing has raised at least two types of discussion on value judgements regarding the relevance of the validation experiments for the (un)expected occasion:

a) Are choice and number of viruses used for validation adequate?

b) Have the tests been performed with sufficient care (statistics in accuracy and precision)? Both questions are leading to a concluding statement that may be worded as follows: The proposed production method can be shown to clear the model virus XY with a reduction factor of 10x. This represents an acceptable level of viral clearance (and safety margin).

The acceptance as safe of such a biological product can only be based on such a «value judgement» with regard to the level of the remaining probability for presence of virus, and for its unknown and potentially pathogenic effect on man. This situation requires a balance between the endless (time- and cost-consuming) effort to reduce the probability of viral hazards, and a practical and affordable way to reach an acceptable low level of risk.

These questions regarding the translation of regulatory requirements into adequate experimental evidence have great impact on the burden of costs for product development. In comparison, biologicals produced from banked (recombinant) cell lines represent by far the best defined and controlled biological production processes to date (Table 6). An international consensus on a scientifically based safety level provides comfort to all involved parties: the patient as non-expert, the manufacturers and the different national regulators because it sets a standard in fairness and safety. The ICH, pursues very actively the goal to provide this required guidance in the form of a document on viral safety.

The final result should permit manufacturers to fulfil global requirements for the demonstration of viral safety with a minimum of experiments and costs or time lost i.e. avoiding redundancy without any major impact on the safety level reached caused by slight differences in national requirements or different emphasis on assay performance or type of assay. It is hoped that national authorities participating in

Table 6: Source of product associated virus.

Endogenous virus: Source*	**Properties**	**Products** Examples
Animal or human tissue	Single individual donors	**hu growth factor,** pancreatic enzymes, (Transgenic) milk proteins
Blood animal or man	Multi-donor pools	**Factor VIII, HSA,** Antibodies (anti Hepatitis)
Primary cells in vitro	Single donor Media additives	Whole **viral vaccines** **Active proteins:** IFN Extracorporal devices
Cell lines	Single donor Subcloned, media additives	Whole **viral vaccines** **Active proteins:** IFN, MAbs
Recombinant cell lines	Single donor Subcloned Media additives	**Active rec. proteins:** rIFN, r Monoclonal antibodies, **subcomp. viral vaccines;** Vectors, Rec cells for **gene therapy**

* All sources are replication competent, because they consist of mammalian cells.

this effort will adhere to this consensus and grant the mutual recognition of experimental evidence generated by manufacturers under these guidelines in the future.

References

References to ICH (International Conference on Harmonisation of Thechnical Requirements for Registration of Pharmaceuticals for Human Use) Section Quality 5 Biotechnology.

ICH Q5A Viral Validation: Viral safety evaluation of biotechnology products derived from cell lines of human or animal origin.

ICH Q5B Genetic Stability: Analysis of the expression construct in cells used for production of r-DNA deroved protein products.

ICH Q5C Stability of Product.

ICH Q5D Cell Substrate: Human and animal cell substrate.

Dr. W. Berthold, Hoffmann-La-Roche, Inc., Bldg 123/3, 340 Kingsland Street, Nutley, NJ 07110-1199, USA.

Brown F, Lubiniecki AS: Viral Safety and Evaluation of Viral Clearance from Biopharmaceutical Products. Dev Biol Stand. Basel, Karger, 1996, vol 88, pp 73-79

Strategies for Viral Removal and Inactivation

D.G. Burstyn[1], *T.C. Hageman*[2]

[1] Alkermes, Inc., Cambridge, MA USA
[2] Biogen, Inc., Cambridge, MA USA

Key words: Viral removal; viral inactivation.

Abstract: This paper provides an overview of the fundamentals of viral inactivation and clearance studies required for mammalian cell culture-derived biotechnological products. Strategies and considerations for evaluating the potential of purification processes to inactivate or remove adventitious or endogenous viruses are also discussed. The principles are illustrated by presenting data derived from the evaluation of clinical manufacturing processes for recombinant human interferon-ß and recombinant soluble CD4.

The information in this paper is representative of what manufacturers have done for a number of years to address on-going concerns regarding the potential for contamination of pharmaceutical products by adventitious or endogenous viruses or viral-like particles. The potential of a process to inactivate viruses is assessed by spiking process buffers or process intermediates with a model virus, holding the test suspension under the physical conditions for the minimum time used during clinical production, and then assaying the suspension for infectious virus.

The potential of a process to clear or remove viruses is most often assessed by spiking process intermediates with a model virus, performing the process step or steps, and then determining the amount of live virus that co-purifies with the product. Although clearance studies can also be performed using radiolabelled viruses, this approach should be avoided whenever possible since the ensuing experiments tend to be difficult to interpret and the background noise of the detection system results in a relatively low, maximum achievable clearance factor. What usually happens is that some percentage of label dissociates from the virus during the study, lowering sensitivity due to the failure to differentiate between label alone and intact labelled virus particles. Additionally, the difficulty in controlling the site and degree of labelling precisely makes interpretation of the data problematic. Nonetheless, radiolabelling can be advantageous if it becomes necessary to study a specific virus with known human pathogenicity or viruses whose laboratory usage is rigidly controlled by human or animal health authorities.

Decisions made at the earliest stage of product development often have an impact on viral clearance and inactivation potential. Therefore, strategic decisions relative to viral inactivation and clearance need to be made at the very beginning of product development, when the design of the clinical manufacturing process first begins to take shape. It is relatively easy, before large-scale clinical manufacturing starts, to include in the process steps or conditions that are known to inactivate viruses generally. Shown in Table 1 are a list of some of these conditions as they affect selected viruses. When possible, conditions shown to inactivate viruses generally should be incorporated directly into the purification process, even if they serve no useful «purification» purpose, but simply exist as a specifically designated virucidal-hold step. Since these same conditions may also have an adverse effect on the molecule of interest, the stability of the protein product should be determined as early as possible in the development process, over a wide range of physical and chemical conditions. Since protein concentration can modulate the effect of these conditions on a molecule, the stability of the molecule should be assessed in both the purified state and as a partially purified intermediate.

As shown in Table 1, filtration can be a powerful tool to remove virus particles non-specifically. In recent years, tremendous strides have been made by the manufacturers of filtration devices to enable the validated use of microporous membranes for the physical removal of viruses. The use of filtration to remove particles is the subject of papers later in this volume.

To illustrate how purification processes are evaluated for their potential to remove and inactivate endogenous or advantageous viruses, data for two recombinant products, human interferon beta (rhIFN-ß)and soluble CD4 (rsCD4), will be used. Although the data shown specifically concern retroviral inactivation and clearance for products produced by recombinant Chinese hamster ovary (CHO)

Table 1: Generalized conditions that inactivate or eliminate selected viruses. Adapted from Hazleton Laboratories technical literature, circa 1988.

	Inactivated by				Eliminated by	
	Low pH	**High pH**	**Lipid solvents**	**30 min @ > 56°C**	**0.2 µm filtration**	**0.025 µm filtration**
DNA viruses						
Adenoviruses	pH ≤ 3	pH ≥ 11	–	–	–	+
Herpesviruses	pH < 4	pH > 9.5	+	+	–	+
Hepadnaviruses	pH < 4	pH > 9.5	+	+	–	+
Parvoviruses	pH ≤ 3	pH ≥ 11	–	–	–	+
Papovaviruses	pH ≤ 3	pH ≥ 11	–	–	–	+
Poxviruses	pH < 4	pH > 9.5	+	+	+	+
RNA viruses						
Bunyaviruses	pH < 4	pH > 9.5	+	+	–	+
Enteroviruses	pH ≤ 3	± pH ≥ 11	–	–	–	+
Flaviviruses	pH < 4	pH > 9.5	+	+	–	+
Orbiviruses	pH < 4	pH > 9.5	–	–	–	+
Orthomyxoviruses	pH < 4	pH > 9.5	+	+	–	+
Paramyxoviruses	pH < 4	pH > 9.5	+	+	–	+
Retroviruses	pH < 4	pH > 9.5	+	–	–	+

cells, the basic principles incorporated in these studies apply equally to studies using other types of viruses and mammalian cell-production systems.

One of the most important considerations in performing viral inactivation and clearance studies is where to perform the studies. Often, the choice is to contract the work to one of the commercial cell-testing laboratories. When contracting studies to another laboratory, they should be audited and evaluated before study initiation to ensure that it is capable of performing the work required. Whether performing the studies internally or externally, it is extremely important to ensure that good basic scientific practices are already in place in the laboratory and that the documentation and training systems are adequate. At the very least: (a) there must be approved, written procedures for assay methodology; (b) the assays should be shown to be precise, accurate and reproducible over the range of conditions expected during their use, and (c) operators must be trained and/or certified in the performance of the assays and equipment.

The first experimental step, before the formal process evaluation studies are performed, is to confirm the validity of the proposed virus quantification methods for use with the actual molecule, buffers, and process intermediates to be evaluated. This is simply an evaluation of the effect on the assays of the test materials alone.

The first control is one of generalized toxicity, to ensure that any indicator cell cytopathology observed during the evaluation experiments is due to the virus alone. In this case the cells are exposed to the process intermediates (in the absence of virus) for the length of time that the test material will be in contact with the cells, which is usually one to two hours for most assays. The cells are incubated as prescribed in the assay protocol and then examined. If a cytopathic or morphological effect is observed relative to the unexposed control cells, the procedure is repeated using dilutions of unspiked solutions until one is found that is not cytotoxic for the indicator cells (see below). Alternatively, it may be necessary to neutralize or otherwise adjust the test solutions before analysis.

It is also possible that the process intermediates alone may reduce the "plating" efficiency of the model virus on the indicator cells, resulting in an inflated calculation of viral inactivation. Therefore, this possibility is evaluated by a second required control. In this case, after exposing the indicator cells to unspiked test solutions, the indicator cells are gently washed and inoculated with amounts of virus at the extremes of the assay range. After incubation, the resulting virus titres are compared with that of the untreated control cells to determine whether there is an effect. If an effect on «plating» efficiency is observed, the procedure is repeated until a dilution is found that does not affect virus quantification.

The first example is the evaluation of an early clinical purification process for rhIFN-ß. For inactivation studies, actual process intermediates are normally spiked with test viruses. For rhIFN-ß this could not be done since even the lowest concentrations of interferon present in any of the process intermediates had a measurable, dose-dependent inhibitory effect on the ability of all the model viruses assessed to form plaques or foci on the indicator cells. Although several methods were evaluated to neutralize the rhIFN-ß present in the process intermediates, including dilution, physical inactivation, and precipitation using monoclonal antibodies, none proves sufficiently effective to the degree required for these studies. Therefore, after consulting the regulatory authorities, the studies were performed using mock process intermediates which did not contain any interferon. The results of the inactivation study are shown in Table 2.

Table 2: Retrovirus inactivation potential of a clinical manufacturing process for recombinant human interferon-ß.

Sample[a]	Active virus per mL	Log inactivation[b]
Medium control	1.0×10^6	
Intermediate A	1.6×10^4	1.8
Intermediate B	4.0×10^5	0.4
Intermediate C	< 20	≥ 5.7
Intermediate D	< 2	≥ 4.7
Total		**≥ 12.6**

[a] Spiked to 5% (v/v) with 2×10^7 Murine Ecotropic Retrovirus.

[b] Inactivation = Medium control ÷ Active virus.

The rhIFN-ß mock process intermediates were spiked 5% (v/v) with a 2×10^7 stock of murine ecotropic retrovirus (MuEco). The generally accepted supposition is that a maximum spike of 5% by volume will not significantly affect the validity of the experiment either due to a physical or chemical change in the process intermediate. This can, and often should, be tested ahead of time in controlled studies.

The amount of virus spiked into each intermediate was approximately 10^6 virus particles per millilitre. The spiked intermediates were held under the processing conditions for the minimum amount of time specified by the manufacturing process and then assayed for biologically active virus using the XC plaque assay.

Since viral inactivation is not a simple first order reaction and is usually composed of a «fast» phase and a «slow» phase [1], inactivation studies may be performed using a series of time points to determine the kinetics of virus killing. As a practical consideration, provided the process time specifications are sufficiently long to ensure that evaluation is well into the slow phase of inactivation, then kinetic studies are unnecessary.

For rhIFN-ß mock process Intermediate A, 1.6×10^4 active virus particles remained after the hold period. Defining the degree of inactivation as the virus spike divided by the active virus remaining, resulted in a calculated log inactivation factor of 1.8. For Intermediate C where total inactivation was observed, the limit of detection was < 20 plaque forming units (pfu). In contrast, for Intermediate D where the model virus was totally inactivated, the limit of detection was 2 pfu. The difference in the detection limit is due to the pre-study finding that undiluted Intermediate C was cytotoxic for the indicator cells. At a 1:10 dilution no cytotoxicity was observed; therefore, this was the lowest dilution at which the sample could be tested in this particular cell-based assay.

The total process inactivation factor is calculated by multiplying the individual step-wise factors together. There has been some discussion as to whether this is a valid calculation. In our judgment, it is valid since the virus in the spike can be considered a homogeneous population. Additionally, the rhIFN-ß process steps evaluated are orthogonal. It would not be valid to combine individually assessed process steps that inactivate based on the same principle, for example, low pH. This calculated total inactivation factor represents an underestimate of the true potential of the rhIFN-ß process to inactivate retroviruses since not every process buffer was evaluated. In general, it is usually sufficient to assess only the number of steps necessary to reach the target inactivation value, and not to evaluate exhaustively the process in its entirety. Additionally, only steps where potential viral contaminants could occur in suspension should be evaluated for their inactivation potential. It is usually incorrect to assess column wash buffers in isolation for their inactivation potential since the viruses, in the presence of the column resin, may behave differently from the way they do in suspension. Inactivation of viruses by column washes can only be evaluated as part of an overall column-clearance study (see below).

These data show that the purification process used for rhIFN-ß was capable of inactivating retroviruses to levels that assured the safety of the target population. Additionally, no single step was relied upon for inactivation; significant retroviral killing occurred at multiple steps. Therefore, no companion retroviral clearance study was performed. This is in contrast to the rsCD4 situation, where initial studies showed that there would be insufficient retroviral inactivation by the process buffers. Therefore, a clearance study was performed to evaluate the ability of the rsCD4 purification process to separate a model retrovirus from the product physically.

Since manufacturers obviously do not want to introduce virus into clinical production equipment or facilities, or to purchase or produce the vast quantity of virus required to spike full scale columns, the first step in a clearance study is to design and validate a small-scale model of the manufacturing process. The design of the model is based on standard process-engineering considerations [2]. These include maintaining: (a) resin bed height while reducing column diameter; (b) flow rates; and (c) ratios of column loads and buffer volumes to bed volume. To ensure the appropriateness of the model scale, the accompanying validation package should include a direct comparison of the resulting product with authentic production batches, and also a qualitative assessment of background proteins expected in the process intermediates. Parameters evaluated should include product yield, product quality, "background" proteins, elution volumes (expressed as column volumes), relative chromatographic separation and possibly the effect of the virus spike on column performance. As regards the scale of the model process, theoretically it should not matter (i.e. 1%, 0.1% or 0.01%) as long as it can be supported by the accompanying validation package. Shown in Table 3 is a summary of the evaluation of the rsCD4 purification process to clear murine xenotrophic retrovirus (MuXeno), a model retrovirus.

The generally accepted supposition is that a maximum spike of 5% by volume will not significantly affect the validity of the experiment due to a physical or chemical change in the process intermediate or, in this case, column performance due to a change in the protein load. This can be tested in control studies and may become important for evaluation of the later process steps, where the total protein content of the intermediates decreases and product purity should increase dramatically.

The amount of MuXeno spiked into the various column loads differs slightly from one another because we calculated the maximum clearance potential of the steps, using fraction volume considerations and minimum dilutions required to avoid cytotoxicity in the assays. In other words, the cost of the experiments was greatly reduced by evaluations that allowed for lowering the amount of concentrated virus purchased and the number of dilutions for assay.

Table 3: Retrovirus clearance-inactivation potential of a clinical manufacturing process for recombinant soluble CD4.

Step	Murine Xenotropic retrovirus (log ffu)		Log CIF[b]
	Total spike	Product fraction[a]	
Column 1	7.3	4.8	2.5
Column 2	6.6	5.4	1.2
Column 3	6.6	< 2.1	≥ 4.5
Column 4	5.3	4.0	1.3
Total			**≥ 9.5**

[a] Virus titre X volume product fraction
[b] Clearance-Inactivation Factor (CIF) = Spike ÷ product fraction

In contrast to the rhIFN-ß situation, the control studies showed that there was no effect of the actual process intermediates on the indicator cells or the ability of the model retrovirus to form foci on the indicator cells. Therefore, process intermediates generated from full-scale clinical manufacturing runs were used to perform these clearance studies.

Shown in Table 3 are the step-wise clearance factors, calculated by dividing the virus spike by the amount of virus in the product fraction. Again it is appropriate to multiply the individual clearance factors (or add log clearance factors) since the virus in the spikes behaves as a homogeneous population. This calculation would not be meaningful if the population were heterogeneous, for example, if a mixture of viruses was being studied, and certainly in cases where clearing host cell proteins was being attempted.

Since the cell-based assay used in this study only detects live virus and is refractory to inactive virus, what is being calculated is a combination clearance and inactivation factor (CIF). Therefore, in this instance, it is not proper to combine factors from separate inactivation and clearance evaluations of the same process step. For example, a value for the inactivation potential of a column load buffer could not be combined with the clearance potential of that column, since it can be argued that the results of the clearance study already take into account both inactivation by the load buffer as well as column clearance. It also takes into account inactivation by column wash buffers. It would be proper to combine separately evaluated inactivation and clearance data only if the steps evaluated are orthogonal. For example, if an inactivation study demonstrated that a column load buffer totally inactivated a virus, there would be no rationale for evaluating the clearance potential of that column. On the other hand, this inactivation factor could be combined with the clearance potential of the other columns to calculate a combined clearance-inactivation factor.

It would be proper to combine factors from separate inactivation and clearance evaluations of the same process step only if the virus quantification method does not rely on cell infectivity assays. For example, a quantitative polymerase chain reaction (PCR), which would enumerate both live and dead virus particles, could be used to provide an independent assessment of clearance that could be combined with inactivation data (generated by assay of «live» virus alone) from the same column step. Although not specifically discussed, ultrafiltration process steps can also be evaluated for their ability to clear virus particles using small-scale apparatus obtained from the manufacturers of most commercial scale equipment.

The remaining and clearly critical question is how much inactivation or clearance is required? In a rational world, it would depend on relative risk. Over the years, in other forums, values of three to six logs over the potential maximum viral load of the starting material have been suggested. What does this mean in practical terms? The limit of quantification of virus-specific assays can be as low as 2 pfu/ml starting material, in this case conditioned media (CM). Assuming that the assay of CM is negative, then the maximum potential contamination is calculated as;

$$2 \text{ pfu/mL} \times 10^3 \text{ mL CM/dose} = 2 \times 10^3 \text{ pfu/dose}$$

and the corresponding required combined CIF is calculated as:

$$2 \text{ x } 10^3 \text{ pfu} \times 10^6 \text{ comfort factor} = 2 \times 10^9 \text{ CIF}$$

Performing the same type of calculation based on non-specific viral detection methods, for example electron microscopy with a limit of detection of approximately 10^5 live or dead particles per millilitre of CM, the maximum potential contamination is calculated as:

$$10^5 \text{ particles/mL} \times 10^3 \text{ mL CM/dose} = 10^8 \text{ particles/dose}$$

and the corresponding required combined CIF is calculated as:

$$10^8 \text{ particles} \times 10^6 \text{ comfort factor} = 10^{14} \text{ CIF}$$

Clearly, one of the most important factors affecting the overall process CIF required is the amount of conditioned medium required to produce a single dose of product.

In summary, by incorporating conditions known generally to inactivate (e.g. low pH) and clear (e.g. microporous filtration) viruses into the process design, coupled with the assessment of actual process steps, including process intermediates, columns, and ultrafiltration, manufacturing processes can be demonstrated to provide sufficient viral clearance and/or inactivation to ensure the safety of pharmaceutical products in the event of contamination of cell cultures with adventitious or endogenous viruses. Required levels of inactivation and clearance can only be assessed in relationship to perceived hazard to the target population and in consultation with regulatory authorities.

References

1 Commission of the European Communities. Validation of Virus removal and Inactivation. Document III/8115/89-EN FINAL, 1991.

2 Sofer GK, Nyström LE: Process Chromatography, A Practical Guide. Academic Press, 1989, ch 7, pp 55-66.

Dr. D.G. Burstyn, Alkermes, Inc., 64 Sidney Street, Cambridge, MA 02139, USA.

Brown F, Lubiniecki AS: Viral Safety and Evaluation of Viral Clearance from Biopharmaceutical Products. Dev Biol Stand. Basel, Karger, 1996, vol 88, pp 81-90

Removal of Virus Through Novel Membrane Filtration Method

S.-I. Manabe

Fukuoka Women's University, Fukuoka 813, Japan

Key words: Virus, Planova, sieving effect, filtration, hollow fibre, membrane, integrity test.

Abstract: To protect against virus infections, the new technology of the virus removal membrane has been developed. The novel virus removal filter PLANOVA™ is validated for virus removal, that is, reproducible and predictable. The removal is based on the sieving mechanism. The removal of viruses by PLANOVA with a given mean pore size ranging between 10 and 75 nm is predictable when the geometrical size and the deformability of the virus are known in advance under the given filtration conditions. The high removability of viruses originates from the unique pore structure, that is the structure of the multi-layer (or multi-screen) each of wich is composed of two types of pores, the capillary and void pores. The good recovery rate of proteins is attained by using material with a good hydroscopic property such as regenerated cellulose. Designing the worst case of the process and selecting model viruses in the process validation experiment are important in the practical application of PLANOVA. The integrity tests by the membrane producer and by the user have been established. Althrough PLANOVA has already worked well in the field of drug manufacture, the combination of virus removal using PLANOVA and virus inactivation treatment gives potentially wide application.

INTRODUCTION

The new appearance of infectious particles such as viruses and prion proteins gives rise to new epidemics and generates social demands against these infections. The progress in biotechnology accelerates the development of novel bio-drugs and is socially acceptable only under free from the potential infections caused by the injection of the drugs. The guidelines or guidance for this strategy have been prepared in many countries such as the USA, Japan, and the European Community [1-3].

The strategy can be realized more completely by the development of new technology for reducing infectious based on well-established methods. The virus removal filter PLANOVA™, which is composed of the hollow fibre membrane BMM™ made of a cuprammonium regenerated cellulose, is able to remove particles effectively, mainly based on their size.

The characteristic feature of the virus removal technologies such as filtration, compared with inactivation technologies such as heat-treatment, is lack of degradation of co-existing proteins and lack of side effects due to residual virus after the inactivation. In general, viruses tend to aggregate and then have higher resistance to inactivation. In contrast to inactivation, filtration works more effectively for viral aggregates than for monodispersed virus.

The virus removal filter is expected to be more reliable than chromatography for the removal of viruses since: (i) the mechanism of removal is established; (ii) integrity testing is carried out, and (iii) usually the filter is used once. When chromatography is applied for virus removal then single use or alternatively the integrity test of regeneration should be developed in advance.

The virus removal filter applicable to the purification process of bio-drugs should be validatable for its ability to remove high levels of virus under conditions where the recovery rate of the protein in question is also high. Here, the validatable filter is defined as the one which shows reproductible and predictable removal of viruses under given filtration conditions [1]. A virion lacks the ability to replicate in the absence of its host cells and can be treated as a lifeless particle. Then the virus removability of a filter can be represented by the logarithmic virus rejection coefficient Φ defined by equation [2],

$$\Phi = \log_{10} N_0/N_f$$

where N_0 and N_f are the infectivity of the virus in the feed solution and filtrate respectively. The value of Φ is different from that of the logarithmic reduction value LRV defined for the sterilization filter since LRV reflects not only the filter but also the filtration system.

I will show the virus removability and protein permeability of PLANOVA under various filtration conditions and also show that PLANOVA is placed at the sublation between the screen type and the depth type filters from the supermolecular structure. The basic concept of the integrity test for the virus removal filter will be discussed.

Materials and Methods

Materials

PLANOVA filters with mean pore sizes of 15, 35, and 75 nm were prepared by the micro-phase separation method [3] from cellulose cuprammonium solution. The PLANOVA series filters are commercially available from Asahi Chemical Ind. Co. Ltd. To prepare gold particles, the reduction of $HAuCl_4$ aqueous solution [4] was used and the distribution of particle size determined by electron microscopy and/or optical spectrometry [5]. Viruses, proteins and DNA were prepared as described [6-8].

Methods

Filtration: Dead end filtration was used unless otherwise specified. To determine the worst case of filtration from the standpoint of virus removal: (i) transmembrane pressure ΔP; (ii) shear rate of the feed solution Υ; (iii) temperature T; (iv) concentration of viruses and/or the total amounts of the virus challenge dose; (v) concentration of proteins and/or the total amounts of the protein particles challenged were used as variables.

$TCID_{50}$, PFU, or animal tests were usually employed to determine virus infectivity. When these methods were impossible, the DNA or RNA concentration was investigated as an alternative. The concentrations of proteins and gold particles were determined by optical spectroscopy. The content of DNA and RNA was determined by spot hybridization and/or PCR. After filtration of the solution containing the viruses, the distribution of the viruses inside the membrane wall was observed using a transmission electron microscope [7].

Results and Discussion

Virus removability of PLANOVA with various mean pore size

Although most of the experimental data have been published [6-15], Table 1 shows typical values of virus removability. As a broad tendency, when the virus size increases, removability increases. On the other hand, when the pore size increases the removability decreases. These results indicate strongly that PLANOVA can remove viruses through the sieving mechanism.

When we plot the logarithmic value of the number of the particles caught at a given location in the membrane wall against the distance from the front surface, the result is a straight line with negative gradient for the cases of HIV [2], JEV [6] and

Table 1: Logarithmic rejection coefficients of microbes for various BMM filters with different mean pore size.

Microbe/medium	BMM15	BMM35	BMM40	BMM50	BMM75
HSV-1/FBS10%D-MEM	(> 6)	> 6	(> 6)	(> 6)	(> 6)
HIV-1/human plasma	(> 5)	> 5	> 5	> 5	> 5
HIV-1/FBS10%D-MEM	(> 6)	> 6	> 6	> 6	> 6
CMV/FBS10%D-MEM	(> 5)	> 5	> 5	> 5	> 5
Xenotrope/5%FBS	(> 8)	> 8	(> 8)	(> 8)	(> 8)
T-4/FBS10%D-MEM	(> 8)	> 8	> 8	7.8	6.5
VSV/FBS10%D-MEM	(> 12)	> 12	> 12	> 12	NT*
Sindbis/FBS10%D-MEM	(> 8)	> 8	> 8	NT*	NT*
SV40/5%FBS	> 8	6.5	(4.5)	(2.5)	(1.5)
DHBV/duck plasma	(> 6)	6	5	NT*	NT*
JEV/FBS10%D-MEM	> 7	5.5	4.5	2.3	1.8
HBV/human plasma	(> 6)	6	5	2.4	1.9
HCV/human plasma	> 4	4	3.5	(2)	1.5
Polio/FBS10%D-MEM	5	0.2	(0)	(0)	(0)
ΦX174/FBS10%D-MEM	〉 9	9	8	4	3
CJD/saline	〉 6	〉 6	4	3.5	2.2
A. laidlawii/FBS7%D-MEM	(〉 5)	〉 5	〉 5	〉 5	〉 5
M. orale/FBS7%D-MEM	(〉 5)	〉 5	〉 5	〉 5	〉 5
M. hyorhinis/FBS7%D-MEM	(〉 5)	〉 5	〉 5	〉 5	〉 5

*NT: not tested; ( ): estimated value.

the gold particle [5] within the region where the pore structure is regarded as uniform. This suggests that PLANOVA removes the particles through multi-step filtration.

Recovery rate of proteins

To attain a high recovery rate, the source material of a filter should be hydroscopic since the absoption of protein should be minimized. PLANOVA is made of cellulose (cuprammonium regenerated cellulose) and shows negligibly small adsorption of proteins [1] resulting in a high level of protein permeability ϕ [7, 8] defined as the ratio between the concentration of protein in the filtrate and that in the feed solution.

Figure 1 shows the molecular weight dependence of ϕ for BMM15 and BMM35, employed in PLANOVA 15 and PLANOVA 35, respectively. It should be noted that BMM shows good permeability of most proteins resulting in nearly complete recovery of proteins. In the case of DNA or larger proteins such as Factor VIII with von Willebrand factor, the value of ϕ is unity. These molecules can deform their shapes to the elongated hair-like shapes along the flow line in a pore [8]. A mycoplasma can also deform in a pore but viruses cannot [16]. Thus, the value of ϕ (= –log ϕ in the case of the monodispersed particles) depends not only on the size effect (sieving mechanism) but also the deformability of the particles.

Another mechanism for penetrant transport through a membrane is diffusion, the driving force of which is the concentration gradient. The particle size dependence of the diffusion coefficient D in water for PLANOVA 75 is shown in Figure 2 [17]. When the particle size increases the values of D decreases monotonously and the difference between the theoretical value calculated through the Stokes-Einstein equations and the experimental value increases with a decrease in the particle size.

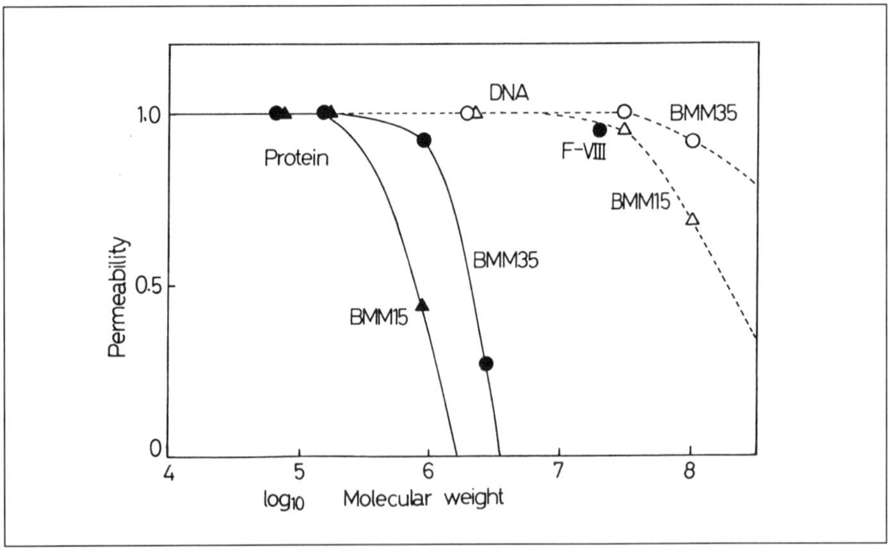

Fig. 1: Molecular weight dependence of permeability of protein and DNA for BMM15 and BMM35.

ERRATUM

Volume 88 of the series Developments in Biological Standardization: in the paper by Kozak, on p. 261, paragraph 3, the word «never» has been omitted. It should read:

(i) The first takes into account the country of origin of the bovine lungs. Lungs have never been sourced from the U.K. and have been obtained from countries free from native cases of BSE which allows a class factor score of 6, or a one-in-a-million chance of a sporadic case of BSE occurring. In addition, all the tissues came from inspected abattoirs and animals inspected by veterinarians.

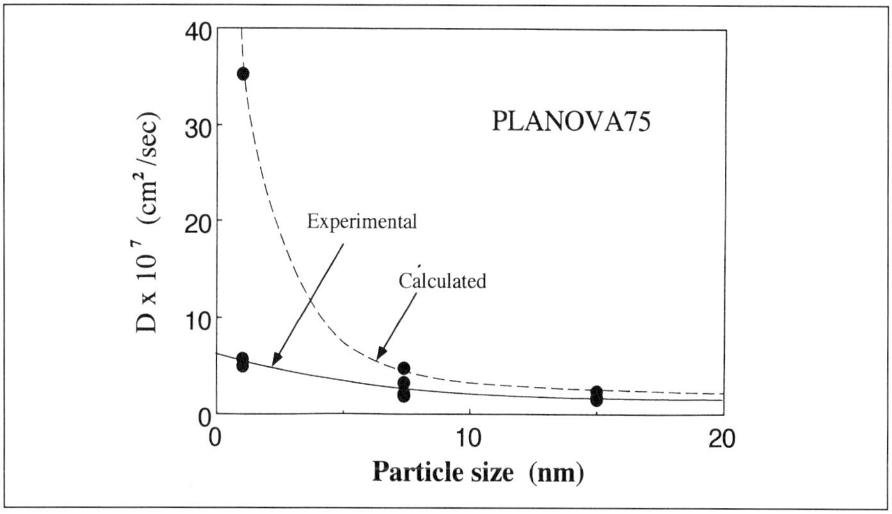

Fig. 2: Particle size dependence of diffusion coefficient D for PLANOVA 75: Broken line is obtained through Stokes-Einstein equation.

This indicates that the contribution of diffusion to the permeability of proteins is more than that expected theoretically.

Characteristics of supermolecular structure of PLANOVA

The specifications of PLANOVA have given in the pamphlet [15] and BMM employed in PLANOVA shows the low crystallinity and negligible small molecular chain orientation [18]. The former is important for good recovery rate of proteins and the latter for virus removability [5].

The high level of virus removability of PLANOVA is realized mainly by the unique pore structure of BMM. The detailed pore structure can be observed through the «negative image of pores» technique [4]. Figure 3 shows the schematic representation of the pore structure of BMM. Figure 3(a) represents the view observed from the inner wall of the hollow fibre and (b) is the cross-sectional view of the plane cut along the radial direction. Figure 3(c) is the model pore structure obtained by simplification of the negative image of pores. These features of the pore structure can be summarized as follows: (i) the membrane wall is composed of many layers; (ii) the pore in a layer is composed of two types of pores of capillary pore and void pore [19]; (iii) the size of the capillary pore ranges between 5 by and 40 nm, and that of the void pore between 50 and 100 nm. It has been confirmed by electronmicroscopy [19] and assessment of dynamic visco-elasticity in water and in air [18] that each layer has the characteristics of the screen filter.

Dependence of virus removability on filtration conditions

The various published experimental data [2-11, 13, 14] and the unpublished data by my collaborators are summarized qualitatively in Table 2. Althrough the mecan-

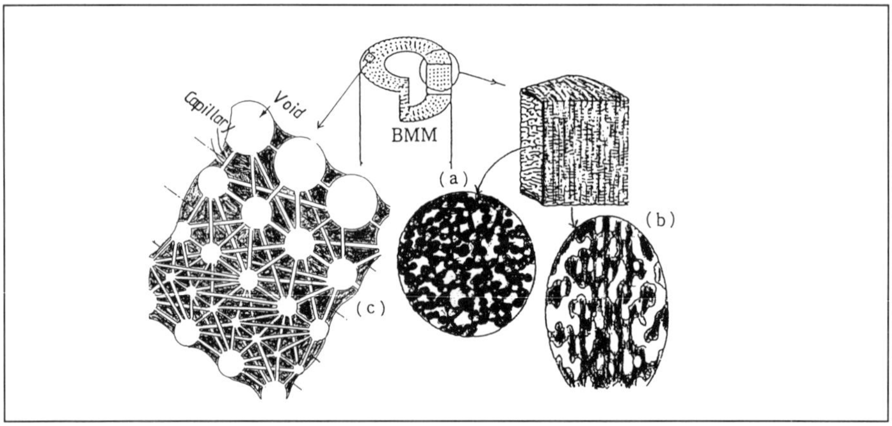

Fig. 3: Schematic representation of pore structure of BMM.

Table 2: Change in virus removability with filtration condition.

Condition	Change	Mechanism	
Transmembrane pressure ΔP	+[a]	Trapping	Clogging in void
Shear rate ϒ	+	Fractionation	Decrease in concentration of virus on surface
Virus			
concentration	–[b]	Trapping	Decrease in capacity of void
challenge dose	–	Trapping Plugging Diffusion	Decrease in capacity of void and capillary Increase in contribution of diffusion
Protein			
concentration	–	Plugging in smaller pore	Shifting flow line to line passing through larger pore
challenge dose	–	Trapping Plugging	Coexisting multimer works
Temperature	X[c]	—	Compensate effect of pore size increase with that of decrease in viscosity

[a] When the variable increases the removability increases,
[b] When the variable increases the removability decreases,
[c] The removability keeps constant independent with the change in the variable.

ism of some of these changes has been clarified. Further study of these mechanisms will be performed.

Based on the experimental results, the worst case of virus removability has been established as follows: temperature, 20°C; ΔP, 200 mmHg; ϒ, 0 sec^{-1} (dead end filtration); concentration of viruses, more than 10^6 $TCID_{50}$/ml; concentration of protein, 10% FBS or its corresponding proteins; filtration volume, 3 l/m^2 (standard). As a precaution, stop processing when the filtration rate decreases to one tenth of the original in the case of constant ΔP and stop when the value of ΔP exceeds 1 bar in the casse of constant filtration rate.

Integrity test

The membrane manufacturer has performed the virus removal validation test and the integrity test to confirm that all the filters show virus removability in excess of a specified value. In this case the filter is the fresh state and at least one test should be done by a non-destructive method.

When we design the validation test, the selection of the model virus is very important. Table 3 summarizes the ideal requirements for the model viruses. Experimental conditions are designed to minimize damage to the virus, that is, ΔP is lowered and the dead-end filtration is carried out. JEV and polio viruses are employed for PLANOVA 35 and PLANOVA 15, respectively.

The membrane user must design the process validation test by referring the worst case of the filter used. In general, when the dimer, trimer, and/or multimers of the proteins coexist, with the monomeric form, actual process fluids showed be used for the process validation test, since the change in the pore size distribution during filtration depends on the particle size distribution of the feed solution. The integrity test that may be employed by the user should be designed to be suitable from theoretical and practical viewpoints. For example, the integrity test before filtration should be the test that possesses sensitivity to detect accidental changes in the pore structure of fresh filters. On the other hand, after the filtration, the pore structure of the membrane changes from the fresh state. Since the complete regeneration of the filter may be impossible, we must apply an integrity test different from the test used with the fresh filter.

Table 3: Requisite for model viruses: Filter validation.

(1) Size; close to mean pore size
(2) Shape; spherical
(3) Assay method; infectivity
(4) Concentration; more than 10^6 $TCID_{50}$
(5) Aggregation; negligibly small (pass double step filtration test)
(6) Stability; heat resistant
shear stress resistant
(condition of $\Delta P \rightarrow 0, \gamma \rightarrow 0$)

Althrough the ideal integrity test after filtration may be performed using the particles having the same size as those of the viruses used, the process validation test (referred to as the absolute method), and the sensitivity of the detection of the particle, is not so high as expected from the virus removal [20]. The alternative proposed is the method where the particle size is smaller than that of the virus in question. When the assumption holds that all the pore size changes during the filtration are continuous and not discontinuous (such as generating extremely large pores, for example, pinholes), the interrelationship between virus removability and its size can be applied. In this case the high sensitivity of the detection of the particle is not necessary because the removability of the particle is lower than that of virus. This integrity test method is called the direct method. To confirm the validity of the above assumption, the leakage test or the corresponding test is essential.

More conveniently, we can employ an indirect method. The indirect method of the integrity test indicates that method is based on the correlation between virus removability and the value of the integrity test although this relationship is not causal. The relation holds empirically only in the case of the fresh filter under the same test condition used in establishing the empirical relation. An example of this method is the fluid permeability and bubble point method [21] (also termed the liquid forward flow method). As the sensitivity of the test is lower than the level of virus removability to be confirmed, the possibility of the existence of a large pore, such as a few μms greater, should be considered. It is clear that when we use the indirect method we must ascertain the following three points: (i) the larger pore does not generate during the filtration; (ii) the filter is generated after the filtration; (iii) the test is performed under the same scale and condition as those of the manufacturer.

Concluding Remarks

PLANOVA has been used in plasma products, such as coagulation factor VIII and IX, and ϒ-globulin, and the products have recieved governmental approval. It has been about four years since commercializing the drugs and no-problems have been reported so far. The virus removability in addition to the results of biochemical effect were reported by the user and the values are comparable or better than those reported by us [22]. Consequently, it can be concluded that the reliability of the virus removal filter has been established.

The practical application of PLANOVA is proceeding. To increase virus safety more effectively, the combination between virus removal and the other inactivation methods should be performed. For example, after filtration using PLANOVA, the heat treatment and/or ultraviolet treatment work more effectively than is the case with the reverse order [23].

References

1 Manabe S: Virus removal and inactivation in process validation, in Murakami H (ed): Basic of Applied Aspect. Kluwer Academic Publishers, Amsterdam, 1992, pp 15-30.

2 Manabe S, Tsurumi T, Ishikawa G, Satani M, Yamashiki S, Hamamoto Y, Kobayashi S, Yamamoto N, Yamaguchi K: Mechanism of human immunodeficiency virus (HIV) removal by regenerated cellulose hollow fiber (BMM). Membrane 1989;14:77-83.

3 Manabe S, Satani M: US patent 4'808'315, 1989.

4 Tsurumi T, Osawa N, Hirasaki T, Yamaguchi K, Manabe S: Mechanism of removing monodisperse gold suspension using cuprammonium regenerated cellulose hollow fiber (BMM hollow fiber). Polymer J 1990;22:304-311.

5 Tsurumi T, Sato T, Osawa N, Hitaka H, Hirasaki T, Yamaguchi K, Hamamoto Y, Manabe S, Yamashiki T, Yamamoto N: Structure and filtration performance of improved cuprammonium regenerated cellulose hollow fiber (improved BMM hollow fiber) for virus removal. Polymer J 1990;22:1085-1100.

6 Hirasaki T, Ishikawa G. Nakano H, Manabe S, Yamamoto N: Mechanism of virus removal by regenerated cellulose hollow fiber (BMM). Membrane 1995;20:135-142.

7 Hamamoto Y, Harada S, Kobayashi S, Yamamoto N, Yamaguchi Y, Iijima H, Manabe S, Tsurumi T, Aizawa H. Vox Sang 1989;56:230-236.

8 Hirasaki T, Sato T, Tsuboi T, Nakano H, Noda T, Kono A, Yamaguchi K, Imada K, Yamamoto N, Murakami H, Manabe S: Permeation mechanism of DNA molecule in solution through cuprammonium regenerated cellulose hollow fiber (BMM™). J Membrane Sci 1995;106:123-129.

9 Ishikawa G, Hirasaki T, Manabe S, Uematsu S, Yamamoto N: Novel determination method of size of virus in solution using cuprammonium regenerated cellulose membrane (BMM). Membrane 1991;16:376-386.

10 Yuasa T, Ishikawa G, Manabe S, Sekiguchi T, Miyamura T: The particle size of hepatitis C virus estimated by filtration through microporous regenerated cellulose fiber. J General Virology 1991;72:2021-2024.

11 Muchmore E, Manabe S, Ishikawa G, Hirasaki T, Alter H, Shih J, Yamaguchi K, Yamamoto N: Chimpanzee inoculation studies confirm removal of hepatitis C virus (HCV) infectivity by microporous membranes and estimate HCV particle size. The 8th Triennial Congress, May 10-13, 1993,Tokyo; Abstract 319b.

12 Tateishi J, Kitamoto T, Ishikawa G, Manabe S: Removal of causative agent of Creutzfeldt-Jakob disease (CJD) through membrane filtration methods. Membrane 1993;18:357-362.

13 Sekiguchi S, Ito K, Kobayashi M, Ikeda H, Tsurumi T, Ishikawa G, Manabe S, Satani M, Yamashiki T: Possibility of hepatitis B virus (HBV) removal from human plasma using regenerated cellulose hollow fiber (BMM). Membrane 1989;14:253-261.

14 Hirasaki T, Noda T, Nakano H, Ishizaki Y, Manabe S, Yamamoto N: Mechanism of removing Japanese encephalitis virus (JEV) and gold particles using cuprammonium regenerated cellulose hollow fiber (i-BMM or BMM) from aqueous solution containing protein. Polymer J 1994;26:1244-1256.

15 Asahi Chemical Ind. Co. Ldt., Commercial catalog «PLANOVA», 1993.

16 Togo I, Manabe S, Fukada S, Kono A: Removal of microorganisms from culture medium using porous cuprammonium regenerated cellulose hollow fiber (BMM). Saibou Baiyo 1991;17:12-25.

17 Fujioka R, Manabe S, Ryutou Y, Kinoshita Y: Diffusion of particle into cellulose solid. Annual meeting of the Society of Fiber Science and Technology, Japan 1995; to be submitted in Polymer J.

18 Fujioka R, Manabe S: Change in fine structure/physical properties of cellulose solid due to adsorption followed by desorption of solvent molecule. Polymer Preprints, Japan (eng. ed.) 1995;44:E243-E243.

19 Manabe S, Kamata Y, Iijima H, Kamide K: Some morphological characteristics of porous polymeric membranes prepared by micro-phase separation method. Polymer J 1987;9:391-404.

20 Noda T, Manabe S, Uragami T, Higuchi A, Sato T: Theoretical approach for integrity test of virus removal filter, in Murakami H (ed): Animal cell technology; Basic & Applied Aspects 1994;6:523-528.

21 Kamide K, Manade S: Characterization technique of straight-through membrane. ACS Symposium, Ultrafiltration Membrane and Application. Plenum Press 1980;173-204.

22 Burnouf-Radosevich M, Appouchaux P, Huart JJ, Burnouf T: Nanofiltration, new specific virus elimination method applied to high-purity Factor IX and Factor XI concentrates. Vox Sang 1994;67:132-138.

23 Tsuboi T, Yamamoto N: Combination of virus removal and inactivation for optimization of recovery date of protein, in Kobayashi T (ed): Animal Cell Technology; Basic and Applied Aspects, 1994, vol 6, pp 9-15.

Dr. S.-I. Manabe, Fukuoka Women's University, Kasumigaoka 1-1-1, Higashi-ku, Fukuoka 813, Japan.

Brown F, Lubiniecki AS: Viral Safety and Evaluation of Viral Clearance from Biopharmaceutical Products. Dev Biol Stand. Basel, Karger, 1996, vol 88, pp 91-98

Evaluation of Anti-Viral Filters

B. Hughes, A. Bradburne, A. Sheppard, D. Young

Glaxo-Wellcome Biotechnology, Beckenham, Kent, UK

Key words: Anti-viral filters.

Abstract: The purification processes used to prepare medicines from human and animal cell lines provide one of the major opportunities for virus risk reduction. The purification process can be validated for its ability to remove a range of viruses and can be shown to provide a significant barrier to the transmission of viruses from cell substrates and raw materials to the patient. Ultrafiltration membranes which are manufactured with extremely tight control on pore size distribution can remove a broad spectrum of viruses with high efficiency. A scheme is described for the evaluation of anti-viral filters and the results from evaluation of two filters are presented. The experiments carried out with ultrafiltration membranes suggest that they can provide significant additional virus risk reduction factors for biopharmaceuticals.

Virus Risk Reduction

Biotechnology has made a significant contribution to human health by providing medicinal products derived from living cells. Some of these medicines have advantages over small molecule pharmaceuticals or may indeed represent the only therapeutic intervention for a particular disease.

Along with the considerable advantages that biopharmaceuticals bring to medicine, there are a number of safety features of the production process that are unique to biotechnology products and require specific consideration. One major area for assessment is the risk of transmission of viral agents from the cell substrate or raw materials, to the patient. In recognition of these risks, and with the examples of transmission in some biological products, manufacturers of biotechnology products together with the regulatory agencies have developed strategies for virus risk reduction which have allowed further safe exploitation of biotechnology for the production of medicinal products.

Virus risk reduction is achieved by application of three major approaches:

- Qualification of cell banks and raw materials
- Design and validation of processes for viral load reduction
- Appropriate testing schemes

This three-pronged approach has been adopted because theoretical considerations and practical experience have shown that no single approach is sufficient to provide an adequate reassurance of risk reduction. Virus assay methodology is too narrow in scope and often too low in sensitivity to detect all possible contaminants in cell banks and raw materials. Process validation is heavily dependent on model systems which may not always give data which fully predict performance at full scale. The information provided by testing schemes is limited by the choice of sampling point and scope and sensitivity of the assay used. This integrated approach of testing and validation is now well established and has contributed to the history of safe use of highly purified biopharmaceuticals.

Role of Process Validation

The design of the process by which a molecule is produced and purified is one point where virus risk reduction can be considered prospectively. This offers the opportunity for planned intervention in which significant barriers to virus contamination can be erected. Processing steps have long been exploited for their intrinsic anti-viral activity. Elevated temperature, extremes of pH, and organic solvents which are used in the purification of proteins can contribute to the inactivation of viruses in process streams. Manufacturers who were aware of the limitations of testing methods assessed the capacity of these steps to inactivate viruses in model systems and showed that significant risk reduction could be achieved during processing. As more experience was obtained in these systems, regulatory authorities and manufacturers developed their understanding of process validation studies and recognised some of the limitations of the data derived from these experiments. The importance of validation of scaled-down processes and the careful preparation of virus spikes was understood [1, 2]. The requirement for inclusion of a number of steps which operate using dissimilar mechanisms of inactivation and removal became clear as well as the need to establish the fate of the spiked virus in order to understand the process fully. Generation of high quality data which address these issues can be complicated with chemical and heat inactivation because of the considerable difficulty in auditing the fate of the added virus spike. These technical difficulties can complicate the application of the validation experiments to full scale. However, heat and chemical inactivation studies have been widely used and there are well established technologies for treating biological products for the inactivation of a wide range of viruses. These treatments are often key steps in the generation of virus risk reduction factors.

Anti-Viral Filters

The inactivation steps have recently been complemented by the advent of anti-viral filters. These filters were specifically developed for separation of virus from protein products and represent a significant change in the design, manufacture and control of membrane porosity. Anti-viral filters bring a completely new mechanism of virus risk reduction and can be fitted into processes to generate large log reduction values (LRV) for a range of viruses. For non-adsorbent filters, there is the potential to audit the fate of added virus spikes which can add confidence to the

LRV obtained during validation studies. Some anti-viral filters operate in a predictable manner showing reproducible size-dependent clearance for a range of virus types indicating that virus size is the key factor influencing clearance.

Evaluation of Anti-Viral Filters

A range of filtration systems has recently become available. These systems have different modes of action, different porosity and require specific installation approaches. To select a system which best fits a particular process, we use a set of selection criteria to evaluate virus-removing filters. Selection criteria are defined by Production staff as well as R&D staff. This approach allows for focussed development work to ensure selection of appropriate systems which can be fitted into routine manufacturing environments.

Selection criteria include:

- LRV data from manufacturers for selected virus types
  LRV required would be influenced by cell substrate, raw materials and LRV from the remainder of the process
- Product recovery through the process
  This will be affected by product molecular weight, concentration and behaviour within the system
- System compatibility
  Filtration systems are required to be compatible with the total process and fit within the manufacturing schedules
- Validatable post-use integrity test
  Confidence in the role of the filter in reducing virus risk can only be gained by adequate confirmation of integrity after each run
- System validation
  The system must be capable of validation at full scale or using validated scaled-down models and demonstrate effectiveness against a range of viruses

In addition to the technical criteria outlined above there are requirements for security of supply, maturity of product and its production process, and the availability of expert support from suppliers.

Millipore Viresolve™ System

The Millipore Viresolve™ system was evaluated by the criteria which were previously established. The Viresolve 70™ system is a composite membrane composed of a pre-formed microporous membrane plus a thin, asymmetric, finely porous retentive layer with unique particle retention characteristics. This membrane achieves extremely effective control of pore-size distribution which can achieve significant LRV even for small, non-enveloped viruses. The protein-sieving characteristics of the membrane approximate to those of a conventional 100,000 Dalton cut-off membrane. The system operates in tangential flow mode with defined and controlled cross-flow and permeate rates. The manufacturers provided data in support of a claimed-virus retention capability of 3 to 8 logs across a size range of 28 nm to 93 nm [3]. The system was claimed to perform in a predictable manner with monotonic increases in LRV across this size range. Significantly, it was claimed that LRV data derived from buffer alone gave worst case conditions and addition of protein always resulted in increased LRV. This implied that validation data performed in buffer alone would underestimate the LRV obtained during routine operation.

Approach to Validation

The Viresolve 70™ met a number of our criteria and was evaluated further. Virus-retention data provided by the manufacturers was acceptable for a range of viruses and product recovery exceeded 95% in early trials. These trials were carried out using conditions which allowed for the complete operation to be completed within the established manufacturing schedules. A pilot-scale system was set up to evaluate the operating parameters of Viresolve 70™ for a process intermediate containing a mixture of proteins with maximum molecular weight of 30,000 Daltons and to establish the LRV for a selection of viruses in our own laboratories. The objective of the validation exercise was to verify the manufacturer's claims for size-dependent LRV and for worst case removal from buffer solutions.

The scale of the production operation would require a one square foot module. This system would not be suitable for virus-spiking studies because of the difficulty in achieving sufficiently high volumes and titres of virus stocks and a 1/3 square foot system was selected for these studies. Conditions of cross-flow and permeate flow were adjusted to establish conditions which emulated the full-scale operation.

Validation: Selection of Viruses

A range of viruses were chosen which covered a wide size range and which included enveloped as well as non-enveloped viruses (Table 1). Viruses which could be grown to high titres without significant processing were chosen over lower titre examples of the same family. Manipulations such as centrifugation, which might be required to generate high titre spikes, might provide high LRV data which would have artefactual components provided by removal of aggregates created during

Table 1: Selection of viruses for evaluation of Viresolve™ membranes

Virus	Diameter (nm)	Enveloped	Assay substrate
ϕX-174	28	N	*E. coli*
EMCV	32	N	MouseL929
Sindbis	54	Y	BHK21
ϕ6	75	Y	*P. syringae*
Maedi-Visna	85	Y	Ovine SCP

ϕX-174 and ϕ6 are bacteriophages.
EMCV is mouse encephalomyocarditis virus.

sample preparation and analysis. Selection was also biased in favour of viruses for which convenient assay systems were available. An additional factor in these experiments was the requirement to choose virus/assay combinations which were resistant to the effects of high titres of interferon, which was the protein in the purification scheme under investigation.

Maedi-Visna virus was the only retrovirus available which could be assayed in the presence of human interferon. The sheep cells in the in vitro assay were less sensitive to interferon than any other substrate which was examined. ϕX-174 bacteriophage was added to all mammalian virus spikes and provided accurate clearance data even when the mammalian virus spike was completely removed by the filter.

Virus Removal Data

Experiments were carried out in buffer before and after addition of protein. In this way, data could be obtained to support the manufacturer's claim that worst-case virus clearance was obtained in the absence of added protein. The first set of experiments was carried out in buffer only. Total protein in the system, contributed by tissue culture protein from the virus spike, was determined to be in the region of 30 μg/ml. Table 2 shows data from passage through a single module (experiments were also carried out with two filters in series). The experiment demonstrates the ability of the system to discriminate between ϕX-174 and encephalomyocarditis virus (EMCV) which have diameters of 28 nm and 32 nm respectively. There is an increase in LRV with particle diameter but the magnitude of this increase is obscured by the limitations in the LRV data from the larger viruses, Sindbis and Maedi-Visna. All infectivity was retained from these two spikes and therefore the starting titre defined the maximum that could be reported. These experiments did however provide evidence that retention was related to size and the reproducible difference between ϕX-174 and EMCV suggested that this was sensitive to relatively small differences in diameter, at least in this size range. The differential filtration of ϕX-174 was monitored in these experiments when complete retention of the mammalian spike virus was achieved. The LRV for phage was maintained, providing strong evidence of the stability of the system during extended operation.

The next experiments were carried out in the presence of protein, to investigate the manufacturer's claims that LRV data from buffer was worst case and higher

Table 2: LRV data from buffer alone.

Virus	Diameter (nm)	LRV
ϕX-174	28	3.0
EMCV	32	3.6
Sindbis	54	*5.7
Maedi-Visna	85	*4.5

* LRV values marked with an asterisk indicate experiments where total retention of the virus spike occurred (Tables 2-4). Values were computed by estimating minimum detection levels in filtrate at 95% confidence limits.

LRV would be obtained in the presence of product. The experimental approach involved repeating the work with ϕX-174, EMCV, Sindbis and Maedi-Visna but adding the product protein from the process step which was identified for location of the filter. This corresponds to approximately 150 μg/ml of interferon protein. The data from these experiments are shown in Table 3. There is a clear and reproducible elevation in LRV for ϕX-174 and EMCV. The discrimination between these two viruses is maintained in the experiment with added protein. The LRV data from Sindbis and Maedi-Visna viruses are limited by input titre in this experiment also and the reported LRV for Maedi-Visna is further depressed by the effect of interferon on assay sensitivity.

Integrity Testing

All Viresolve™ membranes are supplied with values from a system-integrity test, referred to as the Corr-Test™ [3]. This is a modified pressure test developed by the manufacturers. The system uses two immiscible solutions to lower the pressure requirements for the test. An intrusion value (ITV) is obtained for the ratio between flow through all pores in the membrane and those above 28 nm (the diameter of ϕX-174). During validation of the system the manufacturers tested a series of membrane lots with varying porosity. From these data a relationship was drawn between ITV and ϕX-174 LRV. Therefore the intrusion value can be transformed to a Corr-Test™ value which describes the retention of ϕX-174 by the membrane. All membranes are tested by the manufacturer before despatch and by the user after completion of the filtration operation. Membranes are washed to remove any protein build-up before the membrane is tested post-use. Thus, it is possible to assess the integrity of the membrane after the run has finished.

Viresolve 180™

Consideration was also given to the use of virus removing filters in monoclonal antibody processes. The Viresolve 180™ system was chosen because of the higher molecular weight cut-off required for immunoglobulins. Adequate recovery of immunoglobulin protein (> 90%) was demonstrated in early trials. Flux rates were further optimised to establish stable, reproducible recoveries. Experiments were also carried out to determine optimum protein concentration and to identify appropriate points in the process for location of the filters. Acceptable data were obtained

Table 3: Effect of protein on LRV.

Virus	Diameter (nm)	LRV	LRV increase
ϕX-174	28	4	1
EMCV	32	4.4	0.8
Sindbis	54	*5.7	–
Maedi-Visna	85	*3.5	–

for solutions of 2.5 mg/ml IgG and up to 5 mg/ml with later model « high flux » membranes. The filters were installed immediately upstream of a concentration step so that volume reduction could be conveniently achieved after the Viresolve™ step.

The virus retention properties were also investigated. In these experiments, the enveloped bacteriophage ϕ6, and a high titre preparation of Sindbis virus, were used in addition to EMCV and ϕX-174. The LRV data are presented in Table 4. The discrimination between the smaller non-enveloped viruses is lost, as expected from a membrane with these sieving characteristics. However, the retention of Sindbis virus was extremely effective (this is an actual value as virus was detected in the filtrate), and all the added ϕ6 was removed. These evaluations suggest that this membrane system will provide significant removal of the larger, enveloped viruses including retroviruses. Further evaluation of this system focussed on maximising throughput and recovery of product to optimise process compatibility in the manufacturing environment. The availability of « high flux » membranes, capable of higher throughput has improved these aspects of filter performance.

Other Virus Retaining Filters

A number of other ultrafiltration devices were examined in addition to the tangential flow systems described earlier. These filters used the dead-end or tortuous-path mechanism for retention of viruses. The systems which were evaluated were closer to the Viresolve 180™ membrane than the Viresolve 70™ in terms of virus filtration abilities. The systems provided very effective removal of larger enveloped viruses such as Sindbis virus and ϕ6 with LRV similar to the values obtained with the larger pore size Viresolve™ unit. Evaluation of the dead-end systems included a determination of the role of adsorption in the mode of action. Extensive contribution of adsorption to the LRV could affect the predictability of the filters and require extensive validation for each new application. A major advantage of a dead-end filter is the ability to use the modules in convenient cartridge format for direct in-line insertion into process flow.

Summary

An evaluation scheme has been established for the selection of virus-removal filters. The scheme has been applied to the selection of filters for two biotechnology processes. In one process, for purification of a mixture of interferons, a tangential

Table 4: Viresolve 180™ and immunoglobulin.

Virus	Diameter (nm)	LRV
ϕX-174	28	3.05
EMCV	32	3.1
Sindbis	54	7.9
ϕ6	75	*6.6

flow membrane provides reproducible removal of a range of viruses ranging in size from a 28 nm phage to retroviruses of 85 nm diameter. A larger pore size tangential flow membrane was used in the purification process for an immunoglobulin preparation and provided an additional removal capacity of approximately eight logs for enveloped viruses in that process. Other filters operating in flow-through mode also removed high titres of enveloped viruses. The tangential flow systems operate in a predictable manner as LRV increases with virus diameter. These systems can be validated and provide minimum removal factors from buffer solutions ensuring that data from validation experiments will underestimate the removal values obtained in the presence of protein product. Careful selection of anti-viral filters, involving manufacturing teams as well as R&D staff, ensure that membranes will be chosen for process compatibility as well as for anti-viral activity. The availability of these membrane systems provides an opportunity to build powerful anti-viral activity into processes and contribute to overall virus risk reduction for biopharmaceuticals.

Acknowledgements

The Maedi-Visna virus preparation and associated assays were carried out under contract by Dr. Alasdair Shepherd at Q-One Biotech Ltd., Glasgow, Scotland.

References

1 Validation of virus removal and inactivation procedures. (Doc III/8115/89EN). Commission of the European Communities, Brussels, 1991.

2 Points to Consider in the characterisation of cell lines used to produce biologicals. Center for Biologics Evaluation and Research, FDA, 1993.

3 DiLeo AJ, Allegrezza AE Jr, Builder SE: High resolution removal of virus from protein solutions using a membrane of unique structure. Biotechnology 1992;10:182-188.

Viresolve™ and Corr-Test™ are trademarks of the Millipore Corporation.

Dr. B. Hughes, Glaxo-Wellcome Biotechnology, Langley Court, South Eden Park Road, Beckenham, Kent BR3 3BS, U.K.

Brown F, Lubiniecki AS: Viral Safety and Evaluation of Viral Clearance from Biopharmaceutical Products. Dev Biol Stand. Basel, Karger, 1996, vol 88, pp 99-108

Process Scale Considerations in Evaluation Studies and Scale-Up

J.K. Walter, W. Werz, W. Berthold

Process Development, Department Biotech Production, Dr. Karl Thomae GmbH, Biberach, Germany

Key words: Process scale considerations.

Abstract: This paper focusses on the scale of validation studies that are performed in order to demonstrate viral clearance in downstream processing of biopharmaceutical products. A serious concern for rDNA-derived proteins from recombinant mammalian cell cultures and monoclonal antibodies (MAbs) derived from hybridoma cultures is the potential risk from contaminating retroviral particles or adventitious viruses. Accordingly the downstream process has to be designed to reduce a potential virus load significantly: by virus removal and by virus inactivation methods. Such viral clearance is essential for drug safety – and has to be validated. This means that the design of downstream processing has to take into consideration both the capability for viral clearance and the capability for validation respectively.

INTRODUCTION

Downstream processing for the purification and recovery of a pharmaceutical protein on an industrial scale is based on two major separation techniques, chromatography and filtration, which contribute to viral clearance mainly by virus removal; significant support towards drug safety is given by additional viral inactivation methods that are implemented in the downstream processing [1, 2, 3].

A number of proteins are produced on a small scale for several reasons: either they are in an investigational status, e.g. proof of concept at the clinical stage, or as a marketed product with a minor request for high amounts of product. e.g. hormone- or cytokine-type proteins which are highly effective even in a microgram dose.

Typically, any process validation has to be performed using the original equipment located in the prospective environment of the production facility. The equipment should be identical in order to reduce the number of theoretical possibilities that would make the results of the validation work unrepresentative of the process itself; at least the equipment has to be of identical type and of the same size as the commercial scale and must be operated exactly as intended for use in the manufacturing process.

An exception to this stringent rule is the validation of the downstream processing with respect to its capability for virus inactivation and virus removal: the deliberate contamination of the designated manufacturing equipment and facility with different infectious viruses – *some of them might even be human pathogens* – would discriminate against any production because of unacceptable risk factors for the final product. An exception to this exception might be the use of disposable manufacturing equipment or at least the use of disposable parts of the manufacturing equipment that come into direct contact with the product itself.

An alternative for small-scale production can be the use of a replica of the manufacturing equipment. For batch sizes up to several thousand litres it is impossible or at least extremely difficult from a practical point of view to provide the considerable amount of virus that would be required for the corresponding scale validation. An appropriate solution to this difficulty is a truly linear down-scaled model of the unit operation in question provided the process parameters can be kept representative for the manufacturing scale. Both strategies have found acceptance with regard to assuring product safety.

Rationale for the Performance of Validation Studies on Viral Clearance using Downscaled Process Procedures

Relation between manufacturing scale and scale of validation

Before starting the validation studies it has to be considered first what scale of manufacturing is conceivable, and what conclusions for validation have to be drawn. Respective scale considerations are presented in Table 1, where the indicated volumes represent the order of magnitude and are not necessarily absolute numbers.

Table 1: Validation of viral clearance: scale considerations.

Volume of process fluid	Scale of validation	Equipment of validation
< 10 l	**Identical** Identical	**Replica** Identical
< **10 l**	**Downscale** Identical Identical	**Downscale** Replica Identical

Technical feasibility and cost assessment

A typical design for a downstream process is given in Figure 1. Viral clearance should be performed around the capture steps of the process to avoid any potential virus burden for the final purification step. With regard to technical feasibility and cost assessment the fluid volumes that have to be handled are critical. An example for the calculation of a validation study at a manufacturing scale of 1000 L fermentation is given in Tables 2a-5, asssuming a *midstream process volume of 100 L* which is equal to 10% of the harvest volume (Table 1).

Downstream process	Viral clearance	Fluid volume %	Product concentration
Cell-free culture fluid		100	0.2 mg/ml
▼ Capture	› Chromatography	100 – 10	0.2 – 2 mg/ml
▼	› Virus inactivation		
	› Filtration		
Midstream	› Chromatography	10 – 2	2 – 10 mg/ml
▼ Polishing		1 – 0.2	20 – 100 mg/ml
▼ Filling	Total viral clearance 12 – 15 log	20 – 0.2	1 – 100 mg/ml

Fig. 1: Design of downstream processing

Table 2a: Virus inactivation: technical feasibility.

Downscale factor	**1**	**100**	**1000**
Buffer toxicity Carrier buffer	**Cost neutral**		
Virus titration Kinetic study	**Cost neutral**		
Amount of product required for one validation experiment			
volume of product solution **number of fermentation batches**	100 l **1**	1 l **0.01**	0.1 l **0.001**
Amount of virus required for one validation experiment (titre ≅ 10^7)			
virus spike **volume of virus spike**	 **5 – 10 l**	1:20 – 1:10 **0.005 – 0.1 l**	 **0.005 – 0.01 l**
Personnel/laboratory required for one validation experiment			
relation factor	**1**	**0.7**	**0.7**

Table 2b: Virus inactivation: a cost assessment.

Scale of manufacture (Harvest)	1000 l (200 mg/l)		
Process volume (Midstream)	100 l		
Downscale factor	1	100	1000
	Approx. value in US $		
Cytotoxicity of buffer (350 $)	350		
Virus titration (Kinetic Study)	10.000		
Cost of product (1000 $/g)	200.000	2.000	200
Cost of virus (100 $/ml)	500.000	5.000	500
Personnel/Laboratory (1000 $/d)	1.000	700	700
Total cost	711.350	18.050	11.750

Table 2a is a matrix that allows the precise calculation of a validation study on virus inactivation; a respective cost assessment is given in Table 2b.

A significant cost reduction might be achieved for those process steps that have been demonstrated in pre-validation studies or comparable validation experiments to result in a considerably lower viral clearance than the applied virus load, i.e. a virus removal step with a 4-log viral clearance must not be spiked 1:10 using a virus titre of 10^7 viral particles/mL. A 1:100 spike still fits the validation requirements and results in a 90% cost reduction for the virus load.

Table 3a: Virus removal by tangential flow ultrafiltration: technical feasibility.

Downscale Factor	**1**	**100**	**1000**
Buffer toxicity Equilibration		**Cost neutral**	
Virus titration Feed Stock/Load/Permeate/Retentate/ (Regeneration)		**Cost neutral**	
Amount of product required for one validation experiment			
volume of product solution **number of fermentation batches**	100 l **1**	1 l **0.01**	0.1 l **0.001**
Amount of virus required for one validation experiment (titre ≅ 10^7)			
virus spike **volume of virus spike**	 **5 – 10 l**	1:20 – 1:10 **0.05 – 0.1 l**	 **0.005 – 0.01 l**
Personnel/laboratory required for one validation experiment			
relation factor	**1**	**0.7**	**0.7**

Table 3b: Virus removal by tangential flow ultrafiltration: a cost assessment.

Scale of manufacture (Harvest)		1000 l (200 mg/l)	
Process volume (Midstream)		100 l	
Downscale factor	1	100	1000
		Approx. value in US $	
Cytotoxicity of buffer (350 $)		350	
Virus titration of samples (1700 $)		8.500	
Cost of product (1000 $/g)	200.000	2.000	200
Cost of virus (100 $/ml)	500.000	5.000	500
Personnel/Laboratory (1000 $/d)	2.000	1.400	1.400
Total cost	710.850	17.250	10.950

Table 4a: Virus removal by chromatography: technical feasibility.

Downscale factor	1	100	1000
Buffer toxicity Equilibration/Wash/Elution/(Regeneration)		**Cost neutral**	
Virus titration Feed stock/Load/Flowthru/Wash/Elution/(Regeneration)		**Cost neutral**	
Amount of produt required for one validation experiment			
volume of product solution **number of fermentation batches**	100 l **1**	1 l **0.01**	0.1 l **0.001**
Amount of virus required for one validation experiment (titre ≅ 10^7)			
virus spike **volume of virus spike**	 **5 – 10 l**	1:20 – 1:10 **0.05 – 0.1 l**	 **0.005 – 0.01 l**
Personnel/Laboratory required for one validation experiment			
relation factor	**1**	**0.3**	**0.3**

Table 4b: Virus removal by chromatography: a cost assessment.

Scale of manufacture (Harvest)		1000 l (200 mg/l)	
Process volume (Midstream)		100 l	
Downscale factor	1	100	1000
		Approx. value in US $	
Cytotoxicity of buffer (350 $)		1.400	
Virus titration of samples (1700 $)		10.200	
Cost of product (1000 $/g)	200.000	2.000	200
Cost of virus (100 $/ml)	500.000	5.000	500
Personnel/Laboratory (1000 $/d)	3.000	1.000	1.000
Total cost	714.600	19.600	13.300

Matrices for a cost assessment of studies regarding ultrafiltration or chromatography are analogous to the matrix in Table 1a.

Table 5 summarizes the assessments of Table 2-4 and gives an estimate of the average costs related to a duplicate study on viral clearance for *one* virus as well as the benefit of appropriate down-scaled unit operations.

Table 5: Viral clearance: a cost assessment for a duplicate study.

Scale of manufacture (Harvest)	1000 l (200 mg/l)		
Process volume (Midstream)	100 l		
Downscale factor	1	100	1000
	Approx. value in US $		
Inactivation 1/Duplicate	711.350 711.000	18.050 17.700	11.750 11.400
Inactivation 2/Duplicate	711.350 711.000	18.050 17.700	11.750 11.400
Filtration TFUF/Duplicate	710.850 710.500	17.250 16.900	10.950 10.600
Filtration DEF/Duplicate	710.850 710.500	17.250 16.900	10.950 10.600
Chromatography/Duplicate	714.600 713.200	19.600 18.200	13.300 11.900
Total cost	**7.115.200**	**177.600**	**114.600**
Cost reduction factor	1	40	62

Selection of model viruses

The cost assessments clearly indicate that the choice of (model) virus is viral in the economy of validation studies (Table 6). The choice of appropriate model viruses to be used in such validation studies is taken with respect to their individual features which have to cover a broad range of virus species exhibiting different biological, physical and physico-chemical properties. To fit the requirements for an economic downstream processing, i.e. a design of purification procedures reduced in number but enhanced in selectivity and effectiveness, the model virus should be accessible in high titres and detectable in a reproducible and sensitive assay; otherwise high reduction factors of viral clearance for the individual unit operations cannot be demonstrated, and the accumulated virus reduction factor for the complete downstream processing might fail to ensure the required safety assessments.

Table 6: Selection of model viruses.

Virus features	**Feasibility**
– Size of virus – Envelope – Structure and strandedness of genome –Resistance to inactivation	– Relevance for production – Archievable high titre – High sensitivity of detection – Ease of detection

TECHNICAL CONSIDERATIONS FOR THE DOWNSCALING OF UNIT OPERATIONS FOR VALIDATION PURPOSES

Virus inactivation

Virus inactivation can be achieved using harsh environmental conditions such as:

- *high temperatures*, e.g. incubation at 60°C for several hours or ultrashort microwave treatment at temperatures of 85°C or above [4];
- *extreme pH-values*, e.g. incubation at pH 2-4 for 1-2 hours;
- *addition of organic solvents, detergents or chaotropic salts* to dissolve the virus envelope and/or disintegrate the nucleocapsid; or
- *irradiation* by UV-light or γ-irradiation to destroy the virus genome.

The stability of the pharmaceutical protein product determines the inactivation methods that can be applied.

Usually the downscaling of inactivation procedures is feasible assuming that key parameters such as product quality and physical/chemical conditions are strictly followed and monitored. The time course of mixing as well as the method of adjusting the inactivating conditions should be comparable to the original process. To be able to claim for a time range for the virus inactivation, the validation of the inactivation procedure is performed in a kinetic study: samples of the virus-spiked product solution are taken at time intervals to demonstrate the course and slope of inactivation.

Thermal virus inactivation by ultrashort microwave heating

Heat inactivation of virus by using microwave energy is a rare example for the opportunity to run the validation study in the original process equipment, i.e. even the identical equipment and not necessarily a replica, at the manufacturing scale even at highest volumes:

The product solution is flushed in a single path through a disposable coil, which is exposed to the microwave energy; the energy is kept constant and determines the peak temperature while the applied flow rate determines the exposure time which is typically in the range of 2 milliseconds [5, 6]. To save expensive product and virus during the validation study, the required physical conditions – flowrate and temperature – are adjusted by using a carrier solution, which is a buffer identical to the product-containing buffer.

Thereafter a volume of 25 ml of virus-spiked product solution is applied through a shunt and is exposed to the microwave energy under conditions identical to the manufacturing process. After passing the coil, the sample is collected in three fractions: due to dilution the first and the third fractions of approximately 5 ml each are discarded while the middle fraction is used for virus titration.

With regard to the time range of milliseconds, a kinetic study is not feasible, but instead a complete set of samples at around the desired inactivation temperature is analysed to determine the limits of viral inactivation as well as the product integrity. Microwave heat treatment might be of additional use regarding protective measure-

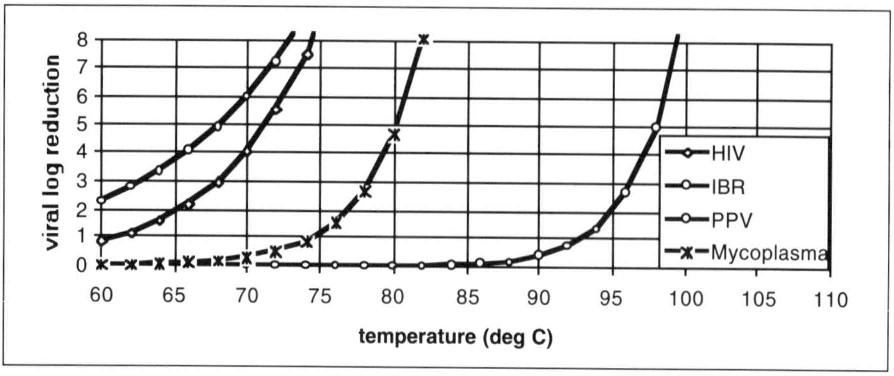

Fig. 2: Inactivation of different viruses.

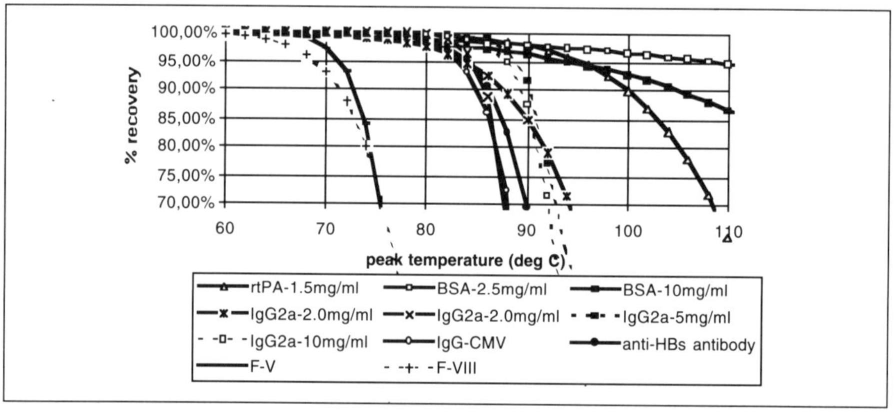

Fig. 3: Stability of proteins.

ments, e.g. the treatment of raw materials such as culture media or their compounds.

Virus removal by filtration

Membrane filtration featuring selective pore sizes is used with regard to the size of the product molecule: an appropriate cut-off of a selected membrane permits the separation of the product from virus particles, e.g. tangential flow filtration through ultrafiltration membranes or dead-end filtration through filter cartridges.

Validation criteria for filtration steps are the membrane chemistry and pore size, geometry of the holder and the membranes, geometry, velocity and pressure of the solution streams and transmembrane pressure, as well as product load per area with respect to volume and product concentration.

Attention has to be given to process fluids with respect to fouling effects to ensure that the viral clearance is due to the membrane's cut-off and not due to a gel layer. In addition, the pumping device (peristaltic, rotary lobe, gear pump, etc.) has to be investigated regarding the shear forces that potentially inactivate the spiked virus.

Virus removal by chromatography

Chromatographic methods exploiting different physico-chemical features of the protein product are designed and optimised for the separation of contaminating proteins as well as for the removal of contaminants such as viruses. The nature of viruses as nucleocapsids or enveloped viruses with a number of surface proteins as an integral part of the envelope makes them accessible to a different separation.

Table 7: Evaluation of downscaled chromatography.

Proof of product quality	Reproducibility of process data
Before and after the chromatographic separation	– yield – capacity – selectivity – resolution – profiles – linear flow rate – pressure – pH-value – conductivity – product load respective matrix volume

Column chromatography using down-scaled equipment requires careful calculation for the process parameters. Chromatographic parameters such as matrix chemistry, bed height of the column, linear flow rate and product load in correlation to matrix volume have to be identical. Another serious concern is the performance of the chromatographic matrix over time – its lifetime.

It is evident that in any case the process buffers, which are used for equilibration, the separation process itself and regeneration processes are identical to those used in the manufacturing process. Contaminating proteins of the virus-spike solution might affect competition and displacement and might have an influence on the chromatographic profile: the separation performance of the column has to be investigated accordingly.

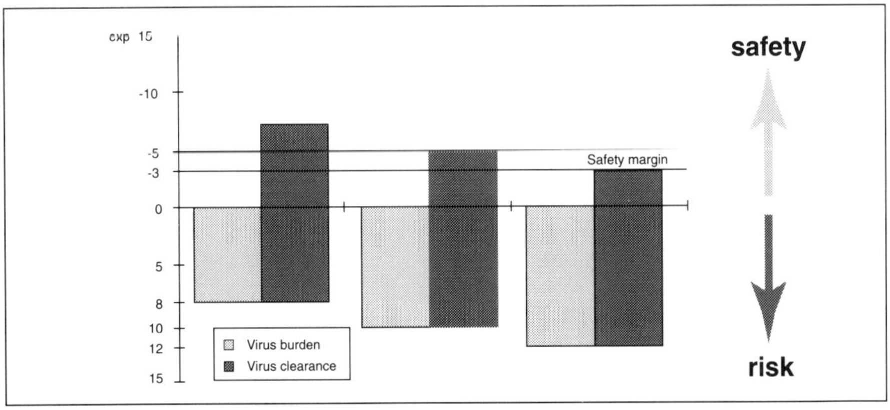

Fig. 4: Virus safety of biotechnological drugs.

Final Remarks

The overall viral clearance is the sum of the individual reduction factors for designated purification steps which have been validated for their capacity for virus inactivation and virus removal. This total clearance factor determines the safety margin of the downstream process.

The maximal potential virus contamination in a fermenter batch should be related to the final dose. The safety margin for virus clearance should be in the range of 10^3 to 10^5 compared to the maximal potential virus contamination. With these or even higher figures of safety derived from reasonably and carefully downscaled unit operations for the recovery of a pharmaceutical protein product, the virus safety can be demonstrated.

Acknowledgement

The data on microwave heat treatment have been elaborated in collaboration with Dr. Stanley Charm and Hossein Zarrin, Charm Bioengineering, Inc., Malden, MA, USA.

References

1 Berthold W, Walter JK, Werz W: Experimental approaches to guarantee minimal risk of potential virus in purified monoclonal antibodies. Cytotechnology 1992;9:189-201.

2 Walter JK, Werz W, Berthold W: Virus removal and inactivation concept and data for process Validation of downstream processing. BFE 1992;9:560-564.

3 Berthold W, Walter JK: Protein purification: aspects of processes for pharmaceutical products. Biologicals 1994;22:135-150.

4 Charm SE, Landau SH: Thermalizer: high-temperature short-time sterilization of heat-sensitive biological materials. Annals of the New York Academy of Sciences, 1987;506:608-612.

5 Charm SE et al: High-temperature short-time heat inactivation of HIV and other viruses in human blood pharma. Vox Sang 1992;62:12-20.

6 Golden RF: Microwave purification for viral inactivation and bioprocess validation. Genetic Engineering News, May 15, 1994;17.

Dr. J.K. Walter, Process Development, Department Biotech Production, Dr. Karl Thomae GmbH, D-88397 Biberach, Germany.

Brown F, Lubiniecki AS: Viral Safety and Evaluation of Viral Clearance from Biopharmaceutical Products. Dev Biol Stand. Basel, Karger, 1996, vol 88, p 109

Risk Assessment and Limitations of Evaluation Studies

J. Löwer

Paul-Ehrlich-Institut, Paul-Ehrlich Strasse 51-59, D-63225 Langen, Germany

Biological products differ from chemical ones in that, during their production, there is at least one step (e.g. living animal, living human, cell culture) which supports the replication of viruses. The risk that such viruses can be transmitted to a recipient of the product does not only depend on their infectivity but also (i) on the nature of their presence in the starting material (whether their presence is theoretical, rare, common or proven) and (ii) on the effectiveness of downstream purification or virus inactivation steps. The contribution of those factors can be judged experimentally. The evaluation of the data should allow a risk assessment that should result in a certain degree of confidence in the safety of a given biological product.

Unfortunately, there is, in practice, no way to prove the absolute freedom from viruses of a given product. Therefore, the risk assessment relies completely on the quality and appropriateness of the available data and on an adequate evaluation. The most pratical limitation is the availability of resources (experienced laboratories, cost). Therefore, it is necessary to design the studies in the most cost effective way. However, to avoid the need for repetitions, the following factors which may influence widely the results should be taken into consideration: the choice of the «model» viruses used, the appropriateness of the scaled-down version, the search for process variables which may alter the efficacy of virus inactivation/elimination steps and the performance of the experiments which should be based on «Good Virological Practice» (correct titrations based on sound statistical grounds, inactivation kinetics, balance for partitioning steps).

It is obvious that all these studies can only be an approximation of the real situation. Therefore, the experimental limitations (e.g. laboratory virus strain versus wildtype strain, summing up of logarithmic reduction factors for several steps) should always be borne in mind. The procedure used reflects our best efforts to come to a conclusive risk assessment, but the reality could be much better – or worse. Even clinical studies, though necessary, cannot give a final answer due to the limited number of patients enrolled and product lots used.

Brown F, Lubiniecki AS : Viral Safety and Evaluation of Viral Clearance from
Biopharmaceutical Products. Dev Biol Stand. Basel, Karger, 1996, vol 88, pp 111-121

Use of a Statistical Strategy to Evaluate Sources of Variability in Viral Safety Experiments for a Recombinant Biopharmaceutical

P.R. McAllister, P.J. Shadle, T.M. Smith, R.G. Scott, A.S. Lubiniecki

SmithKline Beecham Pharmaceuticals, Departments of Statistical Sciences, Quality Operations, Biological Process Sciences and Biopharmaceutical Development, King of Prussia, PA, USA

Key word: Viral safety statistics.

INTRODUCTION

When therapeutic proteins are produced by recombinant expression in animal cell lines, one safety concern is the potential for retroviral contamination caused by expression of endogenous retrovirus genes in the cell line itself [1-3]. Approximately one million humans have been treated with recombinant proteins expressed in animal cell lines and no cases of viral infection have been noted. However, inactivated vaccines as well as traditional biological products and plasma-derived products have given rise to documented cases of Creutzfeld-Jakob disease, poliomyelitis, hepatitis and AIDS [4-6]. Experimentally, it has been shown that infectious retroviruses can cause lymphomas in primates that have been massively irradiated and immunosuppressed [7]. Hence the concern arises that unknown, unsuspected animal retroviruses, if present in a recombinant cell line, might recombine to form an active virus perhaps capable of causing disease in some humans [8-10].

Testing products for retroviruses has many shortcomings. Viral safety therefore depends upon estimating the retroviral risk in the source material and demonstrating that the purification process is capable of removing or inactivating an excess amount of a model retrovirus (or the actual retrovirus isolated, if infectious). Regulatory agencies have suggested that viral evaluation also consider statistical concepts in the experimental design and provide a measure of confidence about the consistency of the purification process [10-12]. In a previous paper, we suggested a method for examining this problem [11]. This paper illustrates the application of an approach to evaluate retroviral risk and clearance for the purification of a recombinant protein which is expressed in CHO cells. The experimental design used permitted estimation of both assay and process variability in the evaluation experiments using Poisson regression and prediction interval analysis to define the limits for the process.

Materials and Methods

Measurement of retrovirus-like particles

A CHO cell line, stably expressing a recombinant human protein, was cultured in suspension. Conditioned medium was harvested by microfiltration and the concentration of retrovirus-like particles was estimated by transmission electron microscopy (TEM). Retrovirus-like particles were isolated by ultracentrifugation of the cell-free culture fluid; the pellets were fixed, embedded, sectioned, mounted, and stained [13]. Risk was calculated based on observed particle levels, concentration factor employed, and volume of conditioned medium equivalent to a dose of final product.

Scaled down models for viral clearance

Purification process steps used to prepare the recombinant protein for clinical trial use were scaled down for use in viral clearance evaluation studies. Column chromatography was scaled down by holding constant column length, linear flow rate, load ratio, buffer compositions and temperature, and by reducing column diameter [1, 12]. The product-containing feedstreams were collected from the full-scale purification process. To the feedstream samples was added 1/20 volume of viral stock containing murine xenotropic leukemia virus (MuLV); chromatography was performed, fractions collected and viral titres assessed by S^+L^- focus formation on mink $MiCl_1S^+L^-$ cells. Buffers were evaluated for cellular toxicity to define the dilution schemes necessary. Similarly, treatments for retroviral inactivation were scaled down by volume, holding constant variables such as pH, buffer composition, protein concentration, reagent addition rate, temperature and duration of treatment.

Data handling

1. Risk estimate

The risk estimate is formed over several full-scale production batches [11]. Throughout this paper we will denote the estimate of a quantity by placing a carat (^) over its symbol. We recommend that this analysis use data from a minimum of three full-scale batches (n_R=3). This choice allows the examination of three levels of variation in determining the risk component of required clearance: (i) assay variance; (ii) sampling variation when selecting the test article from a batch, and (iii) process variation when the set points of critical process variables are not intentionally changed (batch-to-batch consistency). These three components constitute the natural variation of the process and their estimation will allow informed evaluation of the effects of small purposeful changes in the set points of critical process variables (robustness). Our design used two samples from each batch that were independently prepared and counted twice each. This results in a total of 12 risk values, which we designate as R_{ijk}, i = 1, 2, 3; j = 1, 2; k = 1,2. Based on these data, we estimate three variance components for the risk values: (i) $\hat{\sigma}^2_{Assay}$; (ii) $\hat{\sigma}^2_{sample}$, and; (iii) $\hat{\sigma}^2_{Batch}$. The variance components are estimated by treating this experiment as a nested design and equating expected mean squares with their observed values first from the quantities given in Table 1 [14].

The variance estimates are then calculated as

$$\hat{\sigma}^2_{Assay} = SS_A/6;\quad \hat{\sigma}^2_{Sample} = \left(SS_S/3\right) - \left(\hat{\sigma}^2_{Assay}/2\right);\quad \hat{\sigma}^2_{Batch} = \left(SS_B/2\right) - \left(\hat{\sigma}^2_{Sample}/2\right) - \left(\hat{\sigma}^2_{Assay}/4\right).$$

Table 1.

Source	Average	Sum of squares
Assay	$\overline{R}_{ij\cdot} = \frac{R_{ij1}+R_{ij2}}{2}, i=1,2,3; j=1,2$	$SS_A = \sum_{i=1}^{3}\sum_{j=1}^{2}\sum_{k=1}^{2}\left(R_{ijk}-\overline{R}_{ij\cdot}\right)^2$
Sample	$\overline{R}_{i\cdot\cdot} = \frac{\overline{R}_{i1\cdot}+\overline{R}_{i2\cdot}}{2}, i=1,2,3$	$SS_S = \sum_{i=1}^{3}\sum_{j=1}^{2}\left(\overline{R}_{ij\cdot}-\overline{R}_{i\cdot\cdot}\right)^2$
Batch	$\overline{R}_{\cdot\cdot\cdot} = \frac{\overline{R}_{1\cdot\cdot}+\overline{R}_{2\cdot\cdot}+\overline{R}_{3\cdot\cdot}}{3}$	$SS_B = \sum_{i=1}^{3}\left(\overline{R}_{i\cdot\cdot}-\overline{R}_{\cdot\cdot\cdot}\right)^2$

The estimate of variance required for determination of the statistical safety factor is the sum of the three components $\hat{\sigma}^2_R = \hat{\sigma}^2_{Assay} + \hat{\sigma}^2_{Sample} + \hat{\sigma}^2_{Batch}$. To be conservative, we claim only (n_R-1) degrees of freedom for this estimate.

2. Clearance estimate

We assume for this description that the purification process consists of k distinct unit operations. The terminology 'orthogonal operations' is used to denote operations whose modes of either virus inactivation or removal are based on differing chemical or physical mechanisms. The determination is based on scientific grounds and not on any observed clearance capability. The goal of the clearance capacity evaluation is to determine the total clearance capacity of the orthogonal operations in the purification process. We also want to determine the variation of this estimate, to be used in the calculation of the required statistical safety factor.

Let j = 1,2,...k, correspond to orthogonal unit operations in the purification process and C_j to the clearance capacity of unit operation j. C_j is determined experimentally by spiking a test virus, such as murine xenotropic retrovirus, into the input stream of operation j and measuring the virus concentration after operation j. Notationally we let S_I be the spiked virus concentration per unit volume in the input stream, S_o be the virus concentration per unit volume in the output of operation j, v_I be the total volume of the input stream, and v_o be the total volume of the output stream. Using this notation,

$$C_j = \log_{10}\left(\frac{v_I}{v_O}\frac{S_I}{S_O}\right) \quad \textbf{(1).}$$

The total clearance capacity of the entire purification process is determined as:

$$C_T = \sum_{\text{orthogonal operations}} C_j \quad \textbf{(2).}$$

Experimentally, the clearance calculation depends on having estimates for S_I and S_o. These are obtained by performing an infectivity assay and culturing serial dilutions of the test sample with susceptible cells and then manually counting foci or plaques after an incubation period. Such a method provides a set of counts for each dilution in a series. For each sample there will usually be dilutions that yield zero counts and dilutions for which the viral concentration will give more plaques than can be reliably counted. In addition, there are usually other samples, such as non-bound and wash fractions, that are taken during the course of the clearance experiment. Virus concentration estimates are also desired for these samples to allow material balance and stability checks. We have used « Poisson regression » to estimate the virus concentrations of these various samples [15, 16]. Since we are dealing with data that are counted, and we expect the variance of our data to be proportional to its mean, the Poisson distribution is appropriate. For a few samples, the unit operation will remove or inactivate all detectable model virus. In such cases, the virus concentration is estimated as the limit of detection for the assay and the analysis is continued as a minimum estimate rather than an observed value for clearance [1].

In Poisson regression we model the natural logarithm of the plaque count as a function of the dilution factor and the sample type. We assume that the slope of the decrease in plaque count with dilution factor is the same for all samples from the same unit operation. In fact, appropriate dilutions should be experimentally determined for which the solutions do not interfere with the assay. This assumption may be checked through appropriate data analysis techniques such as plotting and examination of residuals.

The Poisson regression model provides estimates of the virus concentration for each sample and an estimated standard deviation for each concentration. In terms of the notation above, we obtain $\hat{\beta}_I = \ln(\hat{S}_I)$ and $\hat{\beta}_O = \ln(\hat{S}_O)$. Also, $\hat{\sigma}_I$ is the estimated standard deviation of the natural logarithm of the input stream concentration and $\hat{\sigma}_O$ is the estimated standard deviation of the natural logarithm of the output stream concentration.

We describe the details of the Poisson regression for a single purification step. Let the subscript s stand for the sample type, s = 1, ..., N_s. Associate s = 1 with the input stream to the purification step and s = 2 with the output stream of the purification step. The other samples, s = 3, ..., N_s. are taken to check material balances and to check stability of the streams.

For each sample we want to estimate the virus concentration per unit volume, $\hat{\beta}_S$. Each test sample is incubated in a serial dilution scheme where d = 0, ..., D stands for $\log_{10}$ of the dilution factor. Usually, three incubation wells are inoculated at each dilution.

We represent the raw plaque counts per incubation well as f, where l = 1, ..., L and L is the number of replicate wells at each dilution. The first dilution yielding countable data is designated d_s. In order to estimate the virus concentration of each sample, only the total counts at each dilution are required and we scale the counts to a convenient unit volume – $f_{sd} = \sum_{l=1}^{L} f_{sdl} / (\text{unit volume})$, s = 1, ..., N_s; d = d_s, ..., D.

We propose the following model:

$$\ln(f_{sd}) = \beta_S + \beta \cdot d \quad \textbf{(3).}$$

The assumption of common slope, ß, allows the data from all samples to be included in one analysis as a generalized linear model (GLIM) for Poisson distributed counts [17]. A set of sample data are plotted in Figure 1 as an illustration.

Briefly, GLIM uses an iterative algorithm to estimate $\underline{\beta}$, the vector of sample concentrations and common slope. The algorithm employs weighted least squares where the Poisson assumption is used to weight the counts appropriately. In addition to parameter estimates, the GLIM analysis provides an estimated variance-covariance matrix for $\underline{\beta}$. Additionally, GLIM allows for «over-dispersion» in the model. That is, the actual data may exhibit more variation than is predicted strictly by the Poisson distribution. An estimate of the over-dispersion variance is obtained and inflates the standard errors of the virus concentration estimates.

Using these components, we can calculate the natural logarithm analog, $\tilde{C}_j$, of clearance for the jth step as $\tilde{C}_j = \ln\left(\frac{v_I}{v_O}\right) + \hat{\beta}_I - \hat{\beta}_O$, with variance $\hat{\sigma}^2_{\tilde{C}_j} = \hat{\sigma}^2_I + \hat{\sigma}^2_O - 2 \cdot \hat{\sigma}_{IO}$, where $\hat{\sigma}_{IO}$ is the estimated covariance between $\hat{\beta}_I$ and $\hat{\beta}_O$. Since we assumed a common slope for the samples from the same purification step, the concentration estimates are not statistically independent and the variance of their sum must account for this correlation. We have also assumed that the volumes of the input and output streams are measured with negligible error relative to the virus concentration estimates.

Our final clearance estimate is found by converting to $\log_{10}$: $\hat{C}_j = \frac{\tilde{C}_j}{\ln(10)}$ and $\hat{\sigma}_{\hat{C}_j} = \frac{\hat{\sigma}_{\tilde{C}_j}}{\ln(10)}$.

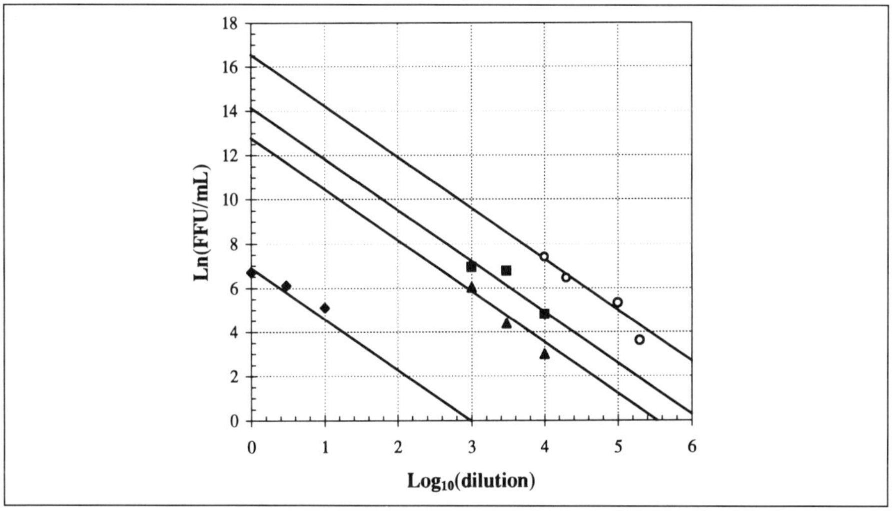

Fig. 1: Poisson regression of MuLV assay data for one purification step. Data from MuLV assays (FFU per plate) were calculated as the natural logarithm of the average virus concentration (FFU/mL), and plotted against the $\log_{10}$ of the dilution factor. Estimation of the common slope and sample intercepts was accomplished using the SAS procedure GENMOD [19]. (■) load sample spiked with virus stock (1:20); (▲) non-bound fraction; (◆) eluate fraction; (○) virus stock; (—) regression lines for each sample.

The total estimated clearance for the purification process is found as the sum over the individual steps as in Eqn (2). The purification steps are taken as independent so that the resulting estimated standard deviation of total clearance is:

$$\hat{\sigma}_{TA} = \sqrt{\sum_{j=1}^{k} \hat{\sigma}^2_{\hat{C}_j}} \cdot \quad \textbf{(4)}$$

The standard deviation calculated in Eqn (4) is based on one run of each purification step. As such it does not include variation due to different feedstocks nor to differences in setup or operating conditions of the purification steps. We have, therefore, labelled it as « assay » variability because the primary source of variation is the viral assay. In order to obtain an overall clearance variation $\hat{\sigma}_T^2 = \hat{\sigma}_{TA}^2 + \hat{\sigma}_{TP}^2$, multiple runs of the clearance experiment need to be performed.

Suppose n_T runs of the purification process are performed, each resulting in an estimate of total clearance, $\hat{C}_{Ti}$. Then the best estimate of process clearance capacity is the average of these individual runs $\hat{C}_T = \frac{\sum_{i=1}^{n_T} \hat{C}_{Ti}}{n_T}$ and the estimated total variance of process clearance is $\hat{\sigma}_T^2 = \frac{\sum_{i=1}^{n_T} (\hat{C}_{Ti} - \hat{C}_T)^2}{n_T - 1}$ with $(n_T - 1)$ degrees of freedom. If one is interested in obtaining an estimate of the additional variation due to the process setup and operation, $\hat{\sigma}_{TP}^2$ can be found by subtraction.

3. Excess clearance

The rationale for creating a statistical safety factor can be understood by considering a quantity called excess clearance (EC). Excess clearance is defined as the difference between the clearance provided by the process (C_T) and the risk (R). This relationship is expressed below:

$$EC = C_T - R \quad \textbf{(5)}$$

EC is a quantity subject to normal variation and, assuming C_T and R are estimated independently, has variance equal to $\sigma_R^2 + \sigma_T^2$. We want to develop a requirement for process total clearance $\hat{C}_T$, such that there will be a high probability that future production batches will have a reasonable safety factor in virus removal. Therefore, the distribution of interest is the distribution of excess clearance (EC) values, where excess clearance is given in Eqn (5).

We define the statistical safety factor, F_S, to be that value of EC such that there is a high degree of confidence that a future production batch will, when purified, exceed the required biological safety factor F_B [11]. To meet this objective we consider a prediction interval for future values of EC. If we form a $(1 - \alpha)\%$ one-sided prediction interval, then future values of EC should be less than the lower bound of this interval only $\alpha\%$ of the time. Further, our requirement is that this lower bound must exceed F_B. We have chosen prediction for a single future observation as the defining concept for F_s. For further discussion of interval estimation, see [18]. A one-sided lower prediction interval for EC is given by:

$$\hat{EC} - t_{(1-\alpha),v}\left(\left(\hat{\sigma}_R^2 + \hat{\sigma}_{\bar{R}}^2\right) + \left(\hat{\sigma}_T^2 + \hat{\sigma}_{\hat{C}_T}^2\right)\right)^{1/2} \quad \textbf{(6)},$$

where $v = \min\{(n_R - 1), (n_T - 1)\}$. We define the statistical safety factor, F_S, as the width of this interval.

Results

A. Estimation of risk and its variance

To estimate retroviral risk, three separate cell culture batches were examined by electron microscopy. A single sample of each was divided and replicate sample preparation included ultracentrifugation and TEM embedding, sectioning, staining, and counting. Assay variability was defined by the variance between replicates,

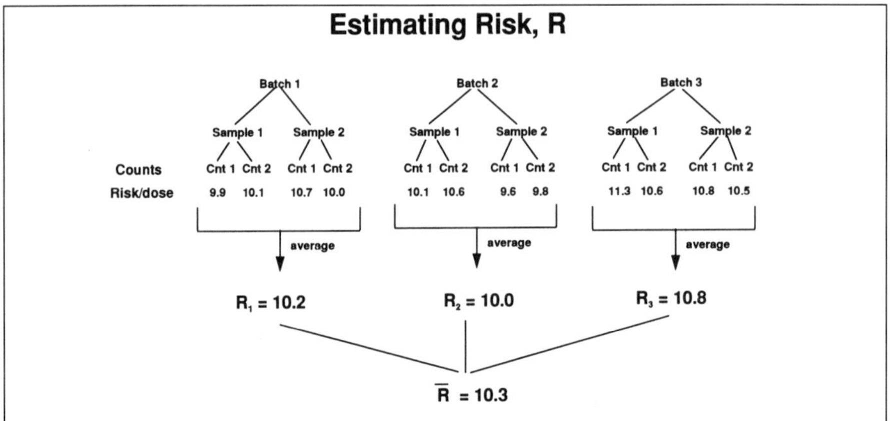

Source of Variation	Estimate of Variance	Percent of Variance	Standard Deviation
Batch	0.12	43	
Sample	0.05	18	
Assay	0.11	39	
Total	0.28	100%	0.53

Fig. 2: The total standard deviation is the square root of the total variance.

while process variability reports the variability that can be ascribed to differences between cell culture batches. Figure 2 shows a summary of the experimental design and the results, reported in terms of $\log_{10}$ of virus-like particles per dose of product. The variance is relatively small, and the data indicate that variability between batches and between assays was relatively similar in magnitude.

B. Estimation of clearance and its variance

To characterize assay and process variability, we performed each of four unit operations three times in a scaled-down model and evaluated viral clearance. Process clearance, the sum (in $\log_{10}$) of the clearances of each unit operation, was calculated as shown in Figure 3. The difference between total variance and assay variance was computed and is reported as process variance. This number was found to be relatively large.

C. Calculating F_s

Figure 4 is a display of how the 95% prediction interval was calculated for excess clearance. Since excess clearance is the difference between clearance and risk, the two sources of variability were summed to estimate the standard deviation of a future process batch. For this data set, the clearance variability dominates the calcu-

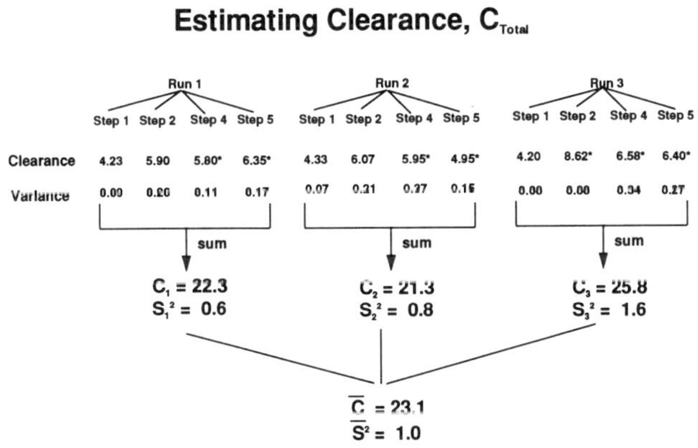

Source of Variation	Estimate of Variance	Percent of Variance	Standard Deviation
Process	4.6	82	
Assay	1.0	18	
Total	5.6	100%	2.4

Fig. 3: The estimate of total variance is obtained by calculating the variance of C_1, C_2, C_3. The estimate of assay variance is obtained by averaging S^2_1, S^2_2, S^2_3. The estimate of process variance is obtained by subtraction. The total standard deviation is the square root of the total variance. * Clearance of the step is limited by the virus stock titre. The reported clearance is a lower limit and is based on the limit of detection of the virus assay.

Statistical Safety Factor, F_s

Excess Clearance : EC = $\overline{C}_{Total} - \overline{R}$

$$= 23.1 - 10.3 = \boxed{12.8}$$

Standard Deviation of a future EC : SD(EC) =

$$\sqrt{\left(\hat{\sigma}_T^2 + \hat{\sigma}^2_{\hat{C}_T}\right) + \left(\hat{\sigma}_R^2 + \hat{\sigma}^2_{\overline{R}_{\ldots}}\right)} =$$

$$\sqrt{\left(5.60 + \frac{5.60}{3}\right) + \left(0.28 + \frac{0.12}{3} + \frac{0.05}{6} + \frac{0.11}{12}\right)} = \boxed{2.79}$$

$$F_s = t_{df,\,(1-\alpha)} * SD$$

Width of 95% lower prediction bound for EC : $\boxed{F_s = 8.2}$

Fig. 4: Detailed calculations for the Statistical Safety Factor.

lation. The data show that, for this process having on average 12.8 logs of excess clearance, the estimated standard deviation of a future excess clearance value is 2.8 logs, and the 95% lower prediction interval has a width of 8.2 logs, with a lower bound of 4.6 logs.

D. Reducing variation

The data in Figure 3 suggest that the third evaluation of the process gave higher clearance values than the first two. However, the viral stock titre was higher in this experiment. Since some of the clearance measurements are limited by the input virus, much of the variability measured as process variabiltiy may in fact be due to differences in stock virus titres. To evaluate this, we re-plotted the clearance estimates against the titre of the virus stock and found a clear correlation (details available from the authors). Normalizing the measured clearance values against the stock titre removed essentially all the variation that had been reported as process variance in Figure 3 and suggests that it is, in fact, related to experimental design. This potential problem could be avoided by performing process replicates on the same day with the same virus stock, or by taking measures to assure a consistent virus titre.

Discussion

In a previous paper [11], we suggested that biological concerns about the type of viral risk and patient population be explicitly separated from statistical concerns about the consistency of a process and the precision of experimental estimates of clearance and risk. These two independent concerns were expressed as F_B and F_S, respectively, and enter in Eqn (7):

$$\text{Clearance required} = \text{Risk} + F_B + F_S \qquad \textbf{(7).}$$

For CHO cells which have retrovirus-like particles but no detectable infectivity in in vitro assays and a plausible basis for genetically determined lack of infectivity [2], we believe F_B can be set equal to zero. A large safety factor is already present because the analysis treats the particles as fully active and as if one particle generated disease.

F_S is defined based on measured data as the width of a one-sided lower 95% prediction interval for excess clearance (EC). The functional meaning of F_S is that it defines the lowest excess clearance that is likely to be observed in a future batch of product. The more variability the process has, or the less precisely the average clearance is estimated, the larger the F_S value and the larger the total amount of clearance required [11].

We performed process clearance experiments three times on each unit operation, to estimate and to separate the contributions of assay and process variability. On initial evaluation, our data suggest that the variation in determining risk is far outweighed by that in measuring clearance. This suggests that additional replicates of counting, sectioning, or sampling are unlikely to affect the outcome or add much confidence about the product's safety. We also note that F_s would be rather insensitive to the inclusion of a unit operation having less than one $\log_{10}$ of clearance, assuming the standard deviation of the clearance estimate for that step was small.

Our data for one purification process showed that with an average excess clearance of 12.8 logs, the width of the one-sided 95% prediction interval (F_s) was about 8 logs. F_s may be as low as 5 logs, based on re-analysis of the data with correction for the viral stock titre. In either case, the process is demonstrated to be highly capable, to remove at least an amount of retrovirus that might maximally be present in these cells. From the patient's perspective the data obtained lead to the conclusion that the average excess clearance is at least 12.8 $\log_{10}$, or that the probability that the average product dose contains a non-inactivated retrovirus (assuming they might be infectious) is less than $10^{12.8}$, or less than about one per 6 trillion doses. With 95% confidence, the lower limit of this probability is $10^{-4.6}$ or less than one per 40,000 doses.

It is cost effective as well as scientifically appropriate to use Poisson regression analysis of plaque or focus formation assay data to maximize the collection of information about experimental variability in the course of performing viral evaluation clearance experiments. Serial dilutions must be assayed since the titre of virus is not known at the time of the experiment. This increases the cost of viral clearance evaluations and motivates the desire to make efficient use of all data collected. The proposed generalized linear models allow data from all samples at all countable dilutions to be used in estimating both the variability of the assay and the effect of dilution. Therefore, more degrees of freedom are available for the estimates and there is less sensitivity to single «unusual» counts. Similarly, risk data collected by counting virus-like particles by electron microscopy are augmented by the use of statistical experimental design and evaluation of replicates.

We took the further step of evaluating process variation by performing each unit operation three times to determine the sources of variation. This added substantially to the cost of performing viral evaluations, yet was necessary for evaluating process variability. Our data suggest that the details of the experimental design, i.e. stock virus titre, can influence the clearance measurements. Our process and product were shown to be safe both for the average batch and for future individual batches based on a prediction interval.

Scientists and regulatory authorities now need to consider how much of this analysis is appropriate for products having starting materials without measurable retroviral infectivity as compared to products which come from starting material with measurable viral titres of known human pathogens. We feel that replicate process clearance experiments are not necessary to establish safety of products provided that excess clearance is high, the source material has no known infectious retroviral risks, and unit operations are orthogonal and scaled down appropriately. However, such experiments may have value in evaluating viral safety of biologics derived from other sources or processes, or to support and justify process changes.

References

1 Lubiniecki AS, Wiebe ME, Builder SE: Process validation for cell culture derived pharmaceutical proteins, in Lubiniecki AS (ed): Large Scale Mammalian Cell Culture Technology. NY, Marcel Dekker, 1990, pp 515-541.

2 Dinowitz M, Lie YS, Low M-AL, Lazar R, Fautz C, Potts B, Sernatinger J, Anderson K: Recent studies on retrovirus-like particles in Chinese hamster ovary cells, in Brown F, Esber E, Williams M (eds): Continuous Cell Lines: An International Workshop on Current Issues. Dev Biol Stand. Basel, Karger, 1992, vol 70, pp 201-207.

3 Lubiniecki AS, Dinowitz M, Nelson E, Wiebe M, May L, Ogez J, Builder S: Endogenous retroviruses of continuous cell substrates, in Hayflick L, Hennessen W (eds): Continuous Cell Lines as Substrates for Biologicals. Dev Biol Stand. Basel, Karger, 1989, vol 70, pp 187-191.

4 Assaad F, Cockburn WC: The relation between acute persisting spinal paralysis and poliomyelitis vaccine – result of a ten year enquiry. Bull WHO 1982;60:231-242.

5 Fradkin JE, Schonberger LB, Mills JI, Gunn WJ, Piper JM, Wysowski DK, Thomson R, Durako S, Brown P: Creutzfeld-Jakob disease in pituitary growth hormone recipients in the United States. JAMA 1991;265:880-884.

6 Sherwood, WC: The significance of the blood-borne viruses: Blood banking and transfusion medicine, in Brown F (ed): Virological Safety Aspects of Plasma Derivatives. Dev Biol Stand. Basel, Karger, 1993, vol 81, pp 25-33.

7 Donahue RE, Kessler SW, Bodine D, McDonagh K, Dunbar C, Goodman S, Agricola B, Byrne E, Raffeld M, Moen JB, Zsebo KM, Nienhuis, AW: Helper virus induced T cell lymphoma in nonhuman primates after retroviral mediated gene transfer. Jnl of Exp Med 1992;176:1125-1135.

8 Office of Biologics Research and Review, «Points to consider in the manufacture and testing of monoclonal antibody products for human use», Food and Drug Administration, MD, 1994.

9 Office of Biologics Research and Review, «Points to consider in the characterization of cell lines used to produce biologicals», Food and Drug Administration, MD, 1993.

10 Commission of the European Communities, Note for Guidance: Validation of virus removal and inactivation procedures. Brussels, Belgium 1991.

11 Shadle PJ, McAllister PR, Smith TM, Lubiniecki AS: Viral validation strategy for recombinant products derived from established animal cell lines, in Beavery EC, Griffiths JB, Zeijlemaker WP (eds): Animal Cell Technology. Developments towards the 21st Century. Dordrecht, Kluwer, 1995, pp 631-635.

12 Levine HL, Tarnowski SJ, Dosmar M, Fenton DM, Gabler R, Gardner JN, Hageman TC, Lu P, Sofer G, Steininger R: Industry perspective on the validation of column-based separation processes for the purification of proteins. J. Parenteral Science & Technol 1991;46:87-97.

13 Poiley JA, Bierly ST, Hillesund T, Nelson RE, Mouticello TM, Raineri R: Methods for estimating retroviral burden. Biopharm 1994;7(4):32-35,57.

14 Snedecor GW, Cochran WG: Statistical Methods, Iowa State University Press, Iowa, 1967.

15 Frome EL, Kutner MH, Beauchamp JJ: Regression analysis of Poisson-distributed data. J American Statis Assoc 1973;68:935-940.

16 Roussel PH, Husson J: Estimating viral concentrations from dilution counts: a reliable computation method programmed on a pocket calculator. J Virological Methods 1991;34:113-140.

17 McCullagh P, Nelder JA: Generalized Linear Models, Chapman and Hall, London, 1989.

18 Hahn GJ, Meeker WQ: Statistical Intervals: A guide for practitioners, NY, John Wiley, 1991.

19 SAS/Stat User's Guide, SAS Institute, Inc., Cary, NC, 1990.

Dr. P. McAllister, SmithKline Beecham Pharmaceuticals, 709 Swedeland Rd., P.O. Box 1539, UE3839, King of Prussia, PA 19406, USA.

Brown F, Lubiniecki AS: Viral Safety and Evaluation of Viral Clearance from Biopharmaceutical Products. Dev Biol Stand. Basel, Karger, 1996, vol 88, p 123

SESSION IV

Manufacturing Considerations

Chairpersons: R. Devine, FDA/CBER, USA
A.S. Lubiniecki,
SmithKline Beecham Pharmaceuticals, USA

Brown F, Lubiniecki AS: Viral Safety and Evaluation of Viral Clearance from Biopharmaceutical Products. Dev Biol Stand. Basel, Karger, 1996, vol 88, pp 125-130

Process Changes and their Effect on Process Evaluation for Viral Clearance

C. Marcus-Sekura

Biotechnology Assessment Services, Inc., Rockville, MD, USA

Abstract: The ability to remove and/or inactivate known and potential viral contaminants during the manufacturing process of a biological product has become an important parameter in assessing the safety of a biological product when it is produced from cell culture or from human or animal tissues. The combination of screening the starting material for infectious virus coupled with a process which will remove and/or inactivate significant levels of infectious virus has been demonstrated to be an effective way to assure product safety. A prime example of the application of this approach was in the manufacture of Factor VIII preparations which did not transmit HIV. Regulatory authorities have subsequently applied this model to a wide spectrum of tissue culture-derived products by requesting examination of the cell line for freedom from contamination with adventitious agents coupled with process evaluation. Difficult issues yet remain concerning agents which should be evaluated and what process changes should trigger a reassessment of the ability of the process to remove/inactivate virus. The broad issues of process changes and their potential effect on process evaluation will be discussed, including changes in the product, cell line, cell culture, manufacturing process, dosage and patient population. An initial process evaluation which is thorough and demonstrates robust removal and inactivation may obviate the need for re-evaluation of minor process changes.

Process evaluation for viral clearance can be defined as evaluation of the manufacturing process for a biological product for its ability to remove and/or inactivate known and potential viral contaminants. Process evaluation is really one component of a multi-pronged approach to assure safety of products from viral contamination; one must examine the cell line for contamination with endogenous or adventitious agents coupled with process evaluation [1]. If there are endogenous agents, an estimate should be made of the particle level and the process should be designed to remove and/or inactivate those agents. Challenge of a process with model viruses to evaluate the ability to clear unanticipated contaminants can provide additional assurance of safety. Routine lot-to-lot testing for potential viral contaminants is another essential component. Furthermore, these are not separate approaches to assure safety; they are inter-dependent. Changes which affect the levels of expression of endogenous agents will in turn have an impact on the process evaluation. Adding or removing tests for virus on the MCB or unprocessed bulk will also have an impact on the importance of process evaluation.

Much has been said about previous experience with viral contamination and its impact on the regulatory process. The impact of experience will be re-visited briefly since it explains some of the reasons for current regulatory requirements for process evaluation. First on the list are vaccines, since experience with incomplete inactivation, contamination with viruses such as SV40, and reversion of attenuated vaccines to virulence, all of which have occured with polio vaccines, has taught us that these problems are not just theoretical risks [2, 3]. Experience with CHO cells, the first widely used continuous cell line used as a cell substrate, indicated that these cells contain endogenous retroviral particles; however, these particles are non-infectious, are not significantly induced during cell culture, and the manufacturing process can be evaluated to remove and inactivate more logs of virus than are present in the starting material [4, 5]. Infectious murine retrovirus is present in hybridomas; however, again the process can be evaluated and shown to remove and inactivate more logs than are present in the starting material. More recently, another infectious murine retrovirus, a recombinant retrovirus (RCR), was detected in a doubly deleted retroviral gene therapy vector. There have been several instances of viral contamination of GMP cell culture facilities. What these incidents indicate is that a clean cell bank and strict adherence to GMPs are not sufficient to avoid viral contamination in very large-scale production and that downstream controls such as process evaluation or lot-to-lot testing for adventitious virus may be helpful in assuring safety. It is also important to recognize that regulatory agencies often suggest approaches but cannot always say exactly why because of proprietary issues.

Several invaluable lessons have been learned from the HIV experience. In the 1980s, licensed blood products believed to be safe transmitted HIV; e.g. Factor VIII. Furthermore, HIV co-purified with Factor VIII during the plasma fractionation process, a process which partitions blood fractions by changes in pH and ethanol concentration. This indicated that purification through multiple steps did not necessarily ensure safety. In addition, for HIV inactivation, small process changes from wet to dry heat, alteration of temperature, or duration of heat treatment have significant effects on the levels of virus inactivated. Interestingly, testing vials of final product for HIV was inadequate: when lots of Factor VIII known to have transmitted HIV infection were examined, even PCR was not sufficient to detect HIV in all lots. However, a combination approach of screening starting material coupled with process evaluation has proved effective in producing current products which do not transmit HIV [6].

Although process evaluation has shown its effectiveness, there are still issues to be decided: these include (i) deciding which agents should be evaluated and (ii) what process changes might warrant re-assessment of the ability of the process to remove/inactivate virus. The second of these issues will be addressed here, and will be based on my experience with products reviewed while at CBER and also some more recent experience as a consultant. Unfortunately, since most of my information is confidential, it will not be possible to discuss specific examples in great detail.

Many sponsors feel their viral clearance studies are sufficient when the Phase I study is approved. However, as a product moves through development to final approval, changes are always made in the process and some of these are significant and may affect virus clearance. First on the list of frequent changes is testing. Addition, elimination or change in sensitivity of specific in-process tests for viruses may affect the need for process evaluation. For example, the elimination of routine lot-to-lot testing for adventitious virus may result in the need for re-evaluation of viral clearance. Other frequently encountered changes are those in the product itself, in the cell line used for production, in cell culture, and in the manufacturing or purification process. Changes in dosage and route of administration, the patient population, and the phase of study are also included in this discussion. These are outside what might be strictly defined as process changes, but are changes which might none-the-less impact on process evaluation. Each of these areas will be discussed individually and some specific examples will be cited. Obviously, all the possibilities cannot be covered but I have tried to include areas where questions are occasionally asked about possible impact when a process is being redesigned. Of course, appropriate decisions about need for re-evaluation of clearance must be based on full information about the specific product.

First on the list are product changes. Frequently manufacturers decide to make a slightly different product using the same fully qualified Master Cell Bank. However, once a cell bank is transformed to express a new recombinant product, it becomes a new line which may have altered growth properties and altered ability to be infected by viruses or to express endogenous retrovirus [7, 8]. Such altered endogenous virus expression has been noted in CHO cells and in hybridomas producted by using the same fusion partners. Another scenario is production of the same product but with a slightly altered amino acid sequence. The impact of this of change may depend on whether the changed sequence alters the purification properties of the modified protein significantly, e.g. it elutes differently from a column providing purification from virus during the evaluation. If the two products are sufficiently similar, one approach might be to postpone re-evaluation until a later phase of product development. The third example is for products in the same product class produced in the same type of cell line, e.g. MAbs produced using the same fusion partner and purified using the same procedure. Here again effects on viral clearance will depend on the actual products; an example can be cited of multiple MAbs reactive with the same epitope produced in the same fusion which purified very differently on a protein A affinity column and required very different elution conditions. Since they eluted at different places, the extent of separation from a given viral contaminant was very different.

Regarding cell line changes, three are common: producing a new MCB, deriving a new MWCB, or expressing the same product in a different cell line. The impact of deriving a new MCB or MWCB would depend on whether the characteristics of the cells were altered, and whether the product is identical and purification unchanged. An example was a case where the «new» MCB was derived from the fermenter during production because it had significantly different properties from the parent MCB. Changing the cell line entirely would probably require re-evaluation of the process.

Another area to consider is implementation of significant cell culture changes. Examples are changes from adherent cells cultured in a T-flask or roller bottle to non-adherent cells cultured in a fermenter, increased passage of the cells, and changing from batch to continuous culture. These changes are likely to result in additional testing to determine the properties of the new cells, but may not require re-evaluation unless other significant changes result, such as in endogenous virus production or purification parameters. Change in serum (type, concentration, substitutes) or other media components such as inducing agents, may have impact on virus removal or inactivation by changing the protein concentration in the unprocessed bulk or by altering the susceptibility of the cells to virus infection [9, 10].

Elimination of serum may reduce concerns regarding initial contamination with adventitious agents and the extent of process evaluation needed. However, serum substitutes may be derived from biological sources, and some cell lines have been shown to be more readily infected with particular viruses in the presence of low serum or in the absence of serum. Scale-up per se is unlikely to have much impact if other parameters remain unchanged.

Three categories of manufacturing process changes will be considered: harvest, purification, and facility changes. Harvest changes would include special cell treatments, pooling, and changes affecting the specific activity of the initial product. Special cell treatments refer to procedures such as trypsinization which might be required if one changed from a non-adherent to adherent cell, in order to populate a medical device with cells expressing a protein of interest. Another example is shaking the culture which is known to increase the titre of infectious retrovirus in the unprocessed bulk greatly. Pooling might have impact if it changes the sensitivity of in-process tests for adventitious agents so that more weight in terms of safety would be focused on viral clearance studies.

Purification changes include scale-up which is probably the most common process change. If all the steps in the process are easily «scalable», then scale-up may have little impact on viral clearance. However, often steps critical to removal or inactivation need to be altered to effect scale-up and it may be necessary to evaluate whether they are as efficient at virus removal or inactivation. Frequently, the inactivation step used in the previous small-scale process cannot be applied to the huge increase in volume accompanying scale-up. Concentration of material into smaller volumes before performing the inactivation may not yield equivalent inactivation so re-evaluation may be needed. A change in the method of inactivation would similarly trigger re-evaluation. A change in the order of steps may also be significant, especially if the steps are placed before a critical inactivation or removal step. An example is a manufacturer who proposed interchanging two column steps; one of which produced the critical removal of virus. The material applied onto the critical column would be entirely different if the order in which the columns were run was reversed. An example of altered placement of a critical step was insertion of an antibody affinity column, using an antibody preparation which had not been evaluated for clearance of murine retrovirus, after the last step providing significant inactivation or removal. Alteration of steps before a critical inactivation or removal step could alter the process intermediate and jeopardize the effectiveness of the step. As for addition of steps using potentially contaminating reagents, one example was a manufacturer who wanted to add a step to remove cell protein and introduced an animal antiserum affinity column as the final step in purification. This provided no opportunity for removal or inactivation of contaminants which might be introduced by the affinity column. Another possible change is elimination of a step providing significant inactivation/removal; the downside of this appears obvious, but such proposals are submitted to CBER. Other purification changes which might have an impact on virus clearance involve changes in protein concentration, pH, ionic strength, solvents, and buffers. The significance of such changes will be variable and will differ for each case. For example, an increase in protein concentration or addition of agents which might stabilize the product during a purification step might simultaneously stabilize a virus contaminant. Other examples would be the variability in retroviral inactivation on protein A columns which depends significantly on pH, and the partitioning of HIV in the plasma fractionation process which is dependent on pH and ethanol concentration. On the other hand, alterations can be used to advantage in getting optimal clearance, but changes need to be monitored. Changes in specific procedures are also potentially important; e.g. column chromatography/HPLC, ammonium sulphate precipitation and filtration. Chromatography changes which result in different elution patterns, or changes in ammonium sulphate fractionation, may alter process intermediates and in turn affect efficiency of critical inactivation/removal steps. Adding a traditional filtration step such as a 0.2μ filtration would have some impact on removal of large viruses such as EBV, but little impact on removal of small viruses unless it removed material for which a virus had affinity. Introduction of a filtration step using a filter specifically designed to remove virus into the process would probably require some re-evaluation using actual process intermediates to substantiate performance in the particular process in which it is used. Changes in storage conditions of intermediates could have an impact e.g. if an intermediate were stored for shorter times because of increased processing capacity in the facility, and critical virus inactivation occurred during storage. Re-processing process intermediates to meet specifications may also have an impact on viral clearance. If a procedure failed to produce the desired product purification to meet specifications and required re-processing, perhaps it also failed to produce the expected viral clearance. Formulation with excipients is rarely an issue; one example which can be cited was a request to use an unlicensed human albumin preparation. Conceivably, use of new biological excipients could be approached by evaluation of their manufacture for viral clearance.

A change in single-use to multi-use is probably the most likely facility change to have an impact on viral evaluation and this might occur if the product(s) being introduced have potential for contamination with adventitious agents, are using viral agents, or use different raw materials. Issues such as use of GMP facilities for research lots of other products are very complex. Concern centres around whether virus could be introduced by these other products. Changes in instrumentation, water, and HVAC may have an impact if they control introduction of contaminants, but are difficult to evaluate in clearance studies.

One additional area encompasses items which are not strictly process changes. When a product is initially evaluated for safety, the evaluation is made based on the presumed indication, dosage and patient population. Changes in dose, frequency or duration of administration might suggest a need for evaluation of model viruses. The demonstrated viral clearance may not be sufficient to accommodate a 100-fold or greater increase in dose. Alteration in the route may also have an impact; for example a product whose manufacture is considered safe for oral administration may not be considered adequately safe for IM or IV administration. One such case involved a vaccine derived by multiple passage in several animal species which had been cured of a viral infection by a particular method. This vaccine was considered safe for oral administration. Product manufactured by that procedure might not have been acceptable for IM or IV administration without changes in manufacture. Similarly a switch from use of a product in terminally ill patients to less ill or normal patients or from adult use to an infant population may require additional evaluation, such as a study of model viruses. Finally the Phase of study may have an impact on the extent of evaluation needed as cited in the 1993 MAb PTC document which suggests evaluation of model viruses be carried out before Phase III studies.

There are a few additional factors which may affect decisions about evaluation of virus clearance. First, what were the results of the evaluation of the initial process? Was it robust? Did it include both inactivation and removal? What was the calculated clearance? Were there several steps using different mechanisms of inactivation or removal? What type of process change is planned? We have seen that some changes are more problematic than others. The product history is also important. Is there a history of contamination of this product or similar products? Is there a known contaminant? Are there endogenous viruses? Was evaluation performed on model viruses? Is there a deliberately added agent such as Sendai virus for IFN production or EBV for human MAb production? Finally, the nature of additional testing performed routinely on the product is critical. Hopefully, sensitive, specific, reliable tests can be developed for use on unprocessed bulk or even final product, such as long-term co-cultivation or PCR assays, which in some instances may modify the need for extensive process evaluation.

In conclusion, it should be emphasized that re-evaluation of the process for virus removal/inactivation may be necessary for significant process changes. For completeness, a wide range of potential process changes has been discussed, many of which will produce insignificant effects on viral clearance. An initial process evaluation which is thorough and demonstrates robust removal and inactivation may obviate the need for re-evaluation for minor process changes.

References

1 Marcus-Sekura CJ, Kozak RW: Continuous cell lines and contaminant testing, in Lubiniecki AS, Vargo SA (eds): Regulatory Practice for Biopharmaceutical Production. Wiley-Liss, 1991, vol 75, pp 3-7.

2 Horaud F: Introductory remarks: Viral safety of biologicals, in Brown F, Horaud F (eds): Virological Aspects of the Safety of Biological products. Dev Biol Stand, Basel, Karger, 1991, vol 75, pp 3-7.

3 Melnick JL: Virus inactivation: Lessons from the past, Virological Aspects of the Safety of Biological products. Dev Biol Stand, Basel, Karger, 1991, vol 75, pp 29-36.

4 Anderson KP, Lie YS, Low ML, Williams SR, Wurm FM, Dinowitz M: Defective endogenous retrovirus-like sequences and particles of Chinese hamster ovary cells, Virological Aspects of the Safety of Biological products. Dev Biol Stand, Basel, Karger, 1991, vol 75, pp 123-132.

5 Dinowitz M, Lie YS, Low ML, Lazar R, Fautz C, Potts B, Sernatinger J, Anderson K: Recent studies on retrovirus-like particles in Chinese hamster ovary cells, in Brown F, Esber E, Williams M (eds): Continuous Cell Lines – An International Workshop on Current Issues. Dev Biol Stand. Basel, Karger, 1992, vol 76, pp 201-207.

6 Marcus-Sekura CJ: Validation of removal of human retroviruses, in Brown F, Esber E, Williams M (eds): Continuous Cell Lines – An International Workshop on Current Issues. Dev Biol Stand. Basel, Karger, 1992, vol 76, pp 215-223.

7 Spector DH: Potential host range expansion of the retroviruses, in Brown F, Esber E, Williams M (eds): Continuous Cell Lines – An International Workshop on Current Issues. Dev Biol Stand. Basel, Karger, 1992, vol 76, pp 153-164.

8 Eglitis MA, Kadan MJ, Wonilowicz E, Gould L: Introduction of human genomic sequences renders CHO-K1 cells susceptible to infection by amphotropic retroviruses. J Virol 1993;67:1100-1104.

9 Silverman L, Cleghon V, Klessig DF: Increased permissivity of monkey cells to human adenovirus multiplication is affected by culturing conditions and correlates with both synthesis of virion fiber protein and altered splicing of its mRNA. Virology 1989;173:109-119.

10 Miller DG, Miller D: Tunicamycin treatment of CHO cells abrogates multiple blocks to retrovirus infection, one of which is due to a secreted inhibitor. J Virol 1992;66:78-84.

Dr. C.J. Marcus-Sekura, Biotechnology Assessment Sevices, Inc., 7413 Ottenbrook Terrace, Rockville, MD 20855, USA.

Brown F, Lubiniecki AS: Viral Safety and Evaluation of Viral Clearance from Biopharmaceutical Products. Dev Biol Stand. Basel, Karger, 1996, vol 88, p 131

Multi-Use Manufacturing Facilities

B.D. Garfinkle

Vaccine Quality Operations, Merck & Co., Inc., West Point PA, USA

Good Manufacturing Practices require the production of biologics to take place under conditions which prevent cross contamination. The Code of Federal Regulations (21 CFR Part 600-800) contains several citations dealing with areas for processing Live Virus Vaccines and spore-forming organisms which cite specific requirements for separate facilities and/or restrictions for the construction of these types of facilities and restricts personnel from working in other areas after working in an area processing infectious agents.

These citations and a lack of engineering controls, which could deal robustly with the questions raised by these issues, have previously discouraged manufacturers from proposing multi-use facilities. However, recent developments in the biotech industry have now made it critical for manufacturers to seek approval again for such facilities. New pressures on the biotechnology industry including shortened product life cycles, the escalation of new facility costs, increased times for design construction and approval of new facilities have again encouraged the industry to desing and construct multi-use manufacturing facilities. Additionally, new engineering controls and use of automated processing and containment systems have made it easier to design and construct such facilities and provide the necessary controls to adhere to GMPs. Specific issues dealing with containment, cleaning between products and segregation of manufacturing process steps will be discussed.

The industry believes that by using contained processing steps, automation, engineering controls and procedures, it is practical to operate a multi-use manufacturing facility successfully. The principles discussed make it possible to manufacture multiple products simultaneously within the same facility while maintaining segregation and preventing cross contamination.

Dr B.D. Garfinkle, Vaccine Quality Operations, Merck & Co., Inc., West Point, PA 19486, USA.

Brown F, Lubiniecki AS: Viral Safety and Evaluation of Viral Clearance from Biopharmaceutical Products. Dev Biol Stand. Basel, Karger, 1996, vol 88, pp 133-137

The Safety of Biopharmaceutical Products – A Personal Viewpoint

N.B. Finter

Tideswell, Sevenoaks, Kent, UK

Key words: Interferon, lymphoblastoid cells, guidelines, manufacturer's responsibilities

Abstract: There are now published guidelines dealing with the production of cytokines, monoclonal antibodies and other therapeutic proteins made by applying biotechnological procedures. Strict conformity with these guidelines will undoubtedly yield safe products, but the question can be asked whether all the tests now specified are really essential in relation to the actual risks involved. During many years of experience with some of these biopharmaceutical products, problems which might on theoretical grounds have occurred have in practice not been encountered. Should biopharmaceuticals still be subjected to a much more intense scrutiny than is given to most other medicinal products? Has the time come to review current control procedures? Unnecessary tests increase the cost of products, and may make them too expensive to be available to millions of people who would benefit from their use.

During the past 20 years, the safety of medicinal products made by biotechnological procedures has been the subject of much discussion and as a result, many Regulatory Agencies have specified what tests they require to be carried out on such a product. A company intending to manufacture a cytokine or monoclonal antibody knows from the start what work it must undertake if it is to produce an acceptable product.

The situation was very different 20 years ago. I was then working in the Beckenham laboratories of the Wellcome Foundation Ltd (now part of Glaxo-Wellcome) and had been involved with the development of interferons for almost two decades. Further progress was greatly hindered by the continuing shortage of interferon for clinical trials. Thus it seemed essential to find some new way of making a human interferon preparation on a much larger scale than hitherto. This meant at that time that large numbers of human cells would be required as the source. I therefore set up a programme to try to make interferon from human cells grown in mass culture; inevitably, this would mean using transformed cells even though there were no precedents for using such cells to make a product for clinical use. The work soon led to the choice of cells of the Namalwa line of lymphoblastoid cells, derived from biopsied tumour tissue from a Burkitt lymphoma patient who died from this disease.

Our proposal to use Namalwa cells for interferon production was subjected, not unexpectedly, to a barrage of criticisms. Many scientists strongly and sometimes vehemently opposed our choice on the grounds that some agents in these cancer cells might contaminate the interferon and eventually lead to cancer or some other disease in the recipients. Some indeed claimed that there were regulations which explicitly forbade the use of transformed cells for the production of pharmaceutical materials (which proved untrue). We felt that such criticisms lacked a sound scientific base. There were already precedents for the manufacture of safe and acceptable pharmaceutical products from a less than ideal starting material. Insulin was one such example: even then it been made for many years by suitable processing and purifying beef and pig pancreases collected from abattoirs. Provided that we could demonstrate beyond reasonable doubt that our final interferon product was safe, we felt that our use of Namalwa cells would be accepted. However, it was obvious that we would need sufficient evidence to justify our choice in the face of all criticisms. With no regulatory guidelines to help us, we had to work out how to proceed from first principles.

The first obvious step was to see what extraneous agents might be present in Namalwa cells and so might contaminate the crude interferon made from them. Using all the standard tests for detecting microbial agents then known, we found nothing except for Epstein- Barr virus, which is present as in almost all human lymphoblastoid cell lines. Each Namalwa cell contains a full copy of the virus DNA, but due to some defect, this never gives rise to infectious virus.

This negative information was reassuring but obviously not adequate. There might be some virus or viruses in Namalwa cells of a type not yet identified, which might cause serious problems if still present in the final product. To deal with potential risks of this type, we were helped by the quite extraordinary stability of the alpha interferons (IFN-α) made by Namalwa cells. These withstand treatments, including exposure to pH 2 (at one time used to define an interferon) and to many organic solvents that would denature or destroy most proteins. My colleague, Dr Karl Fantes, a protein chemist with much experience in handling interferons, was able to devise a procedure for purifying crude Namalwa cell interferon that involved a sequence of four such steps. We felt it highly likely that this procedure would in effect sterilise the final product, irrespective of what microbial contaminants might be present in the crude material. To verify this, we hit on the then novel idea of adding marker substances, one at a time and each at the highest attainable concentration, to samples of our crude interferon, and measuring the amounts in samples taken from the successive stages in the purification process. Our chosen marker substances included representatives of every class of virus, bacterium, bacterial spores, mycoplasmas, and radio-labelled DNA, both whole Namalwa cell DNA and fragments spanning the size of all known oncogenes [1]. Because of the difficulty in detecting small amounts of human and animal viruses in the presence of high concentrations of human interferon, and on the advice of Dr. J. Petricianni, then at the FDA, we included very high titre preparations of various bacterial viruses among these markers substances. The results were clear-cut: only a few marker substances could be detected in the product after even the first purification step, and then only in trace amounts. Markers were similarly added at this stage and were sought in the purified product from the second step. Again, all but a very few were eliminated. No marker added to the crude interferon was ever detected in the second stage purification product.

I still had one great concern, namely that some slow virus such as the agents causing Jacob-Creuzfeldt disease and kuru, which are extraordinarily resistant to inactivation, might be present in Namalwa cells. There are no laboratory tests for most of these slow viruses, but one representative agent, ovine scrapie, has been adapted to grow well in the brains of mice. With the help of experts in the Compton laboratories of the Agricultural Research Council, we showed that when mouse-adapted scrapie virus at the highest attainable concentration was added to crude interferon which was processed as usual, none could be detected after even the first stage in our purification process. This evidence finally convinced me that our interferon could be safely used in man, even though we could not state categorically that there were no other as yet undetected viruses in Namalwa cells. Our purification process provided what in effect was a «chemical autoclave»; it destroyed or eliminated all microbiological agents and DNA, and could be expected similarly to dispose of any that might be detected in the future, whatever their nature [in 1975, the human immunodeficiency virus responsible for AIDS, had not been discovered. When tests for this retrovirus became available, we soon showed that it is not present in Namalwa cells, and also that it is rapidly destroyed under the conditions used in the routine purification of Namalwa cell interferon].

We christened our interferon product, Wellferon, and in due course provided the necessary documentation, including detailed microbiological safety data, to various Regulatory Agencies; with their approval, clinical trials were started with Wellferon in the UK in 1979, in the USA in 1980, and in Japan in 1981. It has since been used in many thousands of patients.

My involvement in the development of Wellferon has made me very conscious of the responsibilities of one who seeks to manufacture a biopharmaceutical product, and Table 1 summarises what I believe these are. I should stress that I retired from Wellcome in 1989, and these are my personal views, which may not be shared by Glaxo-Wellcome.

Let me say then that in my opinion, the safety of the process from the point of view of the product, the staff involved and the environment must be among the main concerns of a manufacturer. Naturally, all the cells, viruses, serum, etc. used in production must be tested for possible microbiological contaminants, and if any are found, their greatest concentration must be established. The chosen purification process must at the very least be capable of eliminating substantially more than these amounts; ideally, it should also be shown capable of inactivating or eliminating any other potentially noxious agent, though this may be difficult to achieve for proteins that are less robust than the IFN-α. Of course, all normal Good Manufacturing procedures must be followed so that no new hazards are introduced during manufacture. The process as a whole must be reviewed regularly, with especial regard for possible consequences

Table 1: A manufacturer's responsibilities.

To develop and justify the production process.
To maintain it and monitor its consistent behaviour.
As the Qualified Person/Responsible Head in the organisation, to accept personal responsibility for the product, including its safety.

resulting from the employment of new batches of reagents, new equipment, or newly-trained operatives, and from the cumulative effects of any permitted minor process changes; however, these are more likely to affect product consistency than its safety. The process should also be reappraised from time to time in the light of scientific advances such as new methods of testing and control, or the discovery of new microbiological agents.

There are now well-established Codes of Practice which govern the production of biopharmaceuticals. In my opinion, mere slavish compliance with such a code may provide a less than ideal method of control and involve tests, including tests in animals etc. that are unnecessary, or carried out too often, or are excessive in other ways. The manufacturer may well be aware of particular problems or particular advantages associated with a product which need to be taken into account. Also, mere rigid compliance with regulations may not make full use of the particular expertise and skills of the staff involved in manufacture and testing, and these may suggest alternative and superior solutions to certain problems.

There is of course almost always a legal requirement that the Regulatory Agency in a country must give approval to a product before it can be used in man. I personally found that the interface with such Agencies was of very great value. It provided an opportunity to explain and justify the whole manufacturing process to an informed and keenly interested group of scientists. In turn, these assessed the process from their fresh and independent points of view, and from their general experience with the manufacture of other biopharmaceutical products. If there are any weak points in a manufacturer's arguments, these are sure to be picked up during such discussions. Regrettably, there are still a few Agencies with a check-list approach, who will not grant approval for a product unless all tests specified in their Guidelines have been carried out, even if some are clearly not relevant. For example, rodents cannot meaningfully be used for a long-term chronic toxicity test on a human IFN-α preparation, as they form specific neutralising antibodies after as few as 10 daily doses of this cytokine. I hope such Agencies will come to accept that exceptions should be made when a manufacturer can give good scientific reasons why a test is not appropriate for his particular product, or can be replaced by a better one.

I feel it is important that the costs involved in carrying out unnecessary or inappropriate tests are not ignored. There are scientific costs in that resources and attention may be dissipated on relatively trivial issues instead of being concentrated on matters of potentially greater concern. Also, because most biopharmaceutical agents are for many years made by only a single manufacturer and his licensees, testing costs are included in the price of the product, and passed on to the general public in the pharmacy price or indirectly through the taxes, insurance premiums etc. that finance supply to the public sector. Furthermore, the initial high cost of most such products puts them beyond the reach of millions of people in the less affluent parts of the world. Contrary to what is popularly supposed, a manufacturer who on good grounds strives to get rid of unnecessary tests, or to test less frequently than hitherto, is working as much in the public interest as his own.

As a final point, I believe one should question why the production of modern biopharmaceuticals is subject to such intense scrutiny. Quite different rules apply to related products such as live virus vaccines and, for example, insulin. Of course these have a long record of microbiological safety, but some biopharmaceuticals such as the recombinant and human cell-derived IFN-α preparations, and recombinant insulin, have now also been used in man for some 12-16 years without, to my knowledge, any problems. Why is there such special emphasis on biotechnology products? In all the testing now required, is there an element of overkill in relation to real risks? If longer field experience is needed before regulations can be relaxed, how much longer must this be? Realistic safety considerations should not be questioned, but are the risks with biopharmaceutical products greatly different from those necessarily involved and accepted with many other pharmaceuticals and indeed other products of all types? In my opinion, these are important questions that need to be addressed by manufacturers and government agencies alike.

Reference

1 Finter NB, Fantes KH, Lockyer MJ, Lewis WG, Ball GD: The DNA content of crude and purified human lymphoblastoid (Namalwa cell) interferon prepared by Wellcome Biotechnology Ltd, in Hopps HE, Petricciani J (eds): In vitro. Cellular and Developmental Biology. Monograph number 6, Abnormal Cells, New Products and Risk, 1985, pp 125-128.

Dr. N.B. Finter, Tideswell, 9 Burntwood Road, Sevenoaks, Kent TN13 1PS, UK.

Brown F, Lubiniecki AS: Viral Safety and Evaluation of Viral Clearance from Biopharmaceutical Products. Dev Biol Stand. Basel, Karger, 1996, vol 88, p 139

SESSION V

Regulatory Expectation for Data on Viral Testing and Process Evaluation for Phase I Studies and Product Licensing in the United States, Europe and Japan

Chairpersons: E. Griffiths, WHO, Switzerland
R. Wolff, FDA/CBER, USA

Brown F, Lubiniecki AS: Viral Safety and Evaluation of Viral Clearance from Biopharmaceutical Products. Dev Biol Stand. Basel, Karger, 1996, vol 88, pp 141-153

Presentation for Panel Discussion 1

Regulatory Expectation for Data on Viral Testing and Process Evaluation for Phase I Studies and Product Licensing in the United States, Europe and Japan

A. Ridgway

Biotechnology Section, Bureau of Biologics, Ottawa, Ont., Canada

In Canada, the system for regulating clinical trials is similar to that in the U.S. We frequently refer manufactures to U.S. «Points to Consider» Documents for guidance in addition to our own guidelines. This morning, the discussions will relate primarily to early Phase-I trials meant to demonstrate safety of the drug. Some fledgling companies, hospital departments or individual clinicians hold a secret hope that they will see an example of efficacy among the handful of patients in such a trial. The unfortunate fact is that it is highly unlikely that the patients in the trial will derive benefit from the drug; their reward derives from the altruism of furthering medical science. It is patients in the later-phase trials who will benefit from any efficacy the drug may hold; these patients, and the trials, will be better served by a well-characterized study drug.

The high cost of viral testing and process evaluation is criticized as a barrier to progress; however, it costs thousands of dollars a day to provide patients with hospital care and to prolong life only a few days. It does not seem unreasonable, therefore, to spend $20,000 to complete a modest series of safety tests on a product lot that can be tested on numerous patients. Confidence in the viral safety of the drug may allow selection of a healthier set of patients for enrolment in the trial and greater confidence in the other aspects of drug safety being investigated.

It has been pointed out to me, at this meeting, by a regulatory affairs person from industry, that when it comes to initial IND submissions, the Bureau of Biologics in Canada seems to place a greater emphasis on the Chemistry and Manufacturing package than on the clinical data. It is true that we require a fairly comprehensive Chemistry and Manufacturing package with the first IND Submission. In Canada, we get proportionally fewer Phase-I trials than in the United States. U.S.-based trials spread into Canada largely at the Phase-II stage so usually U.S. standards have already been met.

We have a growing biotechnology industry and we will face an increasing number of Phase-I trials. Trials involving drugs where some of the viral validation work is incomplete have been permitted. We have no immediate plans to establish any broad policy with respect to this particular issue. Applications will be considered on a case-by-case basis.

A number of factors would be weighed when considering when to show leniency regarding the extent of testing or validation. With respect to products and indications, these include:

- Those without inherent or substantial risk of containing agents pathogenic to humans.
- Those for which pre-clinical or in vitro studies show strong promise of eventual therapeutic value. (This latter point has significant input to the risk/benefit consideration).
- Life-threatening or debilitating conditions for which there are no good alternative treatments.

With respect to routes of administration there would be clearly less concern about a topical agent.

With respect to patient population or numbers:

- Protocol in a hospital setting;
- Numbers limited but sufficient to generate adequate data;
- Perhaps use of terminally ill patients;
- Well-informed patients.

In those areas where the viral testing and validation are recognized as being incomplete, the Chemistry and Manufacturing Package should describe clearly where the process involves equipment and steps equivalent to those for which viral clearance and inactivation have been well demonstrated previously. Manufacturers should therefore make careful choices concerning the methods to be used early in development.

There should be extensive rationales which include references to device master files, published material, etc. Our approach is not to let trial sponsors abrogate their responsibility with respect to the viral safety issue, but to try to have a rational, risk/benefit-based approach that will allow industry and the medical community to proceed with the clinical development by allowing a certain flexibility in completing some of the more difficult aspects of validation of the Chemistry and Manufacturing.

Dr A. Ridgway, Biotechnology Section, Bureau of Biologics, HPB-Bldg. 7, Tunney's Pasture, Ottawa, Ontario K1A 0P2, Canada.

Presentation for Panel Discussion 2

Regulatory Expectation for Data on Viral Testing and Process Evaluation for Phase I Studies and Product Licensing in the United States, Europe and Japan

H. Moriguchi

Research and Development Division, Pharmaceutical Affairs Bureau, Ministry of Health and Welfare, Tokyo, Japan

Key words: Viral testing; product licensing.

My presentation will focus on the regulatory requirement for viral safety and related data of biopharmaceuticals in Japan. I begin my presentation by describing the organization of the Pharmaceutical Affairs Bureau (PAB) that is responsible for the review and registration of biopharmaceuticals.

The legal basis of drug evaluation/registration and review system is described in the Pharmaceutical Affairs Law that regulates the drugs, medical devices and other products to ensure the quality, safety and efficacy of such products. Figure 1 gives a quick overview of the Bureau. PAB is responsible for the pharmaceutical administration in Japan. Its duty is to ensure the provision of safe and effective drugs with high quality standard. Within the Bureau, the Pharmaceuticals and Cosmetics Division (PCD) is in charge of evaluation and registration of new drugs, generics, OTC drugs, in vitro diagnostics and cosmetics. The review of biopharmaceuticals is one of the important tasks of this Division while the research and development aspect of the products is overseen by the Research and Development Division. PCD has 50 staff. There are one medical doctor and twenty-one pharmacists who are responsible for scientific review of drug application. In Japanese bureaucracy, officials including the drug reviewers are rotated to another post every two to three years. This is probably one of the reasons that the Ministry of Health and Welfare (MHW) relies upon outside experts whom the Minister of Health and Welfare appoints as a member of the Central Pharmaceutical Affairs Council (CPAC).

The flow chart in Figure 2 describes the drug review procedure. Usually, generics, OTC drugs and minor changes of new drug approvals are reviewed only by the reviewers of PCD. For new drug applications, the reviewers of PCD are responsible for the preliminary review of the applications and the inspection of non-clinical and clinical data. They also act as the secretariat of CPAC and assist the experts in providing necessary information.

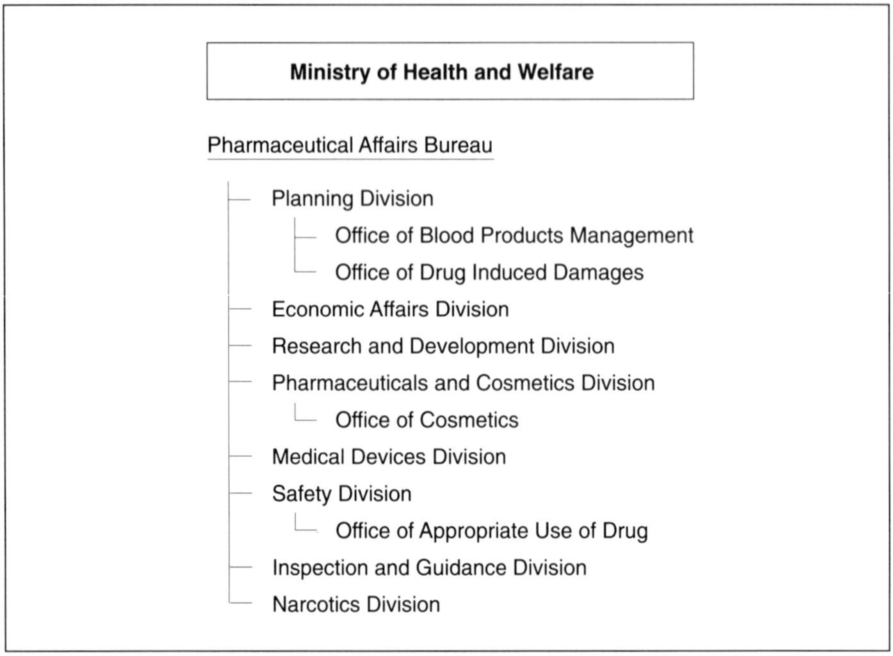

Fig. 1: Organization of Pharmaceutical Affairs Bureau.

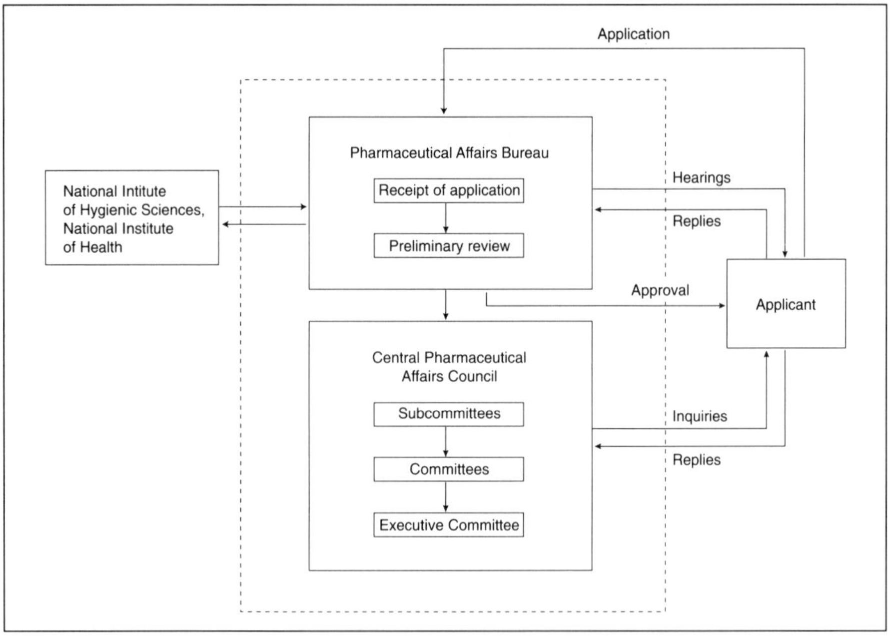

Fig. 2: Flow chart of approval review.

Figure 3 shows a part of the organization of CPAC, which is directly relevant to drug evaluation and approval. CPAC is an advisory body to the Minister of Health and Welfare and consists of 550 experts with medical and pharmaceutical, statistical and other related backgrounds. It has one executive committee, four committees, and twelve subcommittees for the examination of new drugs. Among the four committees, the Committee on Drugs includes nine subcommittees, whereas the Committee on Antibiotic Products, the Committee on Blood Products, and the Committee on Biological Products have one Subcommittee. Figures after the name of each committee show the number of experts it comprises.

CPAC subcommittees and the committees are organized to enable the examination of new drugs according to category and the intended clinical use of the drug for which approval is being sought. For example, the First Subcommittee on New Drugs engages in the examination of hormonal drugs, as well as drugs for diseases of digestive organs, kidney, liver, and so on. Thus, hormones such as insulin or growth hormone, erythropoietins as drugs for renal anaemia caused by dialysis, or interferons as drugs for viral hepatitis, will be the subject of the examination of this subcommittee. Plasminogen activators, such as tPA pro-urokinase, or urokinase, will be examined by the Second Subcommittee on New Drugs, which deals mainly

Name of Committees and Subcommittees	Main items handled
Executive Committee (16)	
Committee on Drugs (24)	
— 1st Subcommittee on New Drugs (15)	Gastrointestinal drugs, urogenital drugs hormones, drugs for topical use (excluding antibacterials)
— 2nd Subcommittee on New Drugs (14)	Cardiovascular drugs
— 3rd Subcommittee on New Drugs (12)	CNS drugs, respiratory tract drugs, anti-allergic drugs, sensory organ drugs
— 4th Subcommittee on New Drugs (13)	Antibiotics, chemotherapeutics
— Subcommittee on Anticancer Drugs (17)	Anticancer drugs
— Subcommittee on Japanese Accepted Names for Drugs (7)	Non-proprietary names
— Subcommittee on Combination Drugs (12)	New combination prescription drugs
— Subcommittee on Radiopharmaceuticals (9)	Radiopharmaceuticals
— Subcommittee on Dental Drugs (6)	Dental drugs
Committee on Antibiotic Products (5)	
— Subcommittee on Antibiotic Products (7)	Standards of antibiotics
Committee on Blood Products (8)	
— Subcommittee on Blood Products (13)	Blood Products
Committee on Biological Products (8)	
— Subcommittee on Biological Products (16)	Biological products

Fig. 3: Organization of the Central Pharmaceutical Affairs Council (relevant to drug evaluation and approval).

with drugs for disease of the circulatory system. If cytokines, such as interferons, colony-stimulating factors, interleukins, or tumor necrosis factor, are submitted for approval as new anticancer drugs, they will be examinated by the Subcommittee on Anticancer Drugs. If their intended clinical uses are other than as anticancer drugs, their safety and efficacy will be reviewed by another appropriate subcommittee.

After the application there is no exclusive or specific committee for reviewing the biotechnology safety and quality aspect of the products. In other words, each subcommittee is obliged to review the products taking account of the state-of-the-art information on biotechnology.

In 1984, MHW issued guidelines concerning the data required for application for approval of drugs manufactured using DNA recombinant tehnology (Table 1). A draft amendment to this notification was proposed in 1992, taking into account the current advances in biotechnology and the harmonization initiative. However, this draft has not yet been finalized.

MHW also announced in 1988 the guidelines concerning the data required for application of approval of drugs manufactured using cell culture technology. The biotechnology products are covered by these two guidelines. However, because scientific advances are occurring at a very rapid pace, the requirements contained in these two guidelines should not be regarded as being all-inclusive nor definitive. In other words, a flexible approach will be adopted when reviewing applications.

In the 1988 guidelines, referring to the methods of propagation, culture and storage of the cells used, it is stipulated that the virus-free status should be verified for cells of the Master Cell Bank and those propagated beyond the condition used for

Table 1: List of guidelines related to biotechnology product.

1	Documents necessary for application for approval of drugs manufactured using recombinant DNA technology
	(Notification N° 243 of PAB, MHW, dated March 30, 1984)
2	Documents necessary for application for approval of drugs produced by microorganisms modified using recombinant DNA technology (excluding in vitro diagnostics) (Draft)
	(Draft revision proposed on August 6, 1992)
3	Documents necessary for application for approval of drugs manufactured using cell culture technology
	(Notification N° 10 of 1st evaluation and registration division, PBA, MHW, dated June 6, 1988)
4	Viral tests and validation of biotechnology products derived from cell lines of human or animal origin
	(Part 1) Hayakawa T, Kenkyu I, 1993;24:1282-1292. (Part 2) Hagakawa T, Kenkyu I, 1994;25:12-24.
5	Guidelines for manufacturing drugs etc. by application of recombinant DNA technology
	(Notification N° 1051 of PAB, MHW, dated December 11, 1986) (Amended by notification N° 434 of PAB, MHW, dated May 21, 1987)

drug production, paying special attention to the viruses that may be found in the animal species from which the cell were originally established. In addition, it is, in principle, prohibited to use cells that carry viruses which are potentially pathogenic to humans. The presence of endogenous viruses such as retroviruses should be investigated for cells of the Master Cell Bank and those propagated for drug production by means of the reverse transcriptase assay and/or electron microscopy.

The report in 1993 from the study group of MHW, shows its position regarding the viral test and validation of biotechnology products, based on state-of-the-art information and the discussions among experts worldwide. The position with regard to viral validation is almost equal to that of the ICH draft guidelines, that is Q5A draft 1. Although this is still « Draft », it is understood that Japanese pharmaceutical companies are following the proposals in these guidelines.

The data requirements on viral safety before starting phase I trials in Japan, there is no approval system for the initiation of clinical trials, even for biopharmaceuticals. In principle, it is the company's responsibility to determine whether the trial can be safely initiated, based on the collected non-clinical data. In reality, data on viral validation are already provided by the company before starting Phase I trials, since the data are necessary for the new drug application.

In 1986, Guidelines for Manufacturing Drugs etc. by Application of Recombinant DNA Technology were established to ensure the quality of the biotechnology products and the safety in the manufacture of such products. According to these Guidelines, companies can ask CPAC for confirmation that the proposed manufacturing process and quality control are beyond the acceptable standard when they wish to initiate manufacturing of the pharmaceutical products using recombinant DNA technology. Although this is not an obligation, almost all companies make use of this confirmation programme before commencing the clinical trials. In these guidelines, there is no specific detailed description concerning viral validation. However, the experts in the Subcommittee and the Committee on Biotechnology in CPAC are well aware of ICH discussions on this issue and review the applications according to the viral validation concept of ICH.

MHW is committed to establishing internationally acceptable standards and requirements for viral validation and viral safety of the biopharmaceuticals in response to the discussions and recommendations of ICH.

Dr. H. Moriguchi, Research and Development Division, Pharmaceutical Affairs Bureau, Ministry of Health and Welfare, Kasumigaseki 1-2-2, Chiyoda-ku, Tokyo, Japan.

Presentation for Panel Discussion 3

Regulatory Expectation for Data on Viral Testing and Process Evaluation for Phase I Studies and Product Licensing in the United States, Europe and Japan

M. Dinowitz, J. Sernatinger, M. Wiebe

Genentech, Inc., South San Francisco CA, USA

Introduction

Product development, as the phrase implies, is an evolving process, and the extent of the analysis and testing that is performed and provided with regulatory submissions will change from the early IND stage to the licence application stage. The stringency and extent of testing that is appropriate at the earliest development stages when the number of patients to be exposed to a new drug is lowest should not be expected to be as extensive as when the potential patient number is much higher, after licensing.

A proper balance must be struck to achieve adequate virus testing and virus removal evaluation which will enable the pursuit of clinical trials during the IND phase without requiring the extensive testing which may be appropriate for licensum. Virus testing for early clinical trials should include appropriate in vitro and in vivo studies on the production cell substrate that would demonstrate that no adventitious viruses are detectable. In addition, clearance of endogenous viruses should be adequately demonstrated such that viruses shown to be present in the cell substrate are cleared to a level that would achieve an excess of removal above the level of potential virus in a dose of the product. It is not necessary to perform virus clearance studies with other viruses not known to be present in the cell substrate at this early stage of product development.

Studies to be provided with a licence application would include those evaluating virus clearance in which the manufacturing process that is evaluated is essentially the same as the actual manufacturing process (although these studies are often performed on a reduced scale), and should demonstrate clearance of endogenous viruses above the level of potential virus in a dose of the product. Additional studies to demonstrate the robustness of the manufacturing process in its ability to remove other virus types may also be provided at the time of the final licence application.

It should be pointed out that we have relied upon guidance documents from the several geographical regions in preparing these comments and, importantly, on the International Conference on Harmonization (ICH) document on viral safety which provides the internationally accepted framework of agreement on these issues.

Comparison Between Testing Appropriate for IND Submission and PLA Submission

The accompanying tables compare what is thought to be an appropriate testing level for IND submission vs. PLA submission. The table outlines testing in three key areas:

- Characterization of the master cell bank (MCB) and/or the working cell bank (WCB) for adventitious viruses and endogenous retroviruses.
- Unprocessed cell culture fluid.
- Virus clearance evaluation studies for endogenous retroviruses and the evaluation of the process for removal of model viruses.

Cell characterization of the production cell substrate (MCB and/or WCB)

The cell substrate (MCB/WCB) should be thoroughly assessed for the presence of adventitious virus by in vivo and in vitro virus testing procedures. The adventitious virus characterization performed on the MCB and WCB should be provided at the IND stage and the same data can be submitted with the PLA provided that the same MCB is being used (Table 1). Some testing may be performed on the WCB in lieu of the MCB. The maximum end of production cell harvest should also be tested at least once by in vitro characterization for adventitious viruses by the time of the PLA submission. It is not necessary to perform maximum end-of-production cell studies at the IND stage, which in many instances will not have been achieved at this point in development. In addition, it is recommended that each unprocessed bulk that is intended for clinical use be tested for the presence of adventitious virus by an appropriate in vitro virus screen before the lot is released.

In general, the cell substrate (MCB or WCB) should be characterized for the presence of endogenous retroviruses by one or more of the infectivity assays that are appropriate for the species of origin of the cells. The data for this characterization should be available for the IND submission. The same information provided with the IND submission is usually sufficient for the PLA submission as well (Table 2). If infectivity is not detected, other studies for the detection of endogenous retroviruses

Table 1: Adventitious virus testing: IND vs. PLA studies.

Procedure/test	IND	PLA
Adventitious virus testing on MCB/WCB*		
• In vitro assay	+	+ (From IND)
• In vivo assay	+	+ (From IND)
• Max. end of production	–	+
Unprocessed Bulk		
Adventitious virus in vitro virus assay	all lots	all lots

* Some testing may be done on WCB in lieu of MCB.

Table 2: Endogenous virus testing: IND vs. PLA studies.

Procedure/test	IND	PLA
Cell characterization	+	+ (From IND)
Endogenous retrovirus • Infectivity	+	+ (From IND)
• Infectivity (CHO; other rodent cells)	May be unnecessary since retrovirus profile of these cell types is well documented	(From IND) May be unnecessary since retrovirus profile of these cell types is well documented
• Max. end of production	+ (Maximum time can be increased during development following appropriate testing)	+ (Using final culture conditions, if different from IND)
• Max. end of production (CHO; other rodent cells)	+ (For particle enumeration only)	+ (Using final culture conditions, if different from IND)

would be appropriate, including electron microscopic (EM) examination of the cells and/or cell culture supernatants. For well characterized cell substrates such as Chinese hamster ovary (CHO) cells from which, historically, no infectious retroviruses have been detected despite extensive testing, the value of additional infectivity studies is questionable and we suggest that it is unnecessary. For CHO cells, EM studies which would enable an estimate of the number of retrovirus particles present per ml (or dose) of cell culture fluid should be sufficient. For some other well characterized cell lines such as some murine cells which have repeatedly demonstrated infectious virus, the value of continuing to demonstrate infectivity is also of questionable value since it may be assumed that infectious and/or non-infectious endogenous retrovirus is present. For these cell substrates, studies that enumerate the level of retrovirus particles, either by infectivity studies (if such viruses are present) or by EM studies should be adequate. This information, available for the IND, can also be used in the PLA submission provided that culture conditions, which may influence the level of retrovirus production, have not changed.

Virus clearance studies

Virus clearance (removal and/or inactivation) evaluation studies should be performed on appropriate steps in the manufacturing process to determine whether there is adequate clearance of endogenous retroviruses (or other relevant virus, i.e. other viruses that could be present in the cell substrate due to prior intentional or unintentional exposure to a virus or for some other specific reason related to the cell history) at the IND stage (Table 3). At this stage, the demonstrated removal should be at or above the potential level per dose at which the virus could be present. At the licence application stage, clearance of endogenous viruses should be demonstrated to be sufficiently above the potential level of virus that could be present in a dose to provide an adequately high safety factor.

Table 3: Virus clearance evaluation: IND vs. PLA studies.

Procedure/test	IND	PLA
Virus clearance evaluation • Endogenous Retrovirus (if known to be present)	+ (at or above potential level per dose, depending on virus)	+ (sufficiently above potential virus level per dose, depending on virus)
• Relevant virus (if known to be present)	+ (at or above potential level per dose, depending on virus)	+ (sufficiently above potential virus level per dose, depending on virus)
• Model viruses for assessment of robustness of process	–	+ (no specific target for clearance needs to be established)

In addition to demonstrating clearance of endogenous and relevant viruses at the licence application stage, the ICH document on viral safety indicates that manufactures should assess the ability of the manufacturing process to remove other viruses to provide a view of the robustness of the process for broader virus clearance. Because these studies are not performed to evaluate a specific safety risk, they should be performed to provide information only and not address specific clearance values.

Dr. M. Dinowitz, Genentech, Inc., 460 Point San Bruno Blvd., South San Francisco, CA 94080, USA.

Presentation for Panel Discussion 4

Regulatory Expectation for Data on Viral Testing and Process Evaluation for Phase I Studies and Product Licensing in the United States, Europe and Japan

J. Koga

Japanese Pharmaceutical Manufacturers Association, Kobe, Japan

Japanese Pharmaceuticals Manufacturers Association will address four brief comments on the issue of clinical trials related to viral safety studies.

1. In Japan, there is no regulatory requirement for clinical trials with regard to viral safety. In other words, manufacturers have to consider this issue totally under our responsibility at this moment. In general, upon starting a clinical trial, a cell banking system is believed to be already established. Viral safety studies should be done. In addition, the product formulation which will be given at a clinical trial should be tested for viral contamination, both in vitro and in vivo. Most probably, process validation studies including spike test, are not completed at this stage.
2. It would be useful to have a guideline dealing with viral safety and clinical trials. This would not only clarify requirements but also, because in many countries there is no IND system as in the UK and USA, which would provide really useful guidance before filing a document. However, it should be done in a very careful manner to set guidelines for example biotechnology-derived products with diverse features. The guidelines should leave many points on a case-by-case basis and also should have detailed, descriptive guidance.
3. Questions on the extent of viral safety studies are important. The exhaustive study on cell banking system is needed to give information on product safety. However, it may be inappropriate to put too much stress on tests such as adventitious virus testing on each lot of unprocessed bulk harvest or spike test with a variety of viruses. On such diverse products, it is very important to have a rationale why such tests should be executed at each step. Dr. Lubiniecki summarized the survey of adventitious virus tests in the USA, EU and Japan. So far, there has been no report on virus contamination in unprocessed bulk harvest. In such instances, it should be carefully assessed what sort of test is needed for this step.
4. There is a guideline dealing with a product produced by cell culture technology. However, after Dr. Hayakawa started his work on editing ICH viral safety guide-

line as rapporteur, he released some guidance, philosophy and other information from the ICH guideline draft in regional scientific and regulatory journals. In other words, manufacturers in Japan have started to prepare for ICH guidelines with an understanding of ICH philosophy over three regions.

Dr. J. Koga, JCR Pharmaceuticals Co. Ldt., Laboratories for Bioengineering and Research, 2-2-10, Murotani, Nishi-ku, Kobe 651-22, Japan.

Brown F, Lubiniecki AS: Viral Safety and Evaluation of Viral Clearance from Biopharmaceutical Products. Dev Biol Stand. Basel, Karger, 1996, vol 88, p 155

BREAKOUT SESSION A

Sensitivity and Specificity

Chairpersons: A.S. Khan, FDA, USA
B. Murphy, NIH, USA

Brown F, Lubiniecki AS: Viral Safety and Evaluation of Viral Clearance from Biopharmaceutical Products. Dev Biol Stand. Basel, Karger, 1996, vol 88, pp 157-162

Retrovirus Screening of Vaccine Cell Substrates

A.S. Khan

Division of Viral Products, Office of Vaccines Research and Review CBER, FDA, Bethesda, MD, USA

Key words: Retrovirus screening of vaccine cell substrates.

Abstract: A strategy for the detection of low levels of infectious retroviruses in vaccine cell substrates is outlined. This strategy involves amplification of possible viral contaminants in cell substrates by co-cultivation of the Cell Bank, with or without prior induction, with potentially susceptible cells followed by extended culturing and subsequent retrovirus testing using general and specific detection assays. In some cases, the virus seed may need to be examined by direct infection and extended culturing of cells that can support the replication of potential retroviral contaminants.

Cell substrates of avian and mammalian species are used in the production of vaccines. In some cases the virus seed stocks, that are used for the production of certain vaccines, have previously been passaged through a variety of cell substrates. The origins of cell substrates used in the production of vaccines currently licensed in the United States are shown in Table 1. These include primary cells or tissues, continuous cell lines and diploid cells obtained from various species including avian, murine, bovine, simian and human. Retroviral sequences are widely found in the genomes of eukaryotic species [1, 2, 3]. These sequences may have been acquired by the host by genetic or horizontal transmission. Although the majority of endogenous retroviral sequences are defective for virus production [4, 5, 3], under certain conditions infectious retroviruses can arise spontaneously from cells by activation of a latent retrovirus or by recombination between endogenous retroviral sequences [1, 6, 7, 8]. These viruses can then stably integrate in the host genome or infect other species via horizontal transmission, integrate and persist in a latent state [1, 9]. Actively replicating retroviruses can be pathogenic in their species of origin (e.g. mink cell focus-forming (MCF) murine leukemia virus [1, 10]) or in other species, via horizontal transmission [11, 12]. A list of known infectious and pathogenic retroviruses that could potentially contaminate the currently used vaccine cell substrates is presented in Table 2. In addition, novel human-tropic retroviruses may arise that could be potentially pathogenic. Thus, it is important to demonstrate the absence of retro-

Table 1: Cell substrates for current U.S. licensed vaccines.

		Live attenuated	Inactivated
Primary cells or tissues	Simian Bovine Murine Avian	OPV Smallpox Measles Mumps Yellow fever	JEV Inluenza
Continuous line	Simian Vero		IPV
Diploid cells	Human WI-38 MRC-5 Simian FRhL-2	Rubella Adenovirus Varicella	IPV Hepatitis A Rabies Rabies

Table 2: Potential retroviral contaminants.

	Infectious	Pathogenic
Type-C	HTLV STLV BLV ALV	+ (human) + (simian) + (bovine) + (avian)
Type-D	SRV (MPMV)	+ (simian)
Type-E	HIV SIV BIV EIAV	+ (human) + (simian) + (bovine) + (equine)
Type-F	FOAMY	–/?
Recombinant retroviruses	?	?
Endogenous retroviruses	?	?

viral contaminants in biological products by testing for retroviruses that may have potentially arisen from the endogenous retroviral sequences present in the host genome or that may have entered the cell by exogenous infection of the host. The most likely source of retrovirus contamination of vaccines is the cell substrate. This may be due to (i) the species of origin, which may be susceptible to known infectious retroviruses; (ii) activation of endogenous replication-competent retroviruses, which may occur during the production process; and (iii) novel recombinant retroviruses, which may arise due to recombination between infectious retroviruses and the large number of defective endogenous retroviral sequences that reside in the host genome. In addition, contamination of the cell substrate may be due to the raw materials used in the production or due to the virus seed, which may have acquired a retroviral contaminant due to its passage history through a variety of different cells (an example is mentioned in the summary of Breakout Session A). Thus, to ensure the safety of vaccines, rigorous retrovirus screening of the cell substrates, virus seeds and the final products, particularly in the case of live vaccines, is recommended.

The development of a retrovirus screening strategy should take into account the cell substrates involved and the potential retroviral contaminants that may be present due to an endogenous or exogenous source of retroviral sequences (discussed above). In the case of primary cells, one should consider the infectious retroviruses which are known to infect the species of origin and include specific assays for their detection. Continuous cell lines become immortalized following a transformation event; thus tumorigenicity tests need to be included in addition to the routine retrovirus testing (Points to Consider for the Characterization of Cell Lines Used to Produce Biologicals, 1993). Certain diploid cells have been well-characterized and extensively used in vaccine production. However, as novel and more sensitive techniques are developed, they may be applied to confirm the safety of currently used cell substrates. If novel procedures are included in the manufacturing process, then their effect upon the generation of retroviruses should be analysed.

To ensure the safety of the final product, retrovirus testing of the potential sources of contamination, e.g. the Master Cell Bank (MCB), the virus seed stock and biological raw materials, is recommended. In certain cases, e.g. live viral vaccines, testing of the End of Production Cells and/or final product may also be needed. A general scheme for the detection of low-levels of infectious retroviruses is shown in Table 3. This involves an amplification step that involves in the case of a virus seed stock infection of potentially susceptible cells and extended culture or in

Table 3: A general strategy for the detection of low levels of infectious retroviruses.

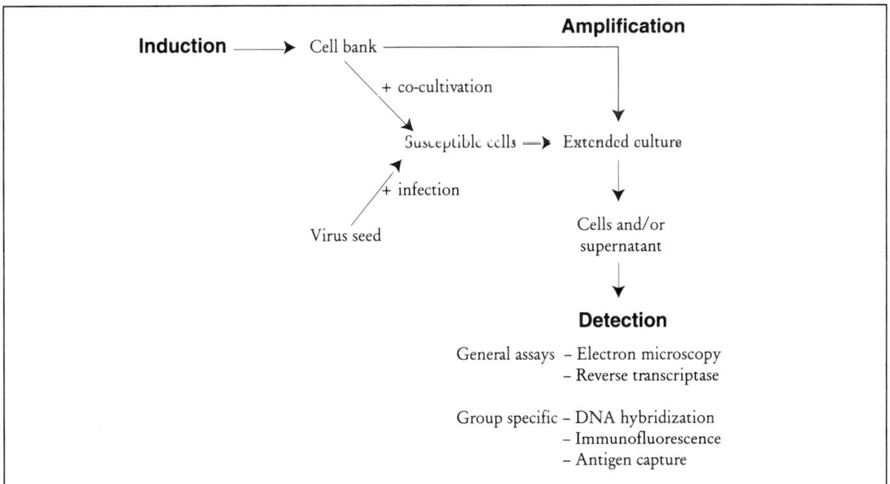

the case of the MCB, co-cultivation of non-induced or induced cells of the MCB with cells that might be permissive for the replication of potential retroviral contaminants. The MCB may be induced with a variety of agents known to activate endogenous retroviruses that reside latently in the host genome; these can include IdU, BrdU, 5-azacytidine, irradiation, or chemical carcinogens [1]. To enable detection of the activated retrovirus(es) the induced cells can be cultured directly for an extended time as well as co-cultured. After extended culture (in general, about four to six weeks), the cells and/or clarified supernatants can be tested for the presence of retroviruses. The use and standardization of retrovirus detection assays are discussed in the summary of Breakout Session A; the general assays, e.g. reverse transcriptase and electron microscopy, are less sensitive but useful for the detection of known and novel retroviruses, whereas a highly sensitive assay, e.g. the polymerase chain reaction assay, is specific for the detection of known retroviruses. In addition, when specific reagents are available, other detection assays such as DNA hybridization, immunofluorescence and antigen capture may also be useful for the general detection of known retroviruses. The cells selected for the co-cultivation should include those of human and non-human primate origin that have previously been shown to be used in the isolation of novel retroviruses or those that are permissive for replication of different types of retroviruses [13]. For example, certain cell lines have been shown to be useful in the isolation of novel primate retroviruses: the human carcinoma cell line A549 was used for the isolation of a retrovirus from an Old World monkey *Colobus polykomos* [14] and is susceptible to some other type C retroviruses; the human rhabdomyosarcoma cell line A204 was used for the isolation of a retrovirus from the Asian primate, *Presbytis obscurus* [15] and is susceptible to other related type D retroviruses. More recently, the human osteosarcoma cell line HOS, human foetal RSb, and human lymphoid Raji cell lines have been shown to express receptors for a variety of different retrovi-ruses [16] and thus may also be useful for the isolation of known and novel retroviruses. In addition, cells known to be sensitive for the detection of known retroviruses may be included, e.g. MOLT 4 clone 8 [17], and human peripheral blood mononuclear cells are susceptible to human and simian lentiviruses. Although no known pathogenicity has been identified with foamy virus infections, these retroviruses have been isolated from a variety of mammalian species and demonstrate a broad host range in vitro. A variety of cell lines of avian and mammalian origins can be used in the detection and isolation of foamy viruses [18-20].

It should be noted that in the case of murine and avian retroviruses, which have been extensively characterized, a variety of biological assays are available for their detection; however, similar sensitive detection assays have not been developed for the majority of primate retroviruses. Highly sensitive and specific polymerase chain reaction assays can be used when the nucleotide sequence of a retrovirus is known; however, to detect a novel retrovirus, general detection assays, which are less sensitive, must be used (discussed in Breakout Session A summary). Since about 1% of the mammalian genome may consist of endogenous retroviral-related sequences, novel viruses may be expected to arise spontaneously in mammalian cell substrates. Recently, highly sensitive detection approaches have been developed that might detect novel retroviruses [21-24]. The application of these assays for the detection of potential retroviral contaminants in cell substrates and biological products needs to be evaluated.

Acknowledgements

I would like to thank Drs. M.C. Hardegree, A. Lewis, R. Sheets, H. Golding, J. Beeler and K. Peden for comments and helpful discussions.

References

1 Coffin J: Endogenous Viruses, in: Weiss R, Teich N, Varmus H, Coffin J (eds): RNA Tumor Viruses, 2nd edn. New York, Cold Spring Harbor Laboratory, Cold Spring Harbor, 1984, vol 1, pp 1109-1203.

2 Kessel M, Khan AS: Nucleotide sequence analysis and enhancer function of long terminal repeats associated with an endogenous African green monkey retroviral DNA. Mol Cell Biol 1985;5:1335-1342.

3 Stoye JP, Coffin JM: Endogenous retroviruses, in Weiss R, Teich N, Varmus H, Coffin J (eds): RNA Tumor 2nd edn. New York, Cold Spring Harbor Laboratory, Cold Spring Harbor, 1985, vol 2, pp 357-404.

4 Amanuma H, Laigret F, Nishi M, Ikawa Y, Khan AS: Identification of putative endogenous proviral templates for progenitor mink cell focus-forming (MCF) MuLV-related RNAs. Virol 1988;164:556-561.

5 Repaske R, Steele PE, O'Neill RR, Rabson AB, Martin MA: Nucleotide sequence of a full-length human endogenous retroviral segment. J Virol 1985;54:764-772.

6 Hayward WS, Hanafusa H: Recombination between endogenous and exogenous RNA tumor virus genes as analyzed by nucleic acid hybridization. J Virol 1975; 15:1367-1377.

7 Khan AS, Laigret F, Rodi CP: Expression of mink cell focus-forming murine leukemia virus-related transcripts in AKR mice. J Virol 1987;61:876-882.

8 Laigret F, Repaske R, Boulukos K, Rabson AB, Khan AS: Potential progenitor sequences of mink cell focus-forming (MCF) murine leukemia viruses: ecotropic, xenotropic, and MCF-related viral RNAs are detected concurrently in thymus tissues of AKR mice. J Virol 1988; 62:376-386.

9 Varmus H, Swanstrom R: Replication of retroviruses, in Weiss R, Teich N, Varmus H, Coffin J (eds): RNA Tumor 2nd edn. New York, Cold Spring Harbor Laboratory, Cold Spring Harbor, 1984, vol 1, pp 369-512.

10 Hartley JW, Wolford NK, Old LJ, Rowe WP: A new class of murine leukemia viruses associated with development of spontaneous lymphomas. Proc Nat Acad Sci USA 1977;74:789-792.

11 Teich N, Wyke J, Mak T, Bernstein A, Hardy W: Pathogenesis of retrovirus-induced disease, in Weiss R, Teich N, Varmus H, Coffin J (eds): RNA Tumor 2nd edn. New York, Cold Spring Harbor Laboratory, Cold Spring Harbor, 1984, vol 1, pp 785-998.

12 Weiss R: Human T-cell retroviruses, in Weiss R, Teich N, Varmus H, Coffin J (eds): RNA Tumor 2nd edn. New York, Cold Spring Harbor Laboratory, Cold Spring Harbor, 1985, vol 2, pp 405-485.

13 Teich N: Taxonomy of Retroviruses,. in Weiss R, Teich N, Varmus H, Coffin J (eds): RNA Tumor 2nd edn. New York, Cold Spring Harbor Laboratory, Cold Spring Harbor, 1984, vol 1, pp 25-207.

14 Sherwin SA, Todaro GJ: A new endogenous primate type C virus isolated from the Old World monkey *Colobus polykomos.* Proc Nat Acad Sci USA 1979;76: 5041-5045.

15 Todaro GJ, Benveniste RE, Sherr CJ, Schlom J, Schidlovshy G, Stephenson JR: Isolation and characterization of a new type D retrovirus from the Asian primate, *Presbytis obscurus* (Spectacled Langur). Virol 1978;84:189-194.

16 Sommerfelt MA, Weiss RA: Receptor interference groups of 20 retroviruses plating on human cells. Virol 1990;176:58-69.

17 Allan JS, Short M, Taylor ME, Su S, Hirsch VM, Johnson PR, Shaw GM, Hahn BH: Species-specific diversity among simian immunodeficiency viruses from African green monkeys. J Virol 1991;65:2816-2828.

18 Hooks JJ, Gibbs CJ Jr.: The foamy viruses. Bact Rev 1975;39:169-185.

19 McClure MO, Bieniasz PD, Schulz TF, Chrystie IL, Simpson G, Aguzzi A, Hoad JG, Cunningham A, Kirkwood J, Weiss R.A: Isolation of a new foamy retrovirus from orangutans. J Virol 1994;68:7124-7130.

20 Rhodes-Feuillette A, Saal F, Lasneret J, Dubouch P, Peries J: Isolation and characterization of a new simian foamy virus serotype from lymphocytes of a *Papio cynocephalus* baboon. J Med Primatol 1979;8:308-320.

21 Heneine W, Yamamoto S, Switzer WM, Spira TJ, Folks TM: Detection of reverse transcriptase by a highly sensitive assay in sera from persons infected with human immunodeficiency virus type 1. J Infect Dis 1995;171:1210-1216.

22 Löwer R, Löwer J, Tondera-Koch C, Kurth R: A general method for the identification of transcribed retrovirus sequences (R-U5 PCR) reveals the expression of the human endogenous retrovirus loci HERV-H and HERV-K in teratocarcinoma cells. Virol 1993;192:501-511.

23 Pyra H, Boni J, Schupbach J: Ultrasensitive retrovirus detection by a reverse transcriptase assay based on product enhancement. Proc Nat Acad Sci 1994;91:1544-1548.

24 Silver J, Maudru T, Fujita K, Repaske R: An RT-PCR assay for the enzyme activity of reverse transcriptase capable of detecting single virions. Nucl Acids Res 1993;21:3593-3594.

Dr. A.S. Khan, FDA/CBER, Office of Vaccines Research and Review, Division of Viral Products, Laboratory of Retrovirus Research, 1401 Rockville Pike, Bldg. 29B, Room 4NN10, HFM-454, Bethesda, MD 20852, USA.

Brown F, Lubiniecki AS: Viral Safety and Evaluation of Viral Clearance from Biopharmaceutical Products. Dev Biol Stand. Basel, Karger, 1996, vol 88, pp 163-165

A Comparison of Methods for the Estimation of Retroviral Burden

S.T. Bierley[1], *R. Raineri*[2], *J.A. Poiley*[3], *E.M. Morgan*[4]

[1] EM Section, Pathology Associates International, Durham NC, USA
[2] TSI Washington Laboratories, Inc., Rockville MD, USA
[3] BRFF, Ijamsville MD, USA
[4] Microbiological Associates, Inc., Rockville MD, USA

Key words: Retroviral burden.

Abstract: The presence of retroviral contamination is of vital concern in the manufacture of cell culture-derived biopharmaceuticals. These cell lines usually have A- or C-type retrovirus-like particles which are visible by transmission electron microscopy (TEM) even when infectivity (IF) or reverse transcriptase activity (RTA) cannot be demonstrated. The supernatant of the post-production cell cultures, therefore, also needs to be evaluated by TEM for viral burden. A major question, however, is how to establish a quantitative viral load estimate for the evaluation of a purification process. The FDA recommends that a purification process for viral contaminants remove or inactivate 3-5 logs over the estimated viral burden. Viral particles are difficult to identify and quantify, however, by conventional negative staining. We present a comparison of infectivity assay, reverse transcriptase assay, negative staining, and thin sectioned TEM. These assays were performed on four samples. Ultracentrifuged sediments of cleared cell-culture media were measured, fixed and processed. Thin sections were evaluated by TEM and the number of viral particles estimated by morphometric derived quantification. Retrovirus particles were easily identified and quantified when examined by TEM as compared to negative staining and correlated with the other viral assays (IF, RTA). These results demonstrate that the TEM thin section method was a superior technique to negative staining for estimating viral particle load in cell-culture supernatant. To validate further the plastic embedding with thin sectioning, we evaluated cell culture supernatants (pellets) for retroviral burden at various dilutions, from two cell lines. Morphometric determinations were made of the number of viral particles present per unit volume and compared to results obtained by infectivity assay. Since the morphometric calculation for viral density assumes even distribution of viral particles, we also evaluated and calculated viral counts on multiple thin sections taken throughout selected pellets.

Materials and Methods

Cell Culture: Three C-type murine retroviruses were used to inoculate either mink lung (MLC) or SC-1 cells. Cultures were passaged twice before seeding cell factories. During log growth, phase cultures were re-fed and supernatants harvested and aliquoted for IF and RT assay.

Infectivity Assay: MLC were infected with supernatant, diluted to proper titre and maintained until foci formed. SC-1 cells were treated with the virus and subsequently lethally UV irradiated. Cells were overlaid with XC cells and, following plaque formation, fixed and stained.

Reverse Transcriptase Assay: Assay included both RT and DNA polymerase templates, controls, and Mn^{++} and Mg^{++} co-factors.

Thin Section TEM: Each viral culture fluid sample was concentrated twice by centrifugation. The remaining supernatant was then centrifuged to obtain a viral culture «pellet» which was fixed in glutaraldehyde-formaldehyde and measured (μm^3). The pellets were processed and embedded in Spurr's epoxy resin. Thin sections (0.08 µm) were cut, mounted on 200 mesh copper grids (grid space = 85 μm^2), and routinely stained. Viral-like particles were quantified from 10 grid squares, i.e. a total volume of 5780 μm^3. The number of viral particles in the entire pellet are calculated as follows: (total volume of pellet/5780 μm^3) (number of viral particles observed). To show reproducibility of viral particles within a pellet, three dilutions of two different viral pellets were evaluated for numbers of virus particles present. Pellets from one dilution of two different viral pellets were sectioned three times with an interval of 15 m between sections to show an even distribution.

Negative Staining: Ten l of 100 nm latex beads were placed on formvar-coated grids (200 mesh), allowed to air dry, and 10 grid spaces were counted. The viral pellets were resuspended with a known quantity of distilled water and a 10 µl aliquot was placed on the grid with the latex beads, drained and stained with 2% phosphotungstic acid. Ten grid spaces were evaluated for viral particles and latex beads. Total viral particles were calculated using the area of the grid and a percent recovery of the latex beads to estimate the viral particles present in the original pellet.

Results and Discussion

Retroviral particles were easily recognized and quantified. Viral particle counts for each pellet at each dilution are presented in Table 1. A direct correlation of viral density existed between the infectivity and the TEM quantitative results. It was found that high titre estimates could be inaccurate, but it was shown that dilution of the sample (Table 2) could give reliable results. Furthermore, even distribution of viral particles in the pellet was demonstrated, evidenced by equal viral counts from each representative section of the pellet (Table 3). An even distribution of viral particles throughout the pellet shows that the morphometric methodology for determination of viral load is a valid method for determining retroviral burden in cell-culture supernatant.

Table 1: Comparison of assays for quantifying virus.

	Infectivity	Reverse transriptase		Transmission electron microscopy	
Retrovirus	(average titre)	(cpm/ml)	(cpm/Infectious particle)	Negative stain	Thin section
Rauscher murine leukemia	50.4×10^5 pfu/ml	6.85×10^5	0.14	0.180×10^5	$>6920 \times 10^5$
Xenotropic murine leukemia	3.71×10^5 ffu/ml	4.39×10^5	1.18	Too clumped to count	$>4190 \times 10^5$
Mink cell focus	15.1×10^5 ffu/ml	5.1×10^5	0.35	0.100×10^5	$>4190 \times 10^5$
P623 AKV Diluted 10-fold	423×10^5	3.31×10^5	0.008	not done	2800×10^5

Table 2: Reproducibility of methodology.

Viral concentration (particles/μl)	Pellet volume (μm^3)	TEM viral density (particles/μm^3)	Total viral count/pellet
Sample A			
1.8×10^4	1×10^9	15730	2.7×10^9
1.8×10^3	1×10^9	871	1.5×10^8
1.8×10^2	2×10^9	136	4.7×10^7
Sample B			
1.2×10^4	1×10^9	4994	8.6×10^8
1.2×10^3	3×10^9	153	7.9×10^7
1.2×10^2	2×10^9	25	8.6×10^6

Table 3: Distribution of particles.

Sample	Total viral count/pellet
A	
First section	4.7×10^7
Second section	4.6×10^7
Third section	4.7×10^7
B	
First section	8.6×10^6
Second section	11×10^6
Third section	11×10^6

A determination of viral titre based on infectivity can differ due to time and condition of the cell culture. RT activity determinations as related to different C-type viruses, yield differing infectivity titres. In the event that a cell line has RT activity and viral particles by TEM but no infectious virus, quantification of titre by TEM would be more dependable and would be more objective. Negative staining estimated the viral titre as 10,000 times lower than the thin sectioned TEM estimate. Pelleted and sectioned TEM would therefore provide an estimate of viral burden allowing an additional margin of safety. Dilutions of the sample provided for greater ease in counting. A determination of whether to dilute a sample can be based on the results of TEM examination of a cell culture.

References

1 Bierley ST, Monticello, TM, Watkins LG, Bond SE, Poiley JA, et al., A comparison of methods for the estimation of retroviral burden. Proc Southeast Electr Microsc Soc, 1992, vol 15, abst.

2 Bierley ST, Monticello, TM, Morgan, EM, Leininger JR: Morphometric Estimation of Viral Burden in Cell Culture Material By Electron Microscopy (TEM), in Bailey GW, Small JA (eds), Proceedings, Electron Microscopy Society of America. Fiftieth Annual Meeting, San Francisco Press, (1992).

3 Poiley JA, Bierley ST, Hillesund T, Nelson RE, Monticello TM, Raineri R: Methods for Estimating Retroviral Burden. Biopharm 1994;7 (4):32-35.

Dr. S.T. Bierley, Pathology Associates International, 4915-D Prospectus Drive, Durham, NC 27713, USA.

Brown F, Lubiniecki AS: Viral Safety and Evaluation of Viral Clearance from Biopharmaceutical Products. Dev Biol Stand. Basel, Karger, 1996, vol 88, pp 167-176

The Reverse Transcriptase Assay: Problems and Practical Solutions

R.A. Metz, L. Mo, S. Per

Quality Biotech® Inc., 1667 Davis Street, Camden, NJ 08104, USA

Key words: Reverse transcriptase, retroviral detection, cellular polymerases.

Abstract: The detection of retroviruses has become critical for addressing the safety concerns associated with biologically derived products, including those derived from blood and cell line substrates, and gene therapy based systems. Most, if not all, retroviruses encode a unique enzyme called reverse transcriptase whose RNA-dependent DNA polymerase activity can be used as a marker for detecting retroviral contamination in test material. In this presentation we document some practical concerns when using the reverse transcriptase assay for detection of retroviruses. We also illustrate important aspects in the assay design by presentation of results that needed supplemental testing to enable accurate assessment of retroviral contamination.

INTRODUCTION

An important safety aspect of biologically derived material is the virological assessment of cell substrates. In particular, the presence of retroviruses is a major source of concern because retroviral infection has been associated with serious disorders such as neoplastic, immunosuppressive, and other degenerative diseases [1]. In many cases, detection of retroviruses by cell culture techniques is difficult because some retroviruses are not cytopathic in vitro [2]. Retroviruses have a unique replication cycle whereby the viral RNA genome is copied into a DNA provirus using a virion associated RNA-dependent DNA polymerase known as reverse transcriptase (RT). The provirus integrates into the host's chromosomal DNA where it may either remain as a latent infection or be activated to produce new progeny. Since RT is found in most, if not all, retroviruses, the enzymatic activity of this marker protein may be measured for the purpose of detecting retroviral contamination in cultured cells and final bulk product [3, 4]. Briefly, the RT assay detects RNA-dependent DNA polymerase activity by quantifying the incorporation of radiolabelled nucleotide precursors into DNA that is copied from a viral template-primer [poly(rA)-oligo(dT)]. Newly synthesized cDNA is separated from the other reaction components by binding to a charged diethylaminoethyl (DEAE) membrane. If retroviral contamination is present in the test material, radiolabelled nucleotides will be incorporated into DNA which is copied from the viral RNA template. If RT is not present in the test material, synthesis will not occur and labelled DNA will not be produced. Although this assay is less sensitive than infectivity assays, it is a rapid, relatively inexpensive, broad spectrum assay which, under the appropriate conditions, can detect the presence of most, if not all, retroviruses.

There are four important considerations when using this assay as a broad-spectrum retrovirus detection method. First, the reverse transcriptase enzymes encoded by different viruses possess varying kinetics and may require special conditions for optimal activity. It is, therefore, difficult to design a single-end-point assay which will detect all RT activity. Secondly, although the assay is typically very sensitive, RT activity associated with low viral titres may be difficult to differentiate from background activity. For this reason, samples are often concentrated by high speed ultracentrifugation. To minimize background from free radiolabelled nucleotides, detection of cDNA production is performed by filtration through a DEAE membrane to capture only newly synthesized cDNA molecules. Thirdly, in our experience, most test samples are cell-free supernatants which may contain cellular polymerases resulting from cell lysis during harvesting. This is particularly troublesome for supernatants obtained from high density cell cultures. Cellular polymerases and other cellular DNA-modifying enzymes may contribute to activities that are difficult to distinguish from that of RT. Finally, samples may contain inhibitors of RT activity (such as nucleases, phosphatases or culture additives) which may lead to false negatives.

To demonstrate the use of an RT detection assay, we describe analyses of four representative samples whose data would be difficult to interpret unless supplemental studies were performed. Based on these findings, we advise that an assay be validated for a particular cell line or protein product using the supplemental studies described below, which are designed to determine sample-assay compatibility before routine analyses on related samples are performed. This pre-testing may indicate a need to alter the method of sample preparation.

Results

The RT assay is based on the detection of RNA-dependent DNA polymerase activity which is present in extracellular retroviral particles [3, 4]. Many factors may influence the level of enzymatic activity measured in this assay, including the time of cell culture before sampling, the type and titre of retrovirus present, and potential enhancing or inhibitory constituents contained in the test material. In general, a linear correlation has been found to exist between the amount of RT activity in a viral stock and the number of viral particles [5-7].

To illustrate the presence of potential factors influencing the detection of RT, we report data from four previously uncharacterized test samples. Supernatants from three test cell lines (Samples A-C) were harvested 24 hrs. after the addition of fresh medium to replicating cells. Sample (D) represents a bulk harvest of a protein product. All samples were first clarified by low-speed centrifugation to minimize the chance of false positive reactions resulting from cellular polymerase activity contained in cellular debris. An aliquot (2 ml) was then ultracentrifuged and the concentrated pellet was solubilized. Aliquots were added to buffered reaction mixtures containing either Mn^{2+} as required for certain Type C viral particles or Mg^{2+} as required for many Type B and D viral particles [8]. Other constituents in the reaction mix included a non-ionic detergent to destroy the viral envelope and release RT, unlabelled and ^{32}P-labelled dTTP, dithiothreitol as a reducing agent, and poly(rA)-oligo(dT) template-primer to discriminate specifically between viral and cellular DNA polymerases [1-3]. For the cellular template assay, which detects cellular polymerases, poly (dA)-oligo (dT) template-primer is used in the reaction mix instead of the RNA template-primer.

Following a two-hour incubation period, each reaction mixture was transferred to a dot-blot manifold and gently filtered through a DEAE membrane. The membrane was washed with a buffered salt solution to remove unincorporated radiolabelled nucleotides. Samples containing radiolabelled DNA were counted and the mean values of triplicate reactions were calculated.

Test Results

For the viral or cellular template-primer assay, when the mean counts per minute (CPM) value for the test sample minus the appropriate negative control (adjusted mean) is less than or equal to 1000 CPM, the sample is considered negative. The dual template assay uses separate reactions, containing either a viral (RNA) or a cellular (DNA) template. DNA polymerases will synthesize DNA from a DNA template at a rate greater than that from an RNA template. Although studies have addressed the differences in rates of DNA synthesis for cellular polymerases using different templates (RNA versus DNA), these studies do little to address the issue of samples containing unknown quantities of cellular polymerases, nucleases, phosphatases, etc. On the other hand, RT will synthesize cDNA from an RNA template at a rate far greater than from a DNA template; it is therefore appropriate to use a DNA template to detect the presence of cellular DNA polymerases. When the difference between the adjusted mean CPM value from the cellular template and viral template (cellular activity being greater than the viral activity) is less than 1000 CPM, the sample is considered to be positive for cellular enzyme activity and negative for RT activity.

The data from the RT assay under Mn^{2+} conditions using a synthetic viral template-primer (rA-dT) combination are shown in Table 1. In this particular set of experiments the reactions performed in the presence of Mg^{2+} were all close to background incorporation and are not reported. Samples A and D have an adjusted mean value greater than 1000 CPM and are scored as positive. Samples B and C have an adjusted mean CPM value less than 1000 CPM and are scored negative.

The data for sample D were particularly difficult to interpret because the triplicate values are not close. This inconsistency suggests either a sampling error or an experimental error (e.g. cross-over contamination). Conversely, the data for sample A are typical for an assay having little variability. The adjusted mean value for both samples suggests RT activity. To determine whether cellular enzymes are contributing to the nucleotide incorporation of samples A and/or D, a separate set of reactions was performed using either Mn^{2+} or Mg^{2+} (Mg^{2+} data are not reported) using a cellular template-primer combination (dA-dT). Table 2 shows the data from this experiment. Both samples A and D have a mean value greater than 1000 CPM and are scored as positive for cellular polymerase. Given the consistency of the triplicate values for sample A, we conclude that the sample is contaminated with cellular

Table 1: Viral template RT assay.

Sample	CPM	Mean	Adjusted mean	Result
AKR-MuLV	100840 99307 103287	101145	101123	*Positive*
VDB	23 28 16	22	0	*Negative*
DMEM	34 52 61	49	27	*Negative*
Sample A	3966 3483 3412	3620	3598	*Positive*
Sample B	120 87 101	103	81	*Negative*
Sample C	28 22 27	26	4	*Negative*
Sample D	8036 105 79	2740	2718	*Positive*

The adjusted mean value is the sample mean minus the mean value of the negative control.

DNA modifying enzymes. Given the high cellular activity (14,537 CPM), compared to viral activity (3,620 CPM), it is likely that the activity reported for sample A on the viral template is also due to cellular polymerase cross-over activity. Sample A is reported as negative for RT activity. In this case, it would be advised that the conditions for sampling be reviewed and altered in order to eliminate or reduce the influence of cellular enzymes.

Sample D, on the other hand, gave the similar inconsistent triplicate CPM values. It would appear likely therefore that this is a sample-related effect rather than an experimental or sampling error. While it would be tempting to discard the outlying datum point, the implications of *US vs. Barr Laboratories* would advise extreme caution instead. Therefore an out-of-limits investigation, parts of which are reported below, was initiated. Our initial hypothesis was that the sample material, which is a bulk harvest protein sample, was trapping free radiolabelled nucleotides either due to protein aggregation or ionic interactions. To explore this possibility, we performed the RT assay in which no viral or cellular template was added to the reaction mix. Similar data were obtained in that triplicate values were disparate (Table 3). In addition, direct radioanalytical analysis indicated that the radiolabel was not homogeneously distributed on the membrane (data not shown). Taken together these data suggest that protein aggregates retarded free radiolabel on the DEAE membrane, resulting in retention and non-homogeneous distribution of free radiolabelled nucleotides.

Table 2: Dual template RT assay.

Sample	Viral template (rA-dT) CPM mean	Cellular template (dA-dT) CPM	Cellular template (dA-dT) mean	Cellular polymerase activity result
E. coli Pol I 1 unit	ND	19591 20623 25263	21826	*Positive*
AKR-MuLV	101145	893 900 959	917	*Negative*
VDB	22	22 25 20	22	*Negative*
DMEM	49	53 61 46	53	*Negative*
Sample A	3620	14264 14899 14447	14537	*Positive*
Sample B	103	18 24 33	25	*Negative*
Sample C	26	45 41 20	35	*Negative*
Sample D	2740	3120 441 57	1206	*Positive*

The adjusted mean value is the sample mean minus the mean value of the negative control.

Table 3: No template RT assay.

Sample	Viral template (rA-dT) CPM mean	No template CPM	No template mean	Result
AKR-MuLV	101145	204 305 191	233	*Negative*
VDB	22	46 46 47	46	*Negative*
Sample D	2740	35 2781 61	959	*Negative/ Positive*

As illustrated in Tables 1 and 2, samples B and C scored negative for both viral and cellular polymerase activities. However, to determine whether other sample-specific factors were influencing these data (such as RT inhibitors), positive and negative control dilution series were performed (shown in Figs. 1 and 2). In this experiment a series of two-fold dilutions was analysed on the controls and samples. In one set of reactions the positive control and samples were diluted with the negative control (viral dilution buffer-VDB). In general, a linear correlation exists between the amount of RT enzyme and the CPM incorporated. The CPM values for sample C remained low. The positive control showed an expected linear fall in activity with each two-fold dilution. Sample B, on the other hand, increased from 103 CPM, when tested undiluted (neat), to 13,000 CPM at a 1:4 dilution (Fig. 1, dilution 3). When the negative control or samples were mixed at an equal ratio with the positive control (Fig. 2, 1:1) and then diluted two-fold in VDB (Fig. 2, 1/2, 1/4, 1/8), values for sample C were similar to the equivalent dilutions of the positive control in VDB. This supports the notion that sample C does not contain substances interfering with RT activity. Conversely, sample B possessed lower activity in dilution 2 (Fig. 2, 1:1 with positive control) and slightly higher activity when diluted further. Taken together these data demonstrate that sample B contains inhibitors of the RT activity. The inhibitors, which mask the presence of viral contamination at lower dilutions, have a more limiting effect on the RT activity when diluted further. This sample is now scored positive. The importance of performing positive control dilution experiments is discussed further below.

The final results for the four samples are summarized in Table 4. The conclusions made from these data emphasize the need to conduct a full series of experiments on previously uncharacterized samples to determine compatibility of the sample and the assay and to ensure that the correct assessment of viral contamination is being made.

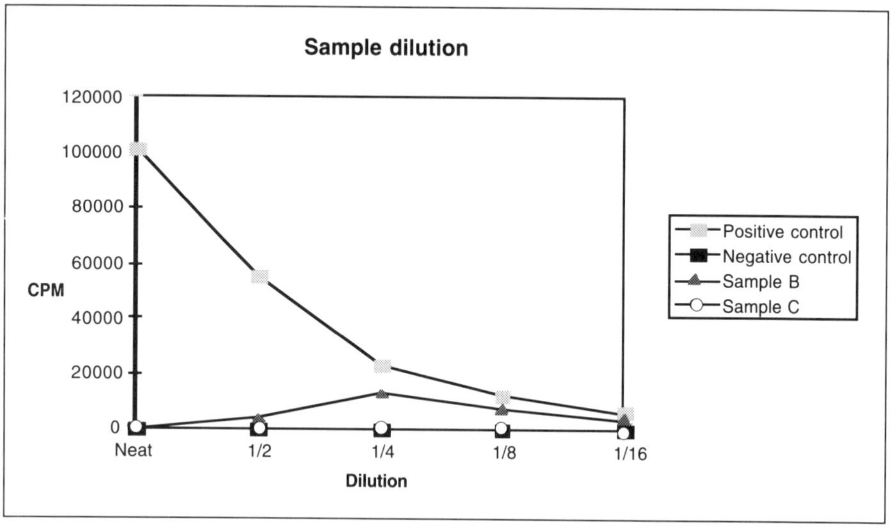

Fig. 1: Test samples B and C, positive and negative controls were assayed undiluted and as two-fold dilutions in VDB.

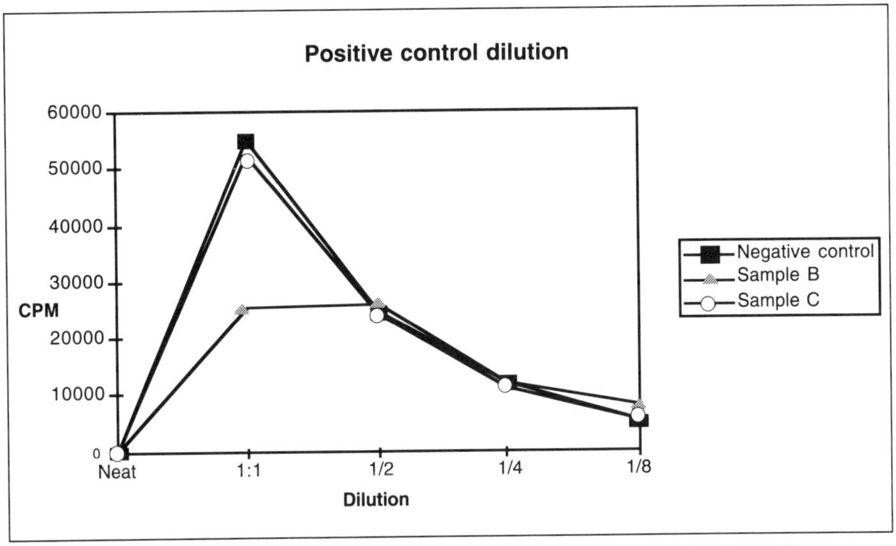

Fig. 2: Test samples B and C, and negative control were assayed undiluted with positive control (1:1), and subsequent two-fold dilutions in VDB.

Table 4: Conclusion.

Sample	Result	
VDB	*Negative*	
DMEM	*Negative*	
Sample A	*Negative*	Contaminating cellular polymerases contribute to positive result on viral template due to cross-over activity.
Sample B	*Positive*	Sample contains retroviral contamination masked by inhibitors of RT activity.
Sample C	*Negative*	Sample is negative for cellular and viral polymerases
Sample D	*Negative*	Sample is a highly concentrated protein solution, which contributed to outlying data points. In these reactions it was unlikely that there was a non-specific retention of free nucleotides. The other reaction samples are negative.

Discussion

Broad spectrum assay

Reverse transcriptase enzymes encoded by different retroviruses are diverse in nature and possess varying kinetics, and may require special conditions for optimal activity [1-4]. It is impractical to perform time-course experiments for each RT under several different conditions due to the added burden of cost and time. Therefore, the RT assay is typically performed using two different divalent cationic conditions (Mn^{2+} and Mg^{2+}) at a single time point. The conditions and reaction time are chosen to maximize detection of most common contaminating retroviruses. It follows then that the assay is not necessarily a quantitative one. The process of determining the suitability of the assay conditions for detection of a variety of retroviruses is an on-going process that will undoubtedly help to establish the range of detection and overall usefulness for monitoring retroviral contamination.

Assay sensitivity

The RT assay, which detects approximately 10^5 p.f.u. per millilitre of virus, is typically less sensitive than other quantitative assays, such as infectivity assays. The criteria for determining a positive result are somewhat arbitrary and are based on a confidence limit obtained from experimental data which determine that the incorporated nucleotides are due to RT activity. Typically, this has been defined as 5-10 times the CPM value of the negative control [1-4]. However, RT activity associated with samples having low viral titres may be difficult to discriminate from background activity. A series of studies conducted at Quality Biotech has established the average background for this assay at 62 CPM with a range of 10-200 CPM. Given the variability in the background CPM, we set the RT activity determination point at 1000 CPM, i.e. values ≤ 1000 are negative and values > 1000 CPM are positive. In those few cases, where values are close to the cut-off value (1000 CPM), we recommend that a more concentrated sample be analysed, along with a two-fold dilution series.

Cellular polymerases

Cellular DNA polymerases, terminal deoxynucleotidyl transferases and other DNA modifying enzymes, when present in the test material, may contribute to activities which are difficult to distinguish from that of reverse transcriptase. Careful sample preparation can often eliminate cell lysis and cellular enzyme contamination. Supernatants from high density cultures are particularly troublesome to harvest free from cellular debris. To reduce contamination with cellular enzymes that may cause nucleotide incorporation, extraneous cellular debris is removed by a low speed centrifugation step. However, soluble cellular enzymes may still persist. To optimize the sample collection procedure and to detect the cellular polymerases, separate reactions which use a DNA template should be performed. Sample A is an example of a test material in which cellular polymerase affected the CPM incorporation in the viral template assay. The presence or absence of cellular DNA modifying enzymes is often indicative of sample handling during collection and treatment before assaying.

Positive control dilution series

An important concern in the RT assay is the potential presence of interfering substances that inhibit or mask enzyme activity. To address this issue, a positive control dilution series is used, whereby the test article is diluted in the presence of positive control virus or VDB. This serves as an essential control for determining whether the sample contains inhibitors of RT activity. Sample B represents a sample containing a retroviral contaminant masked by the presence of a substance which interferes with the detection of RT activity. The data in this particular example can be clearly interpreted. In many cases, when interfering substances are present and retroviral contamination is low, a conclusive result for detection of retroviruses may not always be easily obtained. In such cases, the effect that a sample has on the positive control must be analysed carefully. Any significant inhibitory effect on the positive control-sample dilution, when activity on the VDB-sample dilution is negative, casts doubt on the interpretation of the data. In such cases the final result must be considered inconclusive.

Conclusion

The RT detection assay is a rapid, inexpensive, and rather broad spectrum assay for the detection of contaminating retroviruses. However, there are at least four important issues concerning the use of the RT assay that must be considered carefully. These include the following: 1) the enzyme kinetics and optimal reaction conditions for reverse transcriptase from various retroviral sources can vary significantly; 2) the assay is not quantitative and has somewhat arbitrarily-defined pass or fail criteria; 3) the type of samples commonly analysed are cell free supernatants which may contain DNA modifying activities leading to false positives; 4) samples may contain inhibitors of RT activity, such as nucleases, which may lead to false negatives. Although the last two problems are frequently encountered and are inherent in most biological products, they are not insurmountable and can be assessed directly. In this presentation, we describe data from four test samples, three of which yielded inaccurate results when analysing the data from the RT assay using a synthetic viral template. The purpose in presenting these data was to illustrate the usefulness in performing supplemental control studies which are critically important in assessing the sample-assay suitability.

References

1 Onions DE, Jarrett O: In Onions DE, Jarrett O (eds): Cancer Surveys. Oxford University Press, 1987, vol 6 (1), pp 1661-1680.

2 Coffin JM: Retroviridae and Their Replication, in Coffin JM, Fields BN, Knipe DM et al. (eds): Virology, Second Ed. New York, Raven Press, 1990, pp 1437-1500.

3 Kacian DL, Spiegelman S: Purification and detection of reverse transcriptase in viruses and cells, in: Methods in Enzymology, New York, Academic Press, 1974, vol. 29, pp 150-173.

4 Temin H, Baltimore D: RNA directed DNA synthesis and RNA tumor viruses, in: Advances in Virus Research. New York, Academic Press, 1972, vol. 17, pp. 129-185.

5 Levy JA, Shimabukuro J, McHugh T, Casavant C, Stites D, Oshiro L: AIDS-associated retroviruses can productively infect other cells besides human T helper cells. Virol 1985;147:441-448.

6 Feorino P, Forrester B, Schable C, Warfield D, Schochetman G: Comparison of antigen assay and reverse transcriptase assay for detecting human immunodeficiency virus in culture. J Clin Microbiol 1987;25:2344-2346.

7 Roy-Burman P, Dougherty M, Pal BK, Charman HP, Klement V, Gardner MB: Assay for type C virus in mouse sera based on particulate reverse transcriptase activity. J Virol 1976;19:1107-1110.

8 Scolnick E, Rands E, Aaronson SA, Todaro GJ: RNA-dependent DNA polymerase activity in five RNA viruses: divalent cation requirements. Proc Nat Acad Sci USA 1970;67:1789-1796.

Dr. R.A. Metz, Quality Biotech Inc., 1667 Davis Street, Camden, NJ 08104, USA.

Brown F, Lubiniecki AS: Viral Safety and Evaluation of Viral Clearance from Biopharmaceutical Products. Dev Biol Stand. Basel, Karger, 1996, vol 88, pp 177-188

Use of the Polymerase Chain Reaction to Screen Cell Banks for Retroviruses

A.M. Deatly, J.M. McAuliffe

Viral Vaccine Research, Lederle-Praxis Biologicals, Pearl River NY, USA

Key words: PCR, cell bank screening.

Abstract: The polymerase chain reaction (PCR) may be used to increase the sensitivity for detecting nucleic acids of specific adventitious agents. PCR can be used in combination with other assays such as infectivity or co-cultivation to increase the sensitivity of detecting adventitious agents or to identify a particular adventitious agent present in the test sample. In addition, PCR is useful for confirming the results of other assays such as reverse transcriptase or electron microscopy. Sources of the test sample, passage history and raw materials added to the test sample are considered when deciding upon the appropriate screening procedures and the adventitious agents for which to screen. In developing a PCR assay, the experiments should be designed to address specificity, sensitivity and reproducibility. Assay development issues include design of primers for the particular retroviruses in question, determination of quantity of sample to test, potential levels of interference, and appropriate positive and negative controls. Validation of a PCR assay is necessary to prove reproducibility for detecting the retrovirus. Data are presented on the qualification of a primate master cell bank to assure the absence of simian immunodeficiency viruses.

Introduction

Points to Consider [1] and WHO Guidelines suggest the use of infectivity assays, reverse transcriptase assays, and transmission electron microscopy to screen cell banks for retroviruses [2]. Newer technology, such as the polymerase chain reaction (PCR), is available and may also be useful in screening cell banks for retroviruses. PCR is a powerful molecular biology tool used to amplify specific nucleic acid sequences in an exponential fashion. PCR offers the advantage of high sensitivity, specificity and speed. It may be used when there is no infectivity assay or the sensitivity of an infectivity assay is too low. PCR may be the assay of choice when the test sample interferes with detection of potential retroviruses in an infectivity assay or when a specific retrovirus needs to be identified.

The RNA genomes of retroviruses replicate via a DNA intermediate, or proviral DNA. Proviral DNA molecules have the capacity to integrate into the host cell DNA genome [3]. Therefore, PCR assays can be designed to screen cell banks for proviral DNA rather than screening for genomic RNA. Screening for genomic

RNA of the retrovirus requires a reverse transcriptase step before PCR to produce cDNA copies of the RNA genome.

PCR assays may be used to screen primate cell banks for specific retroviruses with an increased level of detection relative to biological assays. Knowledge of the nucleic acid sequence of the retrovirus target or retroviral gene is a prerequisite for the PCR assay development. PCR assay results only indicate the presence of nucleic acid related to a particular retrovirus. They do not offer conclusive evidence for the presence of infectious retrovirus particles.

In this article, we present our strategy for developing PCR assays to assure the absence of simian retroviruses in primate cell banks. Specific PCR assays have been developed to detect the proviral DNA of simian retroviruses [4-12] such as simian immunodeficiency viruses, simian retroviruses, simian T lymphotropic viruses and simian foamy viruses (see Table 1). Issues related to optimization of PCR assays are discussed to highlight development of reproducible, specific and highly sensitive PCR assays to target simian retroviral sequences.

Table 1: Simian retroviruses infecting monkeys.

Virus	Designation	Retrovirus family	Type
Simian immunodeficiency virus	SIV	lentivirinae	E
Simian retrovirus (type D)	SRV	oncovirinae	D
Simian T lymphotropic virus	STLV	oncovirinae	C
Simian foamy virus	SFV	spumavirinae	F

Cell Banks

Cell banks derived from primate tissues should be screened for the presence of simian retroviruses. Where possible, it is advantageous to test the serum of the primate donor to ensure it is seronegative for the known simian retroviruses [13-15]. A master cell bank can be produced from a continuous cell line (Fig. 1). Since subsequent working cell banks will be produced from aliquots of the master cell bank, only the latter needs to be screened. Even if the primate donor appears retrovirus-free it is advisable to screen the cell line (master cell bank) derived from the animal's tissue.

Polymerase Chain Reaction Assay

PCR results in the amplification of specific DNA sequences using small oligomers (primers) homologous to the regions surrounding the target sequence (Fig. 2 and ref. 16). Target sequences are amplified exponentially through many cycles of stepwise temperature changes which promote:

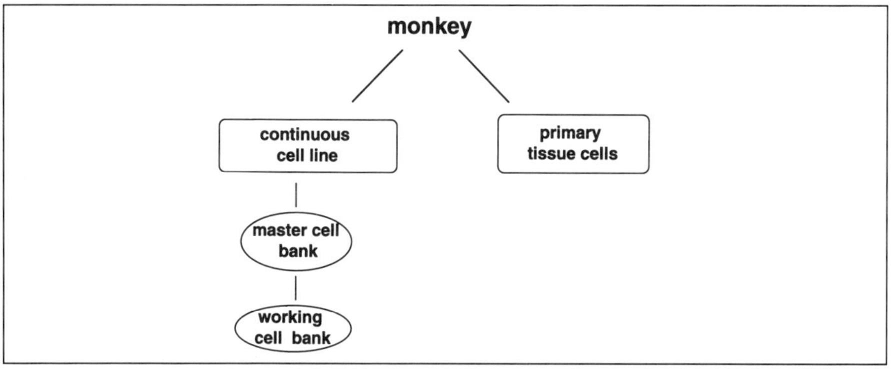

Fig. 1: **Production of a primate master cell bank**. This flow chart represents stages in the production of a continuous or primary cell line derived from monkey tissues. Master cell banks and working cell banks are derived from the continuous cell line.

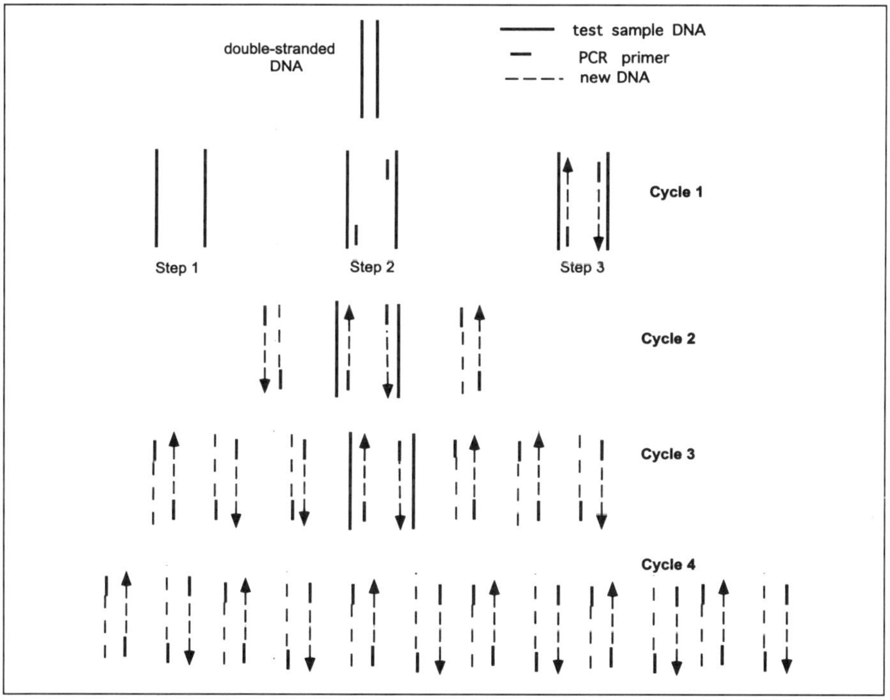

Fig. 2: **The polymerase chain reaction**. This figure illustrates the process of amplifying the target sequence of a test sample using specific primers. In the first cycle, step 1 illustrates denaturation of the double-stranded DNA test sample; step 2 illustrates annealing of the primers on the test sample DNA; and step 3 illustrates elongation from the primers complementary to the test sample DNA sequence. The dashed lines represent elongation from the primers with Taq polymerase. Cycle 2 and 3 represent the annealing of the primers and amplification of both the test sample DNA and the amplified cycle 1 PCR product. Cycle 4 shows that after a few cycles the predominant target sequence is the amplified PCR product. The target sequence is amplified exponentially over the course of 25-35 cycles.

- Step 1 -denaturation of double-stranded DNA molecules
- Step 2 -annealing of the primers to homologous sequences
- Step 3 -elongation of the primers

Elongation of the primers is achieved by polymerization of deoxynucleotides with a heat-stable polymerase. Amplified PCR products are visualized as a band of anticipated size after agarose gel electrophoresis. The PCR bands are stained with ethidium bromide, which intercalates into double-standed DNA and fluoresces under UV light. DNA molecular weight markers and positive controls are electrophoresed to allow determination of the size of the amplified PCR product.

Optimization of PCR Assays

The specific aims in developing a valid PCR assay are high sensitivity, specificity and reproducibility. There are many parameters to consider in the development of a PCR assay which may affect the sensitivity and specificity of the assay [17]. The following parameters are discussed in optimizing a PCR assay to screen cell banks for simian retroviruses:

- DNA purification procedure
- specific primer design
- positive and negative controls
- false positives and false negatives
- potential PCR inhibitors
- test sample volume

DNA purification procedure

In screening primate cell banks for simian retroviruses, the DNA is purified from the master cell bank. PCR assays can be developed which screen for retroviral proviral DNA to avoid using infectious virus particles or retroviral RNA genomes. DNA purified from a master cell bank should be intact, high molecular weight DNA. Proviral DNAs of retroviruses have the potential to integrate into the host cell DNA genome [18]. However, proviral DNA molecules of some retroviruses, such as lentiviruses, may not integrate in host DNA genomes [19]. To increase the possibility of detecting these retroviruses, DNA purification should assure inclusion of potential unintegrated proviral DNAs.

PCR primer design

Design of the primers should correlate with the goal of the particular PCR assay. If the goal is to detect all subspecies of a particular retrovirus, the primers should be homologous to regions of retroviral genes conserved among all the subspecies. With careful design of the primer set, it may be possible to develop pan-specific PCR assays or PCR assays which detect retroviral genes from a broad range of simian retrovirus types or subtypes [20, 21]. If the goal is to design a PCR assay to detect a broad range of retroviruses, which may not be possible with a single primer set, then it may be necessary to design degenerate primers [21-23]. Degenerate primers are mixtures of oligonucleotides that differ only at specific non-conserved nucleotides. If the goal of the assay is to identify the specific retrovirus present in the test sample, then the primer sets should be homologous to sequences that define a specific retrovirus species or serotype [8].

PCR controls

Positive controls should be included to monitor the amplifiability of the retroviral sequences targeted in the assay. Where there are concerns about the use of infectious virus or full-length retroviral clones, a cloned retroviral gene may be sufficient as a control. Negative controls, or controls where no DNA template is added, should be included in the assay to monitor potential introduction of retroviral sequences from PCR reaction components. Potential cross-contamination during the preparation of the PCR reaction samples should also be monitored with no DNA template control tubes open during the template addition step.

To minimize the potential for false-positive results in the PCR reaction, a second primer set designed from an additional simian retroviral gene may be included in a multiplex PCR assay [5, 9, 24-26]. To confirm the presence of a retrovirus in the test sample, both primers should amplify the respective target sequences.

To minimize the potential for false negatives, a primer set specific for sequences such as housekeeping genes may be included in a multiplex PCR assay. This control will help to monitor the integrity or amplifiability of the test sample DNA.

Potential PCR inhibitors

Potential PCR inhibitors present in the test sample may affect amplification of the retroviral target sequences. By testing the positive control DNA in the presence and absence of DNA from the test sample, the presence of PCR inhibitors can be determined.

Nucleic acid present in the test sample may interfere with the detection of simian retroviral sequences. A high concentration of nucleic acid in the sample may affect the ability of the primers to find their homologous sequence counterpart(s). Unrelated but quasi-homologous sequences present in the test sample may also interfere with the primers locating the homologous counterparts. Non-specific amplification of unrelated sequences may be evident by amplification of PCR products of different sizes. More stringent amplification conditions may eliminate non-specific amplification. Evaluation of the level of interference can be determined by comparing the sensitivity of detecting dilutions of the positive control in the presence and absence of the test sample DNA. Assay sensitivity can also be determined by supplementing (spiking) the same dilution of the positive control into different dilutions of test sample DNA.

Test sample size

The amount of sample to test should be similar to a level that could assure that the cell bank is retrovirus-free. For a cell bank, the test sample should be the maximum DNA concentration which gives the highest sensitivity for detecting a supplemented (spiked) positive control. This sensitivity can be calculated by the number of molecules of the positive control present in the sample relative to the number of primate cell genomes.

If the test sample is a cell bank from a continuous cell line originally derived from a single cell, the lowest limit of detection is one copy of the simian retroviral genome per cell. The detection limit of a simian retroviral genome in primary tissue culture, however, should be much lower than one copy per cell [5]. The number of plaque forming units (pfu) for a retroviral infectious unit and the number of pfus produced per infected cell can be estimated. This estimate can be used to determine the range of detection limit needed to assure the absence of the simian retrovirus. It is imperative that reproducible results are obtained each time the same volume of test sample is assayed.

PCR Sensitivity

The absence of a specific retrovirus in the test sample must be qualified by the limit of detection for the particular PCR assay. The sensitivity of the PCR assay must be determined. Dilutions of the positive control are supplemented (spiked) into test sample DNA to determine the lowest concentration of target amplified. The number of positive control molecules is calculated relative to the number of test sample genomes present at the lowest limit of detection. This ratio should correlate with the limit for detection of potential simian retroviruses in the test sample.

PCR Specificity

In optimizing PCR assays, there is usually a balance between sensitivity and specificity. Selected primers may cross-react non-specifically with test sample DNA sequences. Specificity of the PCR reaction is demonstrated by the amplification of an expected size band when using a particular primer set. Liquid [27] or Southern blot hybridization [28] of the amplified PCR products will prove specificity of the PCR amplification. A labelled oligonucleotide homologous to internal sequences of the expected amplified PCR products may be used as a probe. This added step may also improve the sensitivity for detecting specific retroviral sequences.

PCR Reproducibility

To prove that the optimized PCR assay is reproducible, one should validate the assay by testing the samples several times. The same result should be obtained by adding, in duplicate, three replicate sets of different dilutions of the positive control into the test sample.

Examples of PCR Assays to Screen for SIV Proviral DNA

A primate master cell bank had been qualified previously for retroviruses using electron microscopy, reverse transcriptase assay and Southern blot hybridization

[29]. Southern blot analysis was performed with a probe specific for Mason-Pfizer monkey virus, a simian retrovirus type D. PCR assays have recently been applied to screen this master cell bank for other simian retroviruses. The presumed primate origin of this master cell bank are African green monkeys (agm), *Cercopithecus aethiops* subspecies *pygerythrus* [30, 31]. Simian retroviruses associated with these monkeys include SIV_{agm}, a widely diverse genetic group [31]. Therefore, PCR assays designed to target the most conserved retroviral gene, the *pol* gene [31], were selected [4, 5]. One selected PCR assay was originally developed by Allan et al [4] to study the species-specific diversity among simian immunodeficiency viruses associated with African green monkeys. *Pol* gene sequences from DNA purified from uncultured blood of feral West African green monkeys were amplified. The other selected PCR assay was originally developed by Hartung et al [5] to study the viral load in SIV-infected East African green monkeys (Ethiopia and Kenya) which did not demonstrate clinical disease.

Example 1: PCR assay to screen for SIVs associated with West African green monkeys. This assay [4] is a nested PCR reaction. Round one primers (pol A and pol B) and round two primers (pol C and pol D) were designed to target conserved *pol* regions from a widely diverse group of viruses including HIV-2, SIV_{mac}, SIV_{agm}, and SIV_{sm}. Round one primers amplify a 555 bp product and round two primers amplify a 337 bp product. The SIV positive control used in the published assay was SIV_{mac}/BK28. The positive control used in our screen was $pSIV_{mac}$239-5' [32].

Figure 3A shows the agarose gel electrophoresis analysis of round two 337 bp PCR products (see legend for Figure 3A for PCR reaction conditions). No 337 bp product is detectable with test sample DNA only. In addition to the 337 bp product, 465 and 425 bp products are also detectable in samples of highest $pSIV_{mac}$239-5' concentrations. These products are presumably the result of residual amounts of pol A and pol B amplifying in concert with pol C and pol D in the round two reaction. The 425 bp product is SIV_{mac}239-5' specific since it is also detected in the control reaction. Poliovirus receptor gene primers were used as a positive control for test sample DNA amplification. This 226 bp product is amplified in all reactions containing test sample DNA indicating that there are no false negatives.

To determine the sensitivity of this assay, $pSIV_{mac}$239-5' DNA, in concentrations ranging from 3 pg to 0.00001 pg, was added to 3 μg of test sample DNA. The 337 bp product is detectable with $pSIV_{mac}$239-5' concentrations of 3 pg to 0.001 pg in the background of 3 μg of test sample DNA. Since the lowest dilution detected is 0.001 pg, then the sensitivity of this assay is equivalent to detecting one copy of SIV_{mac}*pol* gene per 10^4 test sample cell genomes (Fig. 4).

Figure 3B shows the hybridization results of the experiment shown in 3A. The SIV_{mac}239-5' specific probe hybridizes to the 337 bp PCR amplified products, as well as the 425 bp PCR product (confirming that this product is SIV_{mac}239-5' specific). The SIV_{mac}239-5' specific probe does not hybridize to the 225 bp product amplified with the poliovirus receptor gene primers. This PCR assay is sensitive enough for us to conclude that our test sample does not contain SIV proviral genomes using SIV_{mac}*pol* gene as the positive control.

Example 2: PCR assay to screen for SIVs associated with East African green monkeys. The primers for this assay [5] amplify a 1260 bp of the *pol* gene in one round of PCR. The SIV positive control used in our assay is SIV_{agm}385-*SStI* [33].

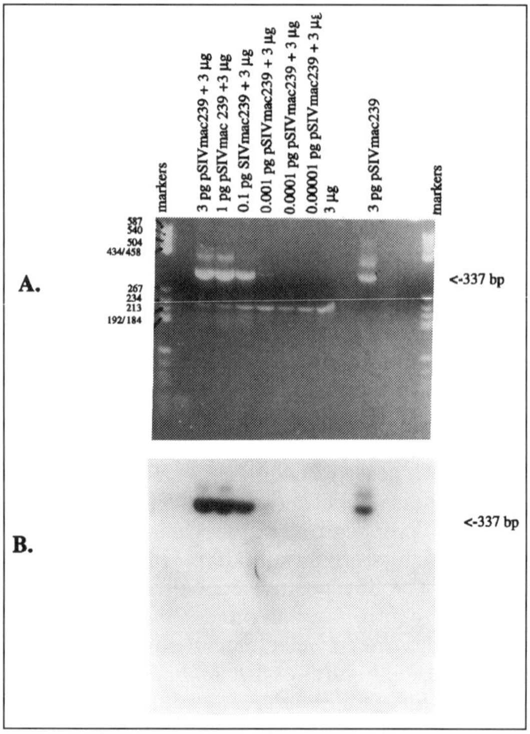

Fig. 3: **Example 1: PCR assay to screen test samples for West African SIV *pol* gene sequences.** This PCR assay is a nested assay with two rounds of PCR amplification.

A. *PCR reaction conditions round 1:* 555 bp of SIV *pol* genes are amplified with the primers pol A (5' AGG GGA GGC TAT ACA TGG GCA AGT AAA TGC 3') and pol B (5' CTG CCT TCT CTG AAA TAG ACC CGA AAA 3'). The PCR reaction mixture consists of 1.5 mM $MgCl_2$, 0.001% gelatin, 50 mM KCl, 10 mM Tris-HCl, pH 8.3, 20 pmoles of each primer, 2.5 units of Perkin-Elmer amplitaq and 0.25 mM dNTPS. The thermal cycler steps include 30 cycles of 94°C, 1 min., 45°C, 1 min., and 68°C, 1 min. *PCR reaction conditions round 2:* 3 µl of round 1 PCR products are added to the same PCR reaction mixture as in round 1 plus 20 pmoles of pol C (5' CAG TAC ATG TTG CAA GTG GAT TTA TAG A 3') and pol D (5' CTT CTT TTA AAA TTC ATG CAA TGA ACT GCC 3') primers. The more stringent round 2 PCR reaction includes 30 cycles of 94°C, 1 min., 60°C, 1 min., and 72°C, 1 min. The final product of round 2 is 337 bp. To control for primate test sample DNA amplifiability, 50 pmoles of SN1 and SN2 primers specific for the human poliovirus receptor gene are also added to the PCR reactions (SN1 is 5' CCA CCC AGG TGC CCG GCT TCT T 3' and SN2 is 5' CTC AGC GAG GCA TTC CGC AGC T 3'; Nawoschik and Deatly, unpublished data). These primers amplify a 226 bp product. 15 µl of each reaction are electrophoresed in 1 × TBE on a 3% agarose gel containing ethidium bromide. Lane 1 and 14 are MWM V (Boehringer Mannheim) DNA markers. Lane 2 is a no DNA template control. Lanes 3-8 are PCR reactions of 3 µg of test sample DNA with the addition of decreasing amounts of $pSIV_{mac}239$. Lane 9 is 3 µg of the test sample DNA alone. Lane 10 is blank. Lane 11 is 3 pg of $pSIV_{mac}239$ alone.

B. Round 2 PCR amplified products were transferred to a nylon membrane according to the method of Southern [28]. The PCR products were hybridized with an oligonucleotide internal to the round two 337 bp PCR $SIV_{mac}239$-5' product (5' GTA AAG ATG GTT GCA TGG TGG GCA GGG ATA GAG CAC ACC TTT GGG GTA CC 3') overnight at 50°C in 5 x SSPE, 10% polyethylene glycol, and 250 µg/ml salmon sperm DNA. The blot was washed twice in 2 x SSC, 0.1% SDS at room temperature, 15 mins, followed by two washes in 2 x SSC, 0.1% SDS at 50°C. The blot was exposed to XAR-5 film for 2 hrs.

Step 1: Determine number of molecules of genomic DNA in 3 μg
size of mammalian genome is 3×10^9 bp
(3×10^9 bp) (660 daltons/bp) = 2×10^{12} daltons (g/mole)
in 3 μg of genomic DNA:
3×10^{-6} g / 2×10^{12} g / mole = 1.5×10^{-18} moles of genomic DNA
(1.5×10^{-18} moles) (6.02×10^{23} molecules /mole) = 9×10^5 molecules

Step 2: Determine number of molecules of SIV_{mac} in 0.001 pg
size of $SIV_{mac}239$ is 1×10^4 bp
(1×10^4 bp) (660 daltons / bp) = 6.6×10^6 daltons (g / mole)
1×10^{-15} g / 6.6×10^6 g / mole = 1.5×10^{-22} moles of SIV_{mac} in 0.001 pg
(1.5×10^{-22} moles) (6.02×10^{23} molecules / mole) = 91.2 molecules

Step 3: Determine molecules of SIV_{mac} / molecules of genomic DNA
$91.2 / 9 \times 10^5 = 1 / 10^4$

Fig. 4.

Figure 5A shows the agarose gel electrophoresis analysis of the PCR products. The 1260 bp product is detectable from the supplemented positive control in lane 1. Because of the apparent non specific amplification of products 1 kb and less in reactions containing test sample DNA, the DNA was transferred to a nylon membrane and hybridized [28] with an SIV_{agm} specific probe (Fig. 5B). There is no detection of SIV_{agm} *pol* gene in the test sample alone. Specific amplification is detectable in the test sample reactions supplemented with 3 pg and 1 pg of SIV_{agm}385-*SStI*. The sensitivity of detecting SIV_{agm}385-*SStI* is equivalent to detecting one copy of SIV_{agm} *pol* gene in 3 test sample genomes. This result indicates that the test sample does not contain SIV_{agm} proviral DNA.

PCR Combined with Other Assays

PCR can be performed in parallel with infectivity [24] or co-cultivation assays (i.e. plus reverse transcriptase [20, 34], electron microscopy, Southern blot hybridization or immunofluorescence antibody staining) to determine if proviral DNA detected by PCR represents an infectious agent. The advantage to using infectivity or co-cultivation assays is that a broad range of retroviruses may replicate under the same conditions. In fact, other retroviruses including unidentified retroviruses may also replicate under the selected conditions and therefore, be detectable. PCR can also be added as a step in other assays such as co-cultivation to increase the sensitivity of detecting the retrovirus in question. PCR can be used in combination with other assays to confirm and identify specific retroviruses detected by an infectivity assay. In addition, other assays may be used in combination with PCR to reduce the number of false positives that may be associated with PCR.

Conclusion

PCR may be used to screen cell banks for proviral DNA of retroviruses. This same approach may be useful to screen viral vaccines. PCR may be more sensitive

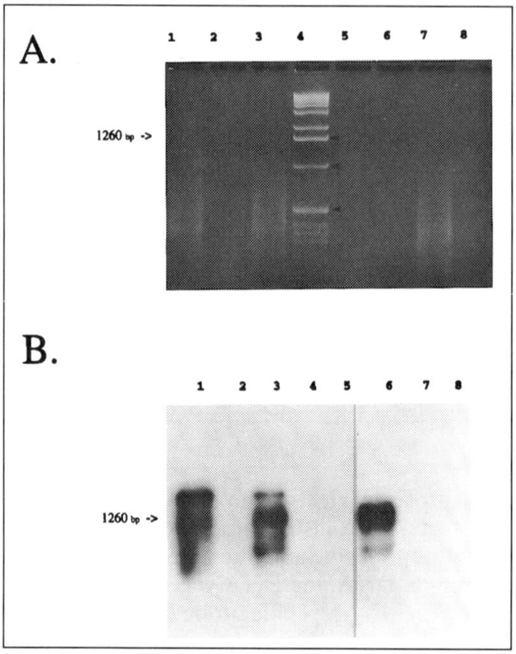

Fig. 5: **Example 2: PCR assay to screen test samples for East African SIV *pol* gene sequences.**

A. *PCR reaction conditions:* 1260 bp of SIV *pol* genes are amplified with the primers SAG*pol*2947+ (5' CCA CCT TAT GAG TGG ATG GGA T 3') and SAG*pol*4207- (5' CTT CTA TGA ATC CAC TGG CTA C 3'). The PCR reaction mixture consists of 2.5 mM $MgCl_2$, 0.01% gelatin, 50 mM KCl, 10 mM Tris, pH 8.3, 50 pmoles of each primer, 2.5 units of Perkin-Elmer amplitaq and 0.2 mM dNTPS. Concentration of the test sample DNA is 1 µg. The thermal cycler steps include 35 cycles of 94°C, 45 sec., 45°C, 90 sec., and 72°C, 90 sec. 10 µl of each reaction are electrophoresed in 1 × TBE on a 1.2% agarose gel containing ethidium bromide. Lane 1 is the amplification of 1 µg of test sample DNA plus 3 pg of SIV_{agm}385-*SSt I*; lane 2 and 5 are empty; lane 3 represents 1 µg of test sample DNA plus 1 pg of SIV_{agm}385-*SSt I*; lane 4 is 1 Kb molecular weight DNA markers (arrows in descending order represent 1.6 Kb, 1 Kb and 0.5 Kb); lane 6 is 100 fg of SIV_{agm}385-*SSt I*; lane 7 is 1 µg of test sample DNA; and lane 8 is a no DNA template control.

B. 5A PCR amplified products were transferred to a nylon membrane according to the method of Southern [28]. The PCR products were hybridized with ^{32}P-labelled 1 kb KpN fragment of SIV_{agm}385-*SSt I* overnight at 65°C in 6 × SSC, 6 × Denhardts, 0.5% SDS, 500 µg/ml salmon sperm DNA. The blot was washed stringently twice in 2 × SSC, 2 x SP at room temperature, 5 min. each and twice at 45°C, 30 min. each, followed by one wash in 2 x SSC, 1 × SP at 45°C, 30 min., and one wash in 2 × SSC, 1 x SP at 65°C, 30 min. Lanes 1-5 were exposed to XAR-5 film for 1 hr. and lanes 6-8 were exposed overnight.

in detecting simian retroviruses relative to infectivity assays or reverse transcriptase assays. PCR may be the assay of choice when the test sample interferes with detection of potential retroviruses in an infectivity assay or when a specific simian retrovirus needs to be identified. In this report, we have outlined strategies to optimize PCR assays to assure specific, reproducible and highly sensitive assays. Two different PCR assays have been presented as examples for screening test samples for SIV targeting the *pol* gene. PCR may also be used in combination with other assays to rule out the possibility of simian retrovirus infectious particles.

ACKNOWLEDGEMENTS

We would like to thank Ms. Maha Kandis for providing us with the cells and Dr. Valerie Randolph for critical assessment of experimental design. We appreciate and thank Drs. Stephen Udem, Barbara Zlotnick, Joan Crowley, Joanne Tatem and Valerie Randolph for careful reading of the manuscript.

REFERENCES

1 Points to Consider in the Characterization of Cell Lines Used to Produce Biologicals, Office of Biologics Research and Review, Center for Drugs and Biologics, FDA (USA) 1993.

2 Losikoff AM, Poiley JA, Raineri R, Nelson RE, Hillesund T: Industrial Experience with the Detection of Retroviruses, Brown F, Esber EE, Williams M (eds): Continuous Cell Lines. Dev Biol Stand. Basel, Karger 1992, vol 76, pp 187-200.

3 Coffin J: Retroviridae and their replication, in Fields BN, Knipe DM et al, (eds): Virology. New York, Raven Press, 1990, vol 2, pp 1437-1500.

4 Allan JS, Short M, Taylor ME, Su S, Hirsch VM, Johnson PR, Shaw GM, Hahn BH: Species-Specific Diversity among Simian Immunodeficiency Viruses from African Green Monkeys. J Virol 1991; 65:2816-2828.

5 Hartung S, Boller K, Cichutek K, Norley SG, Kurth R: Quantitation of a Lentivirus in its Natural Host: Simian Immunodeficiency Virus in African Green Monkeys. J Virol 1992;66:2143-2149.

6 Heneine W, Lerche NW, Woods T, Spira T, Liff JM, Eley W, Yee JL, Kaplan JE, Khabbaz RF: The Search for Human Infection with Simian Type D Retroviruses. J of Acquired Immune Deficiency Syndromes 1993;6(9):1062-1066.

7 Heneine W, Woods TC, Sinha SD, Khan AS, Chapman LE, Schonberger LB, Folks TM: Lack of Evidence for Infection with Known Human and Animal Retroviruses in Patients with Chronic Fatigue Syndrome. Clin Inf Dis 1994;18(Suppl 1):S121-125.

8 Pilcher KY, Avery N, Shiigi SM, Pangares N, Malley A, Axthelm MK, Machida CA: Polymerase Chain Reaction Detection of Type D Simian Retrovirus Proviral DNA from Macaques. J Virol Methods 1994;50:75-86.

9 Dezzuti CS, Lazo A, Yee JY, Blakeslee JR, Mathes LE, Brown BG, Lairmore MD: Detection of Simian T lymphotropic Virus Type 1 using the Polymerase Chain Reaction. Int J Cancer 1992;50:805-810.

10 Schweizer M, Turek R, Hahn H, Schliephake A, Netzer KO, Eder G, Reinhardt M, Rethwilm A, Neumann-Haefelin D: Markers of Foamy Virus Infections in Monkeys, Apes, and Accidentally Infected Humans: Appropriate Testing Fails to Confirm Suspected Foamy Virus Prevalence in Humans. AIDS Res and Human Retroviruses 1995;11:161-170.

11 Schweizer M, Neumann-Haefelin D: Phylogenetic Analysis of Primate Foamy Viruses by Comparison of *pol* Sequences. Virology 1995;207:577-582.

12 Bieniasz PD, Rethwilm A, Pitman R, Daniel MD, Chrystie I, McClure MO: A Comparative Study of Higher Primate Foamy Viruses, Including a New Virus from a Gorilla. Virology 1995;207:217-228.

13 Lerche NW, Marx PA, Gardner MB: Elimination of Type D Retrovirus Infection from Group-Housed Rhesus Monkeys using Serial Testing and Removal. Lab Animal Sci 1991;41(2):123-127.

14 Chen Z, Ben K, Tian B, Zheng Y: Serological Survey of a Captive Macaque Colony in China for Antibodies to Simian Type D Retroviruses. J Med Primatology 1992;21(7-8):377-380.

15 Lerche NW, Lee JL, Jennings MB: Establishing Specific Retrovirus-free Breeding Colonies of Macaques: An Approach to Primary Screening and Surveillance. Lab of Animal Sci 1994;44(3):217-221.

16 Oste C: Polymerase Chain Reaction. Biotechniques 1988;6(2):162-167.

17 Templeton NS: The Polymerase Chain Reaction, History, Methods and Applications. Diagn Mol Pathol 1992;1(1):58-72.

18 Temin HM: The DNA Provirus Hypothesis. The Establishment and Implications of RNA-directed DNA Synthesis. Science 1976;192:1075-1080.

19 Narayan O, Clements JE: Lentiviruses, in Fields BN, Knipe DM et al (eds): Virology. New York, Raven Press, 1990, vol 2, pp 1571-1589.

20 Heneine W, Yamamoto S, Switzer WM, Spira TJ, Folks TM: Detection of Reverse Transcriptase by a Highly Sensitive Assay in Sera from Persons Infected with Human Immunodeficiency Virus Type 1. J Inf Dis 1995;171(5):1210-1216.

21 Gelman IH, Zhang J, Hailman E, Hanafusa H, Morse SS: Identification and Evaluation of New Primer Sets for the Detection of Lentivirus Proviral DNA. Aids Res and Human Retroviruses 1992;8(12):1981-1989.

22 Kirchhoff F, Desrosiers RC: A PCR-derived Library of Random Point Mutations Within the V3 Region of Simian Immunodeficiency Virus. PCR Methods and Appl 1993;2(4):301-304.

23 Morrison HG, Desrosiers RC: A PCR-based Strategy for Extensive Mutagenesis of a Target DNA Sequence. Biotechniques 1993;14(3):454-457.

24 Rose J, Silvera P, Flanagan B, Kitchin P, Almond N: The Development of PCR Based Assays for the Detection and Diffentiation of Simian Immunodeficiency Virus in vivo. J Virological Methods 1995; 51:229-240.

25 Jin MJ, Rogers J, Phillips-Conroy JE, Allan JS, Desrosiers RC, Shaw GM, Sharp PM, Hahn BH: Infection of a Yellow Baboon with Simian Immunodeficiency Virus from African Green Monkeys: Evidence for Cross-Species Transmission in the Wild. J Virol 1994;68(12):8454-8460.

26 Blackbourn DJ, Chuang LF, Sutjipto S, Killam KF, Jr, McCready PM, Doi RH, Li Y, Chuang RY: Detection of Simian Immunodeficiency Virus RNA from Infected Rhesus Macaques by the Polymerase Chain Reaction. J Virol Methods 1992;37(2):109-117.

27 Bruisten SM, Koppelman MH, van der Poel CL, Huisman JG: Enhanced Detection of HIV-1 Sequences Using Polymerase Chain Reaction and a Liquid Hybridization Technique. Application for Individuals with Questionable HIV-1 Infection. Vox Sanguinis 1991;61:24-29.

28 Southern EM: Detection of Specific Sequences Among DNA Fragments Separated by Gel Electrophoresis. J Mol Bio 1975;98:503-517.

29 Swanson SK, Mento SJ, Weeks-Levy C, Brock BD, Kowal KJ, Wallace RE, Ritchey MB, Cano FR: Characterization of Vero Cells. J Biol Standardization 1988;16:311-320.

30 Hayami M, Ido E, Miura T: «Survey of Simian Immunodeficiency Virus Among Nonhuman Primate Populations», in Simian Immunodeficiency Virus, Current Topics in Microbiology and Immunology 1994;188:1-20.

31 Kurth R, Norley S: «Simian Immunodeficiency Viruses of African Green Monkeys», in Simian Immunodeficiency Virus, Current Topics in Microbiology and Immunology 1994;188:21-33.

32 Regier DA, Desrosiers RC: The Complete Nucleotide Sequence of a Pathogenic Molecular Clone of Simian Immunodeficiency Virus. AIDS Res and Human Retroviruses 1990;6(11):1221-1231.

33 Li Y, Naidu YM, Daniel MD, Desrosiers RC: Extensive Genetic Variability of Simian Immunodeficiency Virus from African Green Monkey. J Virol 1989;63:1800-1802.

34 Silver J, Maudru T, Fujita K, Repaske R: An RT-PCR Assay for the Enzyme Activity of Reverse Transcriptase Capable of Detecting Single Virions. Nuc Acids Res 1993;21:3593-3594.

Dr. AM Deatly, Viral Vaccine Research, Lederle-Praxis Biologicals, 401 N. Middletown Road, Pearl River, NY 10965, USA.

Brown F, Lubiniecki AS: Viral Safety and Evaluation of Viral Clearance from Biopharmaceutical Products. Dev Biol Stand. Basel, Karger, 1996, vol 88, pp 189-191

SUMMARY TO BREAKOUT SESSION A

Chairmen: *A.S. Khan, FDA, USA*
B. Murphy, NIH, USA

Participants: *S. Bierley, Pathology Associates, Inc., USA*
M. Kotewicz, Microbiological Associates, Inc., USA
R. Metz, Quality Biotech, USA
A. Deatly, Lederle-Praxis Biologicals, USA
J. Löwer, Paul-Ehrlich-Institut, Germany

The purpose of this session was to discuss the usefulness, limitations and standardization of retroviral detection assays for screening vaccine cell substrates. The session focussed on general and specific assays for the detection of known and novel mammalian retroviruses, especially of primate cell origin. The session was initiated with the awareness that development of a retrovirus-testing scheme should consider the various cell substrates involved in the production of the final product (including those used in the propagation of the virus seed stocks) and the potential retroviruses which might occur during the manufacturing process. Murphy outlined the complex passage history of several vaccines: RSV subgroup A, which had been passaged in primary cells of human and bovine origin (human embryonic kidney and bovine embryonic kidney), human diploid cells (MRC-5), and an African green monkey continuous cell line (VERO), and human para influenza virus type 3 vaccines (PIV3), in which case primary cells of bovine, Rhesus monkey and African green monkey origins were used. In general, biological detection assays which can be highly sensitive and measure infectious virus have not been developed for the majority of primate retroviruses. Thus, less sensitive assays such as Transmission Electron Microscopy (TEM) and Reverse Transcriptase (RT) are used for the general detection of retroviruses. More recently, the highly sensitive Polymerase Chain Reaction (PCR) technique has been applied for the detection of some retroviruses of known nucleotide sequence. To increase the sensitivity of detection, a strategy for the amplification of low levels of infectious, retroviral contaminants in vaccine cell substrates and virus seed stocks was presented by Khan (see paper for details). Briefly, this involves infection of potentially susceptible cells with virus seed or co-cultivation of potentially susceptible cells with non-induced or induced cells of the Master Cell Bank. Following extended culturing, cells and/or supernatants can be tested by general and specific retroviral detection assays.

Transmission Electron Microscopy (TEM) can be used as an initial, general screening procedure for the detection of known and novel retroviruses. It is especially useful for the detection of novel retroviruses for which specific reagents and detection assays have not yet been developed. TEM thin-section method was shown to be superior to negative staining for the identification and estimation of viral count. TEM may not be useful, however, in analysing low levels of retroviral contamination in vaccine products or in cases where the vaccine virus is similar to a retrovirus. Some of the factors which might contribute to inaccuracy in determining the viral count by TEM include the quality of the preparation of the sample and accuracy of visual estimation of the pellet size. The question regarding the reliability of particle numbers obtained by EM was raised and a need was expressed to develop a better assay for quantifying retroviruses. The details of this presentation are given in the paper by Bierley.

A large part of the discussion in the Session was centred around the usefulness and limitations of the **reverse transcriptase (RT) assay** presented by Kotewicz and Metz (see paper). Although the RT assay is not very sensitive, it can be used for the general detection of known and novel retroviruses, based upon their Mg+2 or Mn+2 divalent ion requirements. It was noted that the RT assay was not quantitative; also standardization of the RT assay was critical for meaningful interpretation of the results. The standardization of the assay should demonstrate specificity, sensitivity and reproducibility using known retroviruses, under conditions similar to the test sample analysis. Furthermore, the sensitivity of retrovirus detection in the test sample should be evaluated in spiking experiments. The sensitivity and reactivity of the assay can be improved by varying concentrations or changing the components of the cocktail mix. Efforts need to be made towards increasing the broadness of reactivity of the assay to include detection of a large number of different known and novel retroviruses.

There was considerable discussion of specific examples regarding the interpretation of the results of RT assays. In the case of a test sample which contained cellular debris, detection of RT activity would need further evaluation to determine if the activity was due to a retrovirus or cellular DNA polymerase. If the test sample contains cellular polymerase (double template RT assay), it would not be possible to rule out low levels of retrovirus contamination in the sample; modifications of the assay or other retrovirus detection assays would need to be considered. If the test sample has been processed in any way before the RT assay, e.g. concentration by centrifugation, appropriate spiking experiments should be performed, using known control viruses, to determine virus recovery and the effect of the processing of the test sample on retrovirus detection in the RT assay. In the case of a decrease in the RT activity of the test sample, perhaps caused by some inhibitory factors, a dilution series may need to be analysed. The presence of aggregates in the sample may also interefere in the interpretation of the RT results. Thus in addition to scintillation counting, direct radio-analytical imaging is useful to confirm results. Statistical analysis may be used to establish the positive and negative cut-off values for an RT assay. Since the RT assay is not quantitative, these values can overlap; thus, results of test articles falling in this "grey zone" will be difficult to interpret and may need to be supported by additional data. Limitations of the RT assay must therefore be assessed for each different test sample and its further use based upon the results obtained.

Model viruses for the RT assay were not discussed but the following questions were raised: Which ones should be used? Should all investigators use the same ones? How many different ones should be included in the standardization of the assay? Should everyone be using a universally prepared stock of controls which may allow comparison of results from different laboratories? Should the reporting of the RT activity be changed from cpms to nanomoles? It was clear that further data and discussions between the different groups using the RT assays were needed to determine the standardization criteria for the assay.

The **polymerase chain reaction (PCR) assay** is highly sensitive and can be useful for screening cell substrates for known retroviruses. The sensitivity and specificity of the assay largely depends upon the primers used in the amplification. The development of primers is based upon the available nucleotides sequences of the viruses under investigation. PCR detection of simian immunodeficiency virus (SIV) was described including the sensitivity, specificity and reproducibility of the assay. This assay may be useful for screening cell substrates and vaccine products for potential contamination with SIV and other genetically related viruses. The main limitations of the assay concern the inability to detect viruses that may vary in their nucleotide sequences from the PCR primers used and the reproducibility of the results. To address the latter concern, the assay should be performed in a highly controlled setting to generate reproducible data. False positives as well as false negatives should be confirmed. Several regions of the viral genome should be analysed to increase the level of confidence in the data. Appropriate controls also need to be performed to analyse inhibition of the reaction by the test sample, e.g. a high concentration of DNA can inhibit the reaction. Further details of this assay are presented in the paper by Deatly.

A **novel RT-PCR assay** was described which might potentially be used to detect a number of unknown endogenous and recombinant retroviruses based upon their primer binding sequences (PBS). The assay was presented on research with human endogenous retroviruses in teratocarcinoma cells. The RNA fragments amplified, using a tRNA- derived primer complementary to the retroviral PBS, must be sequenced to distinguish them from amplified cellular sequences (Löwer et al, 1993, Virology, 192:501-511). Although highly contentious, the application of this technique as a screening tool for retrovirus contamination has not been evaluated.

While each assay has its limitations, it was generally agreed that TEM, RT and PCR (and biological assays, when available) are the useful tools which are currently available for the detection of retroviruses that can potentially contaminate vaccines. The sensitivity of detection of the assays may be increased by amplification of low levels of retroviral contaminants in potentially susceptible cells. Regardless of the detection assay used, standardization was critical for meaningful interpretation of the data. This included demonstration of specificity, sensitivity and reproducibility using relevant viruses (when possible) as controls. It was also noted and recognized that the detection assays currently being used have been standardized using known viruses, and novel viruses may or may not be detected under the assay conditions. Thus using more than one approach or detection assay would provide a greater confidence level regarding safety of a product; and efforts towards developing, standardizing and using new, sensitive and broadly reactive detection assays should be pursued.

Dr. A.S. Khan, FDA/CBER, Office of Vaccines Research and Review, 1401 Rockville Pike, Bldg. 29B, Room 4NN10, HFM-454, Bethesda, MD 20852, USA.

Brown F, Lubiniecki AS: Viral Safety and Evaluation of Viral Clearance from Biopharmaceutical Products. Dev Biol Stand. Basel, Karger, 1996, vol 88, p 193

BREAKOUT SESSION B

Adventitious Contaminants Observed during Characterization and QC Testing: Findings and Follow-up

Chairpersons: P. Minor, NIBSC, UK
N. Finter, Tideswell, UK

Brown F, Lubiniecki AS: Viral Safety and Evaluation of Viral Clearance from Biopharmaceutical Products. Dev Biol Stand. Basel, Karger, 1996, vol 88, pp 195-197

Survey of Manufacturers and Developers of Cell Culture–Derived Products for Evidence of Adventitious Contamination

A.S. Lubiniecki

SmithKline Beecham Pharmaceuticals, King of Prussia, PA, USA

At the 1991 FDA Cell Substrate Workshop, Alan Moore of Microbiological Associates presented accumulated results from testing 550 cell lines for the presence of adventitious agents. He reported that less than 1% of the cell lines showed evidence of bacteria, fungi, or adventitious viral contamination, and 1% showed evidence of mycoplasmal contamination [1]. These data argue that adventitious contamination of cell cultures submitted by developers and manufacturers is a real phenomenon, but is not a common occurrence.

As ICH makes progress towards harmonization of guidance concerning assessment and prevention of microbial contamination, it is important to review the current position. This is a major goal of this conference. As part of this overall goal, the three industrial participants in the ICH process wished to assess whether adventitious contamination occurs in GMP cell banks and production cultures. A survey of developers and manufacturers was undertaken jointly by the Pharmaceutical Research and Manufacturers of America, the European Federation of Pharmaceutical Industry Associations, and the Japanese Pharmaceutical Association. The survey asked for information regarding the number of cell banks made and tested for adventitious agents and the number of production runs made and tested for such agents. It also asked for the number of positive test results found for bacteria, fungi, mycoplasma and adventitious viruses. Approximately 100 member firms received the survey questionnaire, and were given approximately four weeks to respond to the trade association. Each trade association compiled confidential responses from its members, and the three pooled their responses for this presentation. I was asked to present the compiled information for all three associations.

A total of 26 responses was obtained which covered 70 cell lines, 88 master cell banks (MCBs), and 126 working cell banks (WCBs). The results of testing these cell banks for adventitious contamination are shown in Table 1. These data indicate that the incidence of bacterial contamination in cell bank preparations is about 2.5%. No evidence of fungal, mycoplasma, or viral contamination was reported.

Table 1: Incidence of microbial contamination during cell bank testing.

Bacteria	5/197	(2.5%)
Fungi	0/197	(<0.5%)
Mycoplasma	0/197	(<0.5%)
Adventitious virus	0/197	(<0.5%)

Quality control testing of production harvests for microbial contamination was also surveyed. Table 2 shows that bacterial contamination was reported to occur with an incidence of 3.7%, while no evidence of fungal contamination was found. A single observation of mycoplasma contamination was reported. No reports of virus isolation were made. Fewer attempts to detect viruses (516) or mycoplasma (1109) were reported compared to those for bacteria and fungi (1818), probably reflecting the common practice of performing viral and mycoplasma testing at the end of an extended multiple-harvest fermentation as an adequate test for contamination of all previous harvests from the series.

The data compiled from this survey indicate that a low incidence (2-4%) of bacterial contamination during GMP cell banking and production operations is found. These results are not surprisingly different from those reported previously by Moore [1], and are expected given the nature of the open manipulations required for many cell culture operations, even in biological safety cabinets under GMP conditions. The lack of detectable fungal contamination events probably reflects the adequacy of current technology for medium preparation and fermentation, use of biological safety cabinets, and environmental controls to prevent entry on 2015 attempts of fungal spores into the cell cultures in a reliable manner. However, this same technology clearly is not completely adequate to prevent entry of bacteria on 73 occasions in 2015 attempts.

The results of this survey also indicate that mycoplasma and viral contamination are quite rare. Observation of a single mycoplasma detection event in 1306 attempts argues that current manufacturing technology and procedures and testing methods are probably adequate to minimize mycoplasma contamination of cell cultures. The lack of reported viral contamination in 705 attempts can be interpreted to support the adequacy of current technology and procedures to prevent frequent viral contamination events. However, other speakers at this conference have reported sporadic

Table 2: Incidence of microbial contamination during quality control testing of harvests.

Bacteria	68/1818	(3.7%)
Fungi	0/1818	(<0.06%)
Mycoplasma	1/1109	(0.09%)
Adventitious virus	0/516	(<0.2%)

contamination events in large scale cell-culture facilities by parvoviruses, orbiviruses, and retroviruses; other virus families are also alleged to cause sporadic contamination.

Garnick and coworkers indicated that they experienced two parvovirus outbreaks in a period where approximately 2000 fermentations were performed [2]. Garnick also hypothesized that such contamination events were rare but inevitable consequences of very large-scale operations, and reflected the ability of low probability events to occur if enough attempts were made. If we accept the figure of 2/2000 or 0.1% as the estimate of incidence of adventitious viral contamination, then it is reasonable that the lack of a positive event in the 705 attempts reported in our survey is a true result. It should also be remembered that a voluntary questionnaire has known defects as an assessment tool, and that these defects may influence the reported outcome. For example, the response rate to this survey was about 25%, which certainly could have allowed sampling error to occur. It could also be concluded that those few organizations which reported viral contamination events at this symposium were among those organizations who did not respond to the survey.

In summary, the industrial trade associations report that a voluntary survey of its member organization indicates that microbial contamination of cell culture banking and cell culture harvests is a rare event. Current analytical and manufacturing technology appear to be adequate to provide a high degree of assurance that the incidence of microbial contamination can be maintained at this low level.

References

1 Moore WA: Experience in cell line testing, in Brown F, Esber E, Williams M (eds): Continuous Cell Lines. Dev Biol Stand. Basel, Karger, 1992, vol 76, pp 51-56.

2 Garnick R: Experience with viral contamination in cell culture, in Brown F, Lubiniecki AS (eds): Viral Safety and Evaluation of Viral Clearance from Biopharmaceutical Products. Dev Biol Stand. Basel, Karger, 1996, vol 88, pp 49-56.

Dr. A.S. Lubiniecki, SmithKline Beecham Pharmaceuticals, King of Prussia, PA, USA.

Brown F, Lubiniecki AS: Viral Safety and Evaluation of Viral Clearance from Biopharmaceutical Products. Dev Biol Stand. Basel, Karger, 1996, vol 88, pp 199-203

Contamination of Genetically Engineered Chinese Hamster Ovary Cells

D.G. Burstyn

Alkermes, Inc., Cambridge MA, USA

Key words: Contamination, CHO cells.

Abstract: In late 1988, during production of a recombinant protein for phase I clinical trials, a failure of the cell culture production system occurred due to contamination of the cells by an orbivirus [1]. The incident occurred at Bioferon GmbH & Co, Laupheim, Germany, a joint venture of Biogen, Inc., Cambridge, MA, and Dr. Rentschler Arzneimittel GmbH & Co (Bioferon is currently a wholly owned subsidiary of Rentschler and is now known as Dr. Rentschler Biotechnologie GmbH). The investigation into, and the subsequent response to, the infection can be divided into three stages: Stage I, Investigation and initial response; Stage II, Secondary response; and Stage III: Continuing response.

BACKGROUND

In the Bioferon cell culture system, cells are expanded from an MWCB vial through a series of T-flask cultures, ultimately to populate polystyrene multitray units (cell factories). Groups of cell factories are placed in a metal rack or pallet and are handled as a segregated unit by semi-automated robots. The units within a pallet are fed and harvested as a unit via a series of interconnecting channels or tubes. The full production culture, consisting of multiple pallets, is designated the production «pyramid». Throughout the propagation and at each harvest of conditioned medium, the cell factories are examined by technicians and supervisors trained in cell culture methods. At the time the infection occurred, Bioferon had been conducting such cell production campaigns for more than ten years and had successfully manufactured licensed products for commercial sale in Germany and for clinical testing in the United States.

STAGE I: Investigation and Initial Response

During the initial production campaign, the following observations were made: (a) loss of viability and cell death during the building of some pyramids; (b) in mature pyramids, there was a decrease in product titre followed by a generalized cytopathic effect (CPE) and cell death; and (c) not all pyramids were affected. The initial response was to quarantine product from all pyramids (including asymptomatic cells) and to launch an intensive investigation into the cause of the phenomenon.

The first indication that the source of the observations was of biological origin was the demonstration that the cytopathic agent could be amplified by passage through cell culture. This led to attempts to isolate the cytopathic agent from cell culture supernatant. The supernatants of lysed cells had a cytopathic effect on the recombinant Chinese hamster ovary (CHO) cells used in production as well as on certain indicator cell lines, particularly WISH cells. The CPE was resistant to chloroform and labile to low pH. Outside laboratories in Hanover, Frankfurt and Muenster/Westfalia were consulted for analytical data to define the cytopathic agent, assumed to be a virus. Electron microscopy of a high-speed centrifugation pellet of cytopathic medium showed virus particles with the structure typical of orbiviruses. Examination of the viral nucleic acid showed it to be double-stranded RNA present in ten individual segments, corresponding to patterns reported in the literature for orbiviruses. At the time of the infection, the virus was identified as a new serotype of bluetongue virus (BTV), based on nucleic acid hybridization and negative serology. Much later, in 1993, the virus was identified as Epizootic Haemorrhagic Disease Virus (EHDV), a close relative of BTV, based on serotyping [1].

BTV and EHDV are widespread members of the reovirus (*r*espiratory *e*nteric *o*rphan *virus*) group [2, 3]. BTV is an arthropod-borne virus, occasionally causing disease in sheep. Inapparent infection also occurs in cattle and deer. EHDV is closely related to BTV, mainly causing a disease indistinguishable from bluetongue in deer. EHDV has also been isolated from cattle, which appear to be its reservoir host.

As part of the investigation, a comprehensive review of the batch production records was initiated in an attempt to define commonality between the affected cells and to differentiate them from the unaffected cells. This analysis indicated that a loss in productivity was first observed 10 to 15 days after foetal bovine serum (FBS) lot 053 (Australian origin) was first used. This loss of productivity was followed 10 - 15 days later by the cells leaving the polystyrene surface and, ultimately, by a total loss of viability of the cells in the affected pallets. While certain pyramids exposed to serum lot 053 lost titre and underwent cell death and lysis, others showed no phenotypic effects and were successfully propagated through all stages of culture.

Because of the scale of the production campaign, it was not possible to obtain and qualify a sufficiently large serum lot to support the entire campaign. In qualifying serum for use in the production, the US Department of Agriculture (USDA) staff veterinarians and the US Code of Federal Regulations Title 9 (9 CFR § 113.53) were consulted concerning the acceptability of serum from certain geographic areas. Early lots of serum used in production were of Australian origin, purchased from a reputable European supplier with an appropriate certificate of analysis indicating that they were free of bacteria, mycoplasma, endotoxins and four specific viruses (foot and mouth disease virus [FMDV], infectious bovine rhinotracheitis [IBR] virus, parainfluenza Type 3 [PI3] and African pig cholera). This certification was based on culture assays (seven days) of test and control serum samples on calf lung cells followed by examination for cytopathic effect and serotyping. Certificates of analysis of the serum lots certified that the serum was collected from foetuses of «cows slaughtered for human consumption inspected ante- and post-mortem and free of FMD and clinical signs of other contagious diseases». There was further certification that they were processed in a European establishment in an area that had been free of FMDV, rinderpest and other diseases infectious to cattle within a 50 km radius during the past twelve months. Additionally the serum was screened for antibodies to IBR and PI3 and shown to be free of them. The individual lots of serum were further tested at Bioferon and shown to support growth of satellite cultures of thc CHO cell line used in production. All serum lots pass the criterion of supporting cell growth.

Attempts to isolate the viral agent directly from FBS lot 053 were unsuccessful, indicating that the virus, if present in the serum, was present at a titre of < 10 TCID50/mL. Nonetheless, FBS as the source of infection is supported by the report that BTV and EHDV are capable of crossing the bovine placenta although it cannot be ruled out that contamination occurred on the slaughter house floor during serum collection. The observation that only some of the cells exposed to the implicated FBS lot became infected could be due either to the low virus titre or to an inefficient adaptation of the virus to CHO cells required for the infection to take place. Lastly, there were no reports of BTV or EHDV outbreaks in Germany at the time of the outbreak, strongly ruling out local farms as the source of the virus.

STAGE II: Secondary Response

Once FBS lot 053 was identified as the common source of viral infection, the quarantine on the material produced by cells that had never been exposed to FBS lot 053 was lifted.

Although the purification process had previously been evaluated for its ability to inactivate and clear model viruses, the ability of the purification process to inactivate and clear the specific viral contaminant was investigated. As part of the process, a pH 5.5 virus inactivation step was included at the recommendation of the USDA and was a stipulation of Biogen's import permit for the partially purified product. The pH 5.5 inactivation kinetics of the contaminant were evaluated and shown to lower viral infectivity by *c.* 10^6. The purification process was shown to clear ^{35}S-methionine – ^{35}S-cysteine radioactively-labelled inactivated virus by a factor of *c.* 10^6. This results in a combined clearance-inactivation factor of *c.* 10^{12}

An independent testing laboratory investigated the potential contamination of serum lot 053, using traditional methods to verify the presence and nature of the virus. Their data demonstrated virus only in the most highly contaminated cell culture harvest (from a fully lysed pyramid) at 2.0-5.0 x 10^3 TCID50/mL. These data demonstrate that if such a virus were present in harvest of conditioned medium from the cell factories at any stage in the production campaign, it would have been inactivated by the pH 5.5 treatment which precedes the initial chromatography step of the purification process. In addition, final traces of the viral material would have been removed from the product. The combination of inactivation factors and the clearance factors for each process step were used to calculate a «comfort factor» of *c.* 10^6, the amount of clearance and inactivation in excess of the maximum viral load.

To assure the safety of the product from exposed but asymptomatic pyramids fed FBS lot 053, an extensive literature search was conducted for reports of human disease associated with BTV. A single report of possible BTV-disease in a laboratory technician was alluded to, but when pursued, the laboratory director indicated his lack of confidence in the diagnosis. Additionally, direct cell-culture assay of intermediates and bulk for virus was conducted with completely negative results.

Lastly, as an added precaution, sheep were injected with 10 mL bulk product (negative for disease and antibodies).

Based on the sum of the evidence, a request was made to the FDA to allow clinical use of material from asymptomatic pyramids fed with FBS lot 053. The FDA eventually rejected the request based on the potential hazard to the target population (AIDS patients).

STAGE III: Continuing Response

As a consequence of the orbivirus contamination, several steps were taken at both Bioferon and Biogen to reduce the likelihood of subsequent contamination of cell cultures by adventitious viruses. These steps focused on materials of biological origin, including increased scrutiny of FBS suppliers through vendor qualification testing and audits of processing facilities. Also, a concerted effort was made to switch from materials of animal origin whenever possible. When not possible, all materials of animal origin were irradiated.

Lastly, throughout the investigation, commercial cell-testing laboratories were used to augment internal expertise and because of the desire to avoid bringing the contaminant into other parts of the facilities. Additionally, agricultural regulations limited the ability to export/import live virus and to work with a suspected ovine or bovine pathogen in unclassified environments. Many of these interactions were scientifically unsatisfying, of low quality, and presented a number of scheduling and logistical problems. This led to a similar, on-going programme, for evaluating and certifying external testing laboratories.

RECOMMENDATIONS

Based on these experiences, the following recommendations are made to others who may be in the unfortunate situation of experiencing a viral contamination in their production cells:

(a) Do not save the product. There is little chance that the regulatory authorities will allow an even remotely implicated product to be used for a clinical trial.

(b) Attempt to develop virus specific monitoring and inactivating techniques.

(c) Attempt to locate source/focus of infection. This will help the clean-up and prevent recurrence.

(d) Focus on immediate clean-up. The priority should be to remove the contaminating agent from the facility so that production can restart.

(e) Do not rush into renewed manufacturing. Short-cuts usually result in some virus escaping clean-up/inactivation. This may result in a renewal of large-scale contamination.

(f) Don't think it can't happen to you. Despite very experienced and sophisticated production and quality control departments, Bioferon had a contamination. All the necessary GMP elements were in place, including pre-use testing and approved incoming raw materials, including the FBS. In addition, at the time of the contamination, Bioferon was producing a marketed cell-culture derived product, natural interferon-ß, under a licence granted by the BGA. The facility was also routinely inspected by representatives of the Federal and local governments.

References

1 Rabenau H, Ohlinger V, Anderson J, Selb B, Cinatl J, Wolf W, Frost J, Mellor P, Doerr HW: Contamination of genetically engineered CHO-cells by Epizootic Haemorrhagic Disease Virus (EHDV). Biologicals 1993;21:207-214.

2 Gorman BM, Taylor J, Walker PJ: Orbiviruses, in Joklik WK (ed): The Reoviridae. New York and London, Plenum Press, 1983, Ch 7, pp 287-359.

3 Fenner F, Bachmann PA, Gibbs EPJ, Murphy FA, Studdert MJ, White DO: Reoviridae, in Veterinary Virology, New York, Academic Press, 1987, Ch 32 pp 577-590.

Dr. D.G. Burstyn, Alkermes, Inc., 64 Sidney Street, Cambridge, MA 02139, USA.

Brown F, Lubiniecki AS: Viral Safety and Evaluation of Viral Clearance from Biopharmaceutical Products. Dev Biol Stand. Basel, Karger, 1996, vol 88, pp 205-206

Key Points from Nettleton and Rweyemamu (1980): Association of Calf Serum with Parvovirus Infection

B. Hughes

GlaxoWellcome, Beckenham, Kent, UK

The paper describes the isolation of a contaminating virus from cultures of BHK21 cells in which persistant cell death was noted. The virus was provisionally designated as a member of the parvovirus family and the contamination was associated with the use of some batches of calf serum.

The agent was isolated from dead cells, cell growth media and batches of calf serum. Clusters of spherical particles of approximately 21 nm. diameter were observed by negative stain electron microscopy. Virus replication was effective only in activity growing cells and large intranuclear inclusion bodies were observed within infected cells. Virus yield was reduced by treatment with BUDR and this inhibition was relieved by addition of thymidine to the medium. These data, together with the resistance of the virus to treatment with heat, acid and lipid solvents, implicated a parvovirus as the causative agent.

Further investigations attempted to determine the source of the infection. Parvoviruses have been associated with porcine sources of trypsin in the past and bovine parvoviruses have also been described. From a wide range, only antisera to minute virus of mice (MVM) was effective in inhibiting infectivity and the structural proteins of the particles were similar to those previously published for MVM.

The discovery that the contaminating parvovirus was not of porcine or bovine origin but that it more probably originated from murine sources was surprising and triggered a search to determine the likely route of infection. The testing of dead cells and medium components was transferred to another, separate laboratory on a distant site. In this round of testing, it was discovered that the agent was associated with batches of complete medium and subsequently that the route of infection was through batches of infected calf serum. The virus was unaffected by heating at 56°C for 30 minutes and 0.2 μ filtration did not remove the infectivity. Chemical inactivation was achieved by treatment with peracetic acid and acetylethyleneimine. The investigations did not identify the primary source of infections, whether from infected animals or during processing operations.

Remedial measures required complete disinfection of all laboratory areas. All cell stocks and serum were destroyed and new cell banks were introduced. All future purchases of serum were made from a new supplier.

Current practice in the company requires serum for growth media to be irradiated and/or tested for the presence of extraneous agents before use. Each batch is fully tested and assessed for its ability to support growth before release for use in manufacturing. All new cell lines are adapted to growth in protein-free media whenever possible.

Reference

Nettleton PF, Rweyemamu MM: Arch Virol 1980;64:359-374.

Dr. B. Hughes, GlaxoWellcome, Biotech Analytical Laboratories Group Development, Langley Court, Beckenman, Kent BR3 3BS, UK.

Brown F, Lubiniecki AS: Viral Safety and Evaluation of Viral Clearance from Biopharmaceutical Products. Dev Biol Stand. Basel, Karger, 1996, vol 88, pp 207-208

SUMMARY TO BREAKOUT SESSION B

Chairmen: *P. Minor, NIBSC, UK*
N. Finter, UK

Participants: *D. Burstyn, Alkermes Inc., USA*
B. Hughes, GlaxoWellcome, UK
A. Lubiniecki, SmithKline Beecham Pharmaceuticals, USA

The issues addressed in the discussion of adventitious contaminants during characterization and QC testing included the following:

1. The frequency of virus contamination of cell banks or production cultures.
2. The type of virus detected.
3. The nature of the cells contaminated, specifically if they are long established cell lines, or like hybridoma lines, closer to whole animals, and therefore possibly more likely to be infected.
4. The stage of production development at which viruses were identified.
5. The source of the virus contamination.
6. The action that should be taken.
7. The assay systems.

Retroviruses were excluded from discussion as they may originate from the cell genome, and be an integral part of the cell used.

A survey of manufacturers has been initiated. Of about 100 manufacturers contacted, 26 have provided information on experience of contamination of harvests or cell banks with bacteria mycoplasma, fungi, or viruses. While bacterial contamination was reported at a frequency of about 3% of tested preparations, no fungi or viral contamination was recorded.

The preliminary findings suggest that the incidence of adventitious virus infection is low in licensed products in developed processes. However, contract testing companies have described virus contamination at a frequency of just under 1% of cells and samples submitted. While some of the samples submitted may be cells at an early stage of development, before the establishment of a manufacturer's cell bank, most are not, although they do not derive from well developed manufacturing processes. Moreover, viral contamination was detected, not only in cells such as hybridoma lines where infection from the species of origin might be thought more likely, but also in cells derived from well-known continuous cell lines used in the production of biotechnology products. While the risk of contamination is difficult to judge with confidence, and may be low, it is not zero.

Three specific examples of virus contamination were presented. Infection of CHO cells by minute virus of mice was described in two separate incidents from Genentech. Infection was detected initially by a general viral screen involving infection of cells in vitro and later by specific assays for the virus. The origin of the virus was not established but was believed to involve imperfect rodent control at some stage in the supply of raw materials. Infection in this case was not detected by a decline in yield. A second example came from the production of foot and mouth disease vaccine in BHK cells by Wellcome some years ago, where contamination by a parvovirus, probably minute virus of mice, resulted in the loss of production cultures. A third example from Bioferon involved contamination by an orbivirus, identified as EHDV that resulted in the failure of cultures. While it is likely that the virus originated in calf serum of Australian origin where orbiviruses such as blue tongue virus and EHDV are well documented, the virus could not be demonstrated in the implicated batch, nor did all production cultures exposed become infected. The level of infection in the serum was therefore probably low.

The action taken on discovering a contaminant in the cases presented was the same. First, contaminated batches were discarded. Second, the virus concerned was identified, and third, the production equipment was decontaminated. The need to identify the virus concerned is illustrated by the fact that decontamination with MVM was accomplished at high pH, while decontamination with the arbovirus used low pH. Identification was therefore needed to identify optimal decontamination procedures.

The studies support the view that a test for viral contamination applied on a routine batch basis is desirable. However, the tests available are not foolproof. General tests include the use of indicator tissue culture systems, usually including the production cell, a human cell line and one other. Because viruses have such diverse properties, it is likely that some will escape detection, either because they do not infect the test cells, or because they produce an inapparent infection.

Detection systems will inevitably focus on the viruses that are perceived to be possible contaminants, and such perceptions may be wrong. Non-infection based assays, such as antigen or antibody production or PCR are extremely specific, although they may have advantages of rapidity.

Infection of cell banks and production cultures by adventitious agents is therefore documented. The frequency is difficult to assess and may be affected among other things by the effectiveness of assay which producers use. Viruses may be introduced by a number of routes including the low frequency contamination of raw materials by rodents. A case can be made for testing each batch for viruses. It is probable that the frequency of contamination is not high in production.

Dr. P. Minor, N.I.B.S.C., Blanche Lane, South Mimms, Potters Bar, Herts ENG 3 QG, U.K.

Brown F, Lubiniecki AS: Viral Safety and Evaluation of Viral Clearance from Biopharmaceutical Products. Dev Biol Stand. Basel, Karger, 1996, vol 88, p 209

BREAKOUT SESSION C

Adenoviral Gene Therapy Issues

Chairpersons: A. Ridgway, Bureau of Biologics, Canada
P. Noguchi, FDA, CBER, USA

Brown F, Lubiniecki AS: Viral Safety and Evaluation of Viral Clearance from Biopharmaceutical Products. Dev Biol Stand. Basel, Karger, 1996, vol 88, pp 211-212

SUMMARY TO BREAKOUT SESSION C

Chairmen: *A. Ridgway, Bureau of Biologics, Canada*
P. Noguchi, FDA, CBER, USA

Participants: *S. Eck, University of Pennsylvania, USA*
K. Hehir, Genzyme Corporation, USA
I. Kovesdi, GenVec, USA
B. Trapnell, Genetic Therapy, Inc., USA
D. Vacante, Microbiological Associates Inc., USA

Replication-defective adenovirus vectors present considerable promise for use in the delivery of therapeutic genes. They can accommodate large gene expression cassettes and deliver them to a wide variety of cell types with high efficiency. One of the most intense efforts to be addressed with adenovirus vectors has been correction of the genetic defect in cystic fibrosis. These phase I studies have shown that the viral capsids themselves may induce a rapid inflammatory response in the lungs of cystic fibrosis patients. The viral mass required to elicit this response has been very similar for each reported instance and suggests that the total viral mass (including both infectious and non-infectious viral particles) will be important when determining dosage. A second important lesson to be learned from these early trials is that current methods for production of replication-defective adenoviruses also result in the generation of replication-competent adenoviruses RCA at a low but significant frequency. These are the dreaded RCA which are thought to be capable of causing disease when administered in high enough doses.

The adenovirus breakout session brought together five groups who have been working on the manufacture and use of these vectors for gene therapy of cystic fibrosis and other applications. The goal was to discuss experiences in two areas: methods for determining bio-availability of adenovirus vectors and methods for the detection of RCA. A secondary, but perhaps more important goal for this session was to determine whether data exist for reconsideration of the FDA current recommendation for less than 1 RCA/dose.

1. There was general agreement that taking 30-year-old laboratory methods into the clinic is a difficult task.
2. Considerable progress has been made towards determining adenovirus bio-availability. This was defined as the ratio of functional particles to total particles which is also the same as the ratio of infectious units to optical density of the

virus stock multiplied by the extinction coefficient for purified adenovirus. First, infectious unit determination by plaque assay seems to be the most widely used method. However, data were presented which showed it to be a complex and somewhat inefficient process, dependent upon many factors including time of virus adsorption and time for development of plaques. Second, determination of total virus particles by multiple methods showed that determination of optical absorbance at 260 nm was reliable in addition to being inexpensive and quick. There was also general agreement with the early determination of the adenovirus extinction coefficient for absorption at 260 nm.

3. Recombination is an inherent problem when manufacturing E1A-deleted adenovirus vectors in 293 producer cells. These cells contain a portion of the adenovirus genome encompassing E1A and additional flanking sequences. Sequence analysis of many independent RCA revealed that each isolate had arisen through homologous recombination between sequences flanking the E1A region, with a structure identical to that of wild-type adenovirus. Validated assays exist which can detect RCA at very low frequency. However, it was acknowledged that large-scale production of adenovirus will be hindered by this propensity for recombination and that different production methods are needed.
4. What level of RCA is safe? Is our current recommendation of <1 RCA/dose still appropriate? More data are needed.
5. Clinical trial of these vectors is an iterative process which will show us which reagents and approaches are safe.

Dr. P. Noguchi, FDA/CBER, 1401 Rockville Pike, HFM-515, Rockville, MD 20852, USA.

Brown F, Lubiniecki AS: Viral Safety and Evaluation of Viral Clearance from Biopharmaceutical Products. Dev Biol Stand. Basel, Karger, 1996, vol 88, p 213

BREAKOUT SESSION D

Lessons Learned from Related Product Categories

Chairperson: J. Finlayson, FDA/CBER/OBRR, USA

Brown F, Lubiniecki AS: Viral Safety and Evaluation of Viral Clearance from Biopharmaceutical Products. Dev Biol Stand. Basel, Karger, 1996, vol 88, pp 215-216

Follow-up Studies of Hepatitis C Association with an Intravenous Immunoglobulin

M.W. Yu

Division of Hematology, Office of Blood Research and Review, CBER/FDA, Bethesda, MD USA

Key words: Hepatitis C virus; immunoglobulin.

On February 23, 1994, Baxter Healthcare Corporation withdrew all lots of its Immune Globulin Intravenous (IGIV), Gammagard®, because of reports of hepatitis C virus (HCV) transmission. Worldwide, approximately 200 suspected transmissions have been reported since October 1993. These were the first such reports received since the licensing of the product in the USA (February 18, 1986).

We tested more than 40 Gammagard lots prepared between 1988 and November 1993 for HCV RNA and assessed the results in relation to the type of anti-HCV screening performed on the starting plasma and the number of case reports received. HCV RNA was rarely found in lots prepared from plasma that had not been screened for anti-HCV or from plasma mostly or partially screened by EIA-1 (single antigen). In contrast, many lots prepared mostly or entirely from plasma screened by EIA-2 (multi-antigen) were found to contain HCV RNA. Although multiple lots are given frequently to a single patient and lot numbers are often not identified, more cases were associated with lots prepared solely from EIA-2 screened plasma. One such lot had the highest level of HCV RNA among all lots tested, and was associated with the most transmissions. Many recipients' serum specimens provided by the Centers for Disease Control were found to contain HCV RNA.

We then investigated whether anti-HCV screening depletes antibodies capable of complexing with and/or neutralizing HCV. Antibodies to HCV envelope proteins, E1 and E2, have recently been suggested to be the neutralizing antibodies for HCV. We found that Gammagard lots manufactured from unscreened plasma had higher titres of both antibodies than those lots prepared from either EIA-1 or EIA-2 screened plasma. Highest titres of both antibodies were detected in two experimental IGIV lots prepared solely from anti-HCV EIA-reactive plasma units.

We also found that partitioning of HCV (as detected by RNA) during Cohn-Oncley fractionation of plasma leading to immune globulin is affected by the presence of anti-HCV in plasma. In a plasma pool with high levels of HCV antibodies including both anti-E1 and anti-E2, we found previously that the majority of HCV was partitioned into the precipitate fractions, such as cryoprecipitate, fraction I, and fraction III, and that the overall reduction in HCV RNA was 4.7 logs. On the contrary, in a look-back plasma unit devoid of both anti-E1 and anti-E2, we found that a greater amount of HCV is partitioned into the early supernatant rather than the precipitate fractions. As a result, the overall reduction in HCV RNA leading to immune globulin was 3.5 logs, i.e., less significantly than that in the presence of anti-HCV. Although studies to follow up this episode are continuing, data obtained thus far indicate that the removal of complexing and/or neutralizing antibodies because of screening was largely responsible for the HCV transmissions by Gammagard.

Lessons learned from the Gammagard incident are that: 1) the viral safety of a US licensed immune globulin product, as previously manufactured, appears to have been marginal and was compromised because of anti-HCV screening of plasma in the absence of compensating factors; and 2) to safeguard the viral safety of immune globulin products, viral inactivation/removal procedures should be incorporated into the manufacturing process.

Dr. M.W. Yu, Division of Hematology, OBRR/CBER/FDA, 8800 Rockville Pike, Bldg. N29, Rm. 305, HFM-345, Bethesda, MD 20892, USA.

Brown F, Lubiniecki AS: Viral Safety and Evaluation of Viral Clearance from Biopharmaceutical Products. Dev Biol Stand. Basel, Karger, 1996, vol 88, p 217

Evaluation of Human Parvovirus B19 DNA Contamination of Factor VIII Products

Y. Matsunaga

Department of Epidemiology, The National Institute of Health, Tokyo, Japan

Key words: Human parvovirus, Factor VIII.

Antibody positive rate to B19 virus in haemophiliacs is almost 100%, indicating a higher infection rate in haemophiliacs than in ordinary population. There are many reports on transmission of B19 virus by Factor VIII concentrates.

We have developed a titration technique for B19 DNA by combined PCR and Southern hybridization. The minimum limit of detection is 0.01 fg cloned B19 DNA (3.5 copies). B19 DNA was detected in about 80% of the products during 1991 to 1993. The concentration of viral DNA in the positive products was 10^0-10^3 copies per 17μl (10^2-10^5/ml) and the DNAs were DNase resistant, suggesting that they are encapsidated.

Screening of 27,400 plasma from Japanese healthy donors by PCR detected eight positive donors (1/3,400). The highest concentration of B19 DNA copies per ml of plasma was $10^{8.8}$ and the lowest was $10^{2.8}$. Furthermore, we assayed IgM antibody in the DNA positive samples. The antibody was detected in six samples containing $10^{3.4}$-$10^{6.4}$ B19 DNA copies/ml.

B19 antigen in plasma can be detected by ELISA or PA (gelatin particle agglutination) with minimum sensitivity of approximately 10^7 virus particles/ml. Therefore, if each plasma sample is screened before making a pool for cryo-precipitate by both B19 antigen and IgM antibody detection methods, and the positive samples are excluded, then B19 virus concentration in the starting material would be lower. Since the present technology for purification of Factor VIII reduces contaminated B19 levels by -5 $\log_{10}$, the final product would be virus free when tested by this PCR method.

These studies were done partially in co-operation with the Japanese Red Cross Plasma Fractionation Center.

Dr. Y. Matsunaga, Department of Epidemiology, The National Institute of Health, 1-23-1 Toyama, Shinjuku-ku, Tokyo, Japan 162.

Brown F, Lubiniecki AS: Viral Safety and Evaluation of Viral Clearance from Biopharmaceutical Products. Dev Biol Stand. Basel, Karger, 1996, vol 88, pp 219-220

Abortion and Death in Pregnant Bitches Associated with a Canine Vaccine Contaminated with Bluetongue Virus

R.L. Levings, L.A. Wilbur, J.F. Evermann, I.R. Stoll, D.E. Starling, C.A. Spillers, G.A. Gustafson, A.J. McKeiman, J.C. Rhyan, D.H. Halverson, R.F. Rosenbusch

DVM MS, Biologics Virology Laboratory, National Veterinary Services Laboratories, USDA, APHIS, Veterinary Services Ames, IA USA

Key words: Abortion, death, canine vaccine, BTV.

Two pregnant bitches were vaccinated with a USDA licensed, four-component canine-modified live virus vaccine. The bitches became febrile and depressed within 48 hours after vaccine administration. Both aborted late term pups and eventually died or were euthanized.

A virus was isolated from the placenta and the mesenteric lymph nodes of the bitches, and was identified as bluetongue virus (BTV). A similar BTV was isolated from the vaccine serial and identified as a domestic strain, serotype 11. Two additional field cases were reported associating abortion and death with pregnant bitches following vaccination with the same manufacturer's modified live vaccine. A BTV was isolated from tissues from each of these cases and from the vaccine associated with these cases. The manufacturer voluntary recalled the serials of vaccine identified as being contaminated with BTV. The virus isolated from vaccine was expanded in cell culture and 6 logs administered intramuscularly to four pregnat bitches. Three bitches were observed to abort between days 5 and 6. One of the aborting bitches died on day 6. On post mortem, the fourth bitch had an involuting uterus and evidence of foetal attachments. BTV was re-isolated from the bitches and the foetal tissues.

In a separate experiment, groups of six-week-old puppies, 12-week-old puppies, non-pregnant adult dogs, and late term pregnant bitches were given either the contaminated vaccine or the isolate. Two of two pregnant bitches receiving the vaccine (5.5 log BTV) died or aborted and died within six days. Two of two pregnant bitches receiving the isolate (5.6 log BTV) aborted by day 6. The pregnant bitches receiving the isolate subsequently developed severe respiratory signs and either died or were euthanized by day 8. None of the six-week-old puppies (n = 6 per treatment and two contacts), 12-week-old puppies (n = 2 per treatment), or the non-pregnant adult dogs (n = 2 per treatment) developed clinical illness. A control group of two late-term pregnant bitches received a single dose of the same canine vaccine that was not contaminated with BTV. Both bitches remained clinically normal and whelped live pups.

Dr. R.L. Levings, DVM, MS, Biologics Virology Laboratory, National Veterinary Services Laboratories, USDA, APHIS, 1800 Dayton Rd, Box 844, Ames, IA 50010, USA.

Brown F, Lubiniecki AS: Viral Safety and Evaluation of Viral Clearance from Biopharmaceutical Products. Dev Biol Stand. Basel, Karger, 1996, vol 88, p 221-222

SUMMARY TO BREAKOUT SESSION D

Chairman: *J. Finlayson, FDA/CBER, OBRR, USA*

Participants: *D. Baker, Baxter Biotech, USA*
F. Horaud, Institut Pasteur, France
M. Savage, Bayer Corporation, USA
H. Willkommen, Paul-Ehrlich-Institut, Germany
M. Yu, FDA/CBER, OBRR, USA

Although important lessons about viral safety can be learned from experience with a variety of biological products (e.g. vaccines, animal sera, purified animal proteins), most of the examples in this session came from human plasma derivatives.

The conclusions emerging from the session can be summarized by answering two questions, namely: (i) what kind of factors underlay previous viral events? (ii) in view of these factors, what are the major lessons learned and the areas requiring close attention in the future?

Important factors underlying past adverse events included: (a) insufficient knowledge of an excipient; (b) over-optimism regarding viral inactivation process; (c) a technical/design flaw; (d) unexpected (and unoticed) characteristics of a product lot; (e) emergence of a new virus and (f) a change in source material.

Major lessons for the future include: (a) the continuing importance of viral validation and the possible need for re-evaluation as laboratory techniques and our understanding improve; (b) the importance of well designed, well thought-out validation studies; (c) the importance of accuracy in scaling down processes for validation purposes; (d) the importance of statistics in validation and (e) the need for many different types of expertise (e.g., chemical, virological, manufacturing) in designing and executing validation studies.

Consistent with these lessons, the session emphasized that when viral validation studies are undertaken, it is essential to understand: (a) each individual product, (b) the virology involved, (c) the mechanism of viral reduction and (d) the details of the viral assay system.

Similarly, for manufacturing, the session emphasized the importance of: (a) knowing as much as possible about the source materials and (b) having robust, high-capacity viral removal/viral inactivation processes.

Finally, it was considered essential to pay close attention in the future to: (a) process changes; (b) equipment changes; (c) source material changes; (d) encouraging clinical colleagues to maintain complete records and to record lot numbers.

Dr. J. Finlayson, FDA/CBER, OBRR, 1401 Rockville Pike, 200 N, HFM-330, Rockville, MD 20852-1448, USA.

Brown F, Lubiniecki AS: Viral Safety and Evaluation of Viral Clearance from Biopharmaceutical Products. Dev Biol Stand. Basel, Karger, 1996, vol 88, p 223

BREAKOUT SESSION E

Adventitious Agent Screening Assays

Chairpersons: A. Lewis, FDA/CBER/OBRR, USA
A.C. Lee, The Salk Institute, USA

Brown F, Lubiniecki AS: Viral Safety and Evaluation of Viral Clearance from Biopharmaceutical Products. Dev Biol Stand. Basel, Karger, 1996, vol 88, pp 225-227

SUMMARY TO BREAKOUT SESSION E

Chairmen: *A. Lewis, FDA/CBER/OBRR, USA*
A.C. Lee, The Salk Institute, USA

Participants: *C. Marcus-Sekura, Biotechnology Assessment Services, Inc., USA*
R. Garnick, Genentech, Inc., USA
Y. Matsunaga, NIH, Japan
J. Beeler, FDA/CBER/OVRR, USA
R. Levings, VSL/USDA, USA

The purpose of the Breakout Session on Adventitious Agent Screening Assays was to attempt to identify those issues related to testing for viral contaminants that are of concern and to consider means of applying screening assays to the maximum level of their effectiveness.

A panel of speakers and the audience had on open discussion of a wide variety of issues related to testing for adventitious viral agents. The Session was introduced by a review of the tissues culture and animal systems in current use to screen for adventitious agents. The introduction provoked discussions of the selection of indicator cells, the volume and dosage of test article to be evaluated, the use and availability of neutralizing antibody to inactivate test articles, FDA admonitions to use scientific judgment and the need to clarify the applicability of GLP documents and develop regulatory mechanisms to update procedures that are being displaced by better technology.

Following the introduction, the issue of developing alternative screening assays based on PCR technology was considered. When properly applied, such assays have the potential to be more sensitive, more specific and provide higher quality data much more rapidity than other screening techniques. By combining PCR assays with cell culture and process evaluation, better data concerning the presence and, what is more important, the absence of adventitious agents might be developed at lower cost than by current methods. An excellent example of the power of PCR technology was presented by an analysis of the contamination of CHO cells in large production vessels by minute virus of mice (MVM). By applying both PCR technology and tissue culture assays for infectious virions, this particular problem was resolved. As a result of this experience, PCR technology is being applied routinely to monitor daily operation at the Genentech facility. In this example, detection of

MVM contamination by PCR technology was 100-fold more sensitive than tissue culture assays. As the PCR data are available in eight hours compared to cell culture that takes 21 days or about 500 hours, this represents a significant decrease in the time required to determine the absence of this virus in manufactured products.

PCR technology has also been applied to the detection of parvovirus B19 DNA in Factor VIII concentrates from human plasma. There have been many reports of transmission of this virus by Factor VIII. In the example presented in the Breakout Session, PCR assays could detect 3.5 copies of the viral genome per ml of concentrate. The PCR technique used detected B19 virus DNA in the plasma of 8 of 27.000 healthy donors. On the basis of this experience, PCR technology appears to provide an opportunity to evaluate manufacturing processes and Factor VIII product for clearance of parvovirus B19. To eliminate the virus from Factor VIII product, screening of donor plasmas before pooling for the presence of both B19 DNA and B19 specific IgM antibody was recommended. This study suggests that PCR technology, together with antibody screening assays, also provide the means to detect and possibly eliminate other adventitious agents from human plasma products.

Pestiviruses are known to be present in bovine sera and in porcine and bovine trypsin, reagents that are routinely used in the manufacture of biologicals. PCR reactions have been developed that detect from 10 TCID50 to approximately one copy of a pestivirus genome per ml. As all domestic cattle herds in the United States are infected with pestiviruses and as current pestivirus vaccines are not preventing the spread of pestivirus diseases, PCR technology provides the most sensitive means of screening tissue-culture reagents, cell lines, cell banks and manufactured products for these viruses. As pestiviruses are inactivated by gamma irradiation and heat (techniques that are used routinely to treat bovine serum) the usefulness of this approach needs to be augmented by evaluating the relationship between the presence of detectable pestivirus DNA and the presence of infectious pestivirions in the reagents used to manufacture biologicals.

Orbiviruses producing infections in ruminants are not usually known to infect canines. However, there has been a recent outbreak of orbivirus infection in dogs in Idaho and Oregon following vaccination. The canine disease was characterized by febrile illnesses, abortion, renal failure and death. The orbivirus isolated from these animals was blue tongue type 11, a strain that is present in cattle. The virus was also recovered from the vaccine and the vaccine strain reproduced the syndrome when inoculated into dogs. The vaccine product was recalled. The source of the contamination has not been identified. Although the lot of bovine serum used in producing the vaccine is suspected, attempts to recover the virus from the aliquots of this lot that remained were unsuccessful. The concern conveyed by this outbreak is the possibility that blue tongue type 11 virus has acquired a new pathogenic phenotype and the ability to cross species barriers.

The data and deliberations of this Session achieved the goals of identifying issues of concern and issues of assay sensitivity in the following manner:

(i) Regulatory agencies need to re-assess long-standing techniques that are being used to assay for adventitious agent contamination, develop methods for comparative testing and when appropriate, replace outdated methods with more sensitive and more cost-effective processes.

(ii) Second, PCR technology appears to offer a very rapid, highly sensitive and highly specific means of detecting viral contaminants in biologicals as well as a means for daily monitoring of manufacturing processes for their presence.

(iii) Large-scale tissue culture operations appear to be susceptible to contamination by small, highly stable, rodent viruses that can infect the CHO and BHK-21 cells used in these operations. Special precautions seem to be needed to ensure the absence of these agents from large-volume manufacturing process.

(iv) Blue tongue virus type 11 may have acquired an expanded host range. The work of our colleagues as they study this canine isolate should be followed closely. In this era of emerging viruses and xenografting, our canine companions could be sending us an important message.

Dr. A. Lewis, FDA/CBER/OVRR, 1401 Rockville Pike, HFM-400, Rockville, MD 20852, USA.

Brown F, Lubiniecki AS: Viral Safety and Evaluation of Viral Clearance from Biopharmaceutical Products. Dev Biol Stand. Basel, Karger, 1996, vol 88, p 229

BREAKOUT SESSION F

Adventitious Agents from Animal-Derived Raw Materials and Production Systems

Chairpersons: *K. Hellman, FDA/CDRH, USA*
C. Vincent, FDA/CDER, USA
J. Honstead, FDA/CVM, USA
R. Rohwer, Medical Research Service, USA

Brown F, Lubiniecki AS: Viral Safety and Evaluation of Viral Clearance from Biopharmaceutical Products. Dev Biol Stand. Basel, Karger, 1996, vol 88, pp 231-234

Adventitious Agents from Animal-Derived Raw Materials and Production Systems

K.B. Hellman[1], *J.P. Honstead*[2], *C.K. Vincent*[3]

[1] Center for Devices and Radiological Health, U.S. Food and Drug Administration, Rockville, Maryland USA
[2] Center for Veterinary Medicine, U.S. Food and Drug Administration, Rockville, Maryland USA
[3] Center for Drug Evaluation and Research, U.S. Food and Drug Administration, Rockville, Maryland USA

Key words: Transmission, adventitious agents, TSE.

Recognizing that the transmission of adventitious agents, including viruses and agents of the transmissible spongiform encephalopathies (TSEs), is an important concern for all animal-derived raw materials and production systems used in the manufacture of medical products, the United States Food and Drug Administration (FDA) convened a session at the June 1995 International Conference on Viral Safety and Evaluation of Viral Clearance from Biopharmaceutical Products held at the National Institutes of Health to address issues associated with the management of adventitious agents. The Session Organizing Committee represented the FDA's Center for Devices and Radiological Health (CDRH), Center for Drug Evaluation and Research (CDER), and Center for Veterinary Medicine (CVM).

The central objective of the session was to exchange current information and discuss management approaches in the development of guiding principles for minimizing/eliminating transmission of adventitious agents via medical products. The Organizing Committee developed and identified the critical elements for this development from the perspective of the government regulatory communities and representative production industries. An important goal was to initiate a dialogue between the scientific communities of the public and private sector towards development of common principles and methods.

The elements identified for this development include:

1. definition and characterization of the animal source and animal maintenance;
2. methods of material procurement and storage;

3. material processing and process-monitoring procedures; and,
4. evaluation of agent clearance with identification of model protocols and the rationale for model selection.

Information recommended for consideration in each of these elements is shown in Table 1.

Evaluation of Agent Clearance

Model protocols

Rationale for selection of model/relevant agents

The session provided the opportunity to discuss these elements and to describe the different approaches developed to ensure that a product is free of biological contaminants. Process considerations for minimizing the transmission of adventitious agents by medical products, and updates of the regulatory considerations for manufacture of medical products from animal sources were the focus of a half-day programme featuring invited presentations and panel discussions. Issues were addressed as they relate to the development of biopharmaceutical products, transgenic animal-derived products, product components and processed materials, and tissue engineered products.

The information presented at this session and treated in the respective papers of this issue will be helpful to the FDA and the international community in fulfilling the regulatory responsibility for evaluating products using animal-derived raw materials and production systems.

Preliminary risk assessment exercises were proposed that could enable users of animal-derived raw materials to make more informed decisions on the need and

Table 1: Development of guiding principles for animal-derived raw materials and production systems in medical products: Factors for consideration

Animal source	**Animals maintenance**
Origin	Medication history, antibiotics
Genetic background	Level/validation of containment
Health history	Housing
Adventitious agents	Care, monitoring
Latent/Endogenous flora	Sentinel animals
Feeding history	Caregivers
Material procurement and storage	**Material processing**
Collection method and documentation	Fresh material
(i.e., slaughter, surgery, tissues, fluids)	Process monitoring
Storage conditions	Pooling

type of processing necessary before product manufacture that would also be useful to the final risk/benefit analysis. Criteria that were identified as basic to any risk assessment strategy include:

a) stringent material sourcing;

b) quantification of risk in the starting materials;

c) validation of agent(s) removal;

d) screening of animals and products;

e) final product testing.

The end-use of the product, and the route(s) of administration and exposure to potential pathogens, should dictate the level of concern. In quantifying the risk in starting materials, the type of tissue(s) used and its likelihood of harbouring or being contaminated by the adventitious agent(s) are important. For complete identification of the animals, all information on regulatory inspection (ante- and post-mortem), transportation, and importation should be available. Finally, the production process should be monitored and should include steps that remove the agent(s) either by inactivation or physical separation. The disposal of agent(s) sequestered but not inactivated is important and should be considered with the manufacturing process(es).

It was generally agreed that clearance studies are valuable in assuring the safety of medical products from contamination by adventitious agents. Clearance protocols, i.e. validation of the removal or inactivation of the agents, provide an added level of assurance that the final product will be free from contamination.

Clearance studies involve spiking high-titre infectious agent into scaled-down manufacturing steps and evaluating the ability of each step to remove or inactivate the agent. The agents to be evaluated should represent the potential contaminants of the starting material or appropriate models. Evaluation of study results should consider the sensitivity of the assay(s) with appropriate adjustments for samples in which the agent is not detected.

The agents of transmissible spongiform encephalopathy (TSE), such as scrapie in sheep and bovine spongiform encephalopathy (BSE) in cattle, are the most resistant adventitious agents and among the most difficult to detect. As such, they constitute a stringent challenge for clearance. Bio-assay in suitable animal models such as the hamster/scrapie or mouse/scrapie models is the only generally available detection method. Since cost and time factors are issues regarding detection by bio-assay, research on other comparable potential detection assays has become critically important.

Since the TSE agents are resistant to inactivation by methods successful with other biological agents, adequately controlled clearance studies including the titration of spiked loading material and output samples become extremely important. Moreover, since the most effective known inactivation processes are not always compatible with maintaining the integrity of the biomedical product of interest, the only approach for non-destructive removal of the TSE agents is through material sourcing and adequate control of tissue/cell collection procedures.

The concern posed by the potential of TSE agents in regulated products also prompted the FDA to issue a December 1993 Letter to Manufacturers of FDA-regulated products which proposed recommendations for identification of animal-derived raw materials, documentation of raw-material sourcing, and maintaining traceable records at the manufacturing site. The FDA has also issued additional letters in this area [see 59 FR 44591-44594, published August 29, 1994]. The FDA further recommends that appropriate screening tests and process validation be performed to ensure that the product is free of biological contaminants.

Available guidance on risk assessment approaches used by different industries include the EEC Regulatory Document Note for Guidance entitled « Guidelines for Minimizing the Risk of Transmitting Agents Causing Spongiform Encephalopathy via Medicinal Products », Biologicals 1992;20:155-158 and the six-parameter risk assessment scheme developed by the Federal Health Office of Germany, « Safety Requirements for Medicinal Products Derived from Body Constituents of Cattle, Sheep, or Goats to avoid Risk of Transmission for BSE or Scrapie », February 1994.

Areas for further consideration by the international community evolved from the session. Among others is the need to develop common language/definitions regarding animal-derived raw materials and systems. For example, when considering animal sourcing, what are the differences between a closed herd and one that is isolated or contained? Is containment feasible for all animal species? With regard to risk assessment strategies, it is apparent that the development and adoption of common guiding principles and general risk assessment strategies are needed for minimizing/eliminating transmission of adventitious agents through biomedical products where animal-derived materials and systems are used as the initial source. These strategies should be developed co-operatively between the public and private sector. Since the TSE agents pose special problems for the safety of biomedical products and since there is a lack of information regarding their origin, detection and inactivation, increased attention must be focussed on these adventitious agents.

Dr. K.B. Hellman, Coordinator for Biotechnology, Center for Devices and Radiological Health, Food and Drug Administration, Rockville, MD 20852, USA.

Brown F, Lubiniecki AS: Viral Safety and Evaluation of Viral Clearance from Biopharmaceutical Products. Dev Biol Stand. Basel, Karger, 1996, vol 88, pp 235-236

Zoonotic Agent Consideration in Xenotransplants

Introductory Comments for Breakout Session F: Adventitious Agents from Animal-Derived Raw Materials and Production Systems

J.P. Honstead

Food and Drug Administration, Center for Veterinary Medicine, Division of Animal Feeds, Rockville, MD, USA

Key words: Xenotransplants, zoonotic agents, TSE; BSE.

This paper describes an issue in the Center for Veterinary Medicine that relates to xenotransplants.

In the August 29, 1994 *Federal Register,* FDA proposed a rule which would prohibit specified offal from adult sheep and goats from use in ruminant feed. Specified offal is defined in this proposal as any tissue from the brain, spinal cord, spleen, thymus, tonsil, lymph nodes, or intestines. The specified offal might possibly contain the agent that causes scrapie, a transmissible spongiform encephalopathy (TSE) of sheep that is epidemiologically associated with the occurrence in the United Kingdom of a TSE of cattle called bovine spongiform encephalopathy (BSE).

Processed tissues from sheep and goats are used as ingredients in animal feeds. These products are derived from inedible slaughter by-products and dead, dying, diseased and disabled (or 4-D) animals. Generally, these slaughter by-products and 4-D animals are processed by rendering which involves cooking to separate fats, proteins and water. Processed animal by-products have a long history of safe use in the United States as a source of nutrients for animals. However, the agents responsible for the transmission of BSE and related TSE diseases are not well characterized, but it is known that they survive some rendering processes.

Even though BSE has not been diagnosed in the United States, scrapie is present in US sheep flocks. Because there may be a causal relationship between scrapie and BSE, FDA believes that the potential implications for animal health require oversight to minimize the possibility of introduction of the disease into US cattle. Although no case of human disease has been associated with an animal TSE, exposure of humans to TSE agents in ruminant tissues through xenotransplants can be prevented with proper precautions.

This meeting is an important step towards an understanding of the potential for zoonotic agents to occur in animal-derived tissues and the responsibilities of both FDA and industry in the safety and oversight of xenotransplants.

Dr. J.P. Honstead, Food and Drug Administration, Center for Veterinary Medicine,
Division of Animal Feeds, 7500 Standish Place, HFV-222, MPN II, Rockville, MD 20855, USA.

Brown F, Lubiniecki AS: Viral Safety and Evaluation of Viral Clearance from Biopharmaceutical Products. Dev Biol Stand. Basel, Karger, 1996, vol 88, pp 237-241

Transmissible Encephalopathies and Biopharmaceutical Production

M.M. Robinson

USDA-ARS Animal Disease Research Unit, Washington State University, Pullman WA, USA

Key words: TEs, biopharmaceuticals.

Summary

The use of post-mortem tissues as sources for the production of biologicals, vaccines and feedstuffs has led to the transmission or generation of transmissible encephalopathies in some recipients. For example, the use of pituitary-derived human growth hormone and gonadotropins has resulted in the transmission of Creutzfeldt-Jakob disease to other humans [1], the use of formalin-inactivated sheep brain as a source for louping ill vaccine led to the transmission of scrapie to over 1,000 sheep from one vaccine lot [2], and the use of rendered products from ruminant carcasses in the domestic animal food chain led to the emergence and epizootic of bovine spongiform encephalopathy in the United Kingdom [3]. Infection with transmissible encephalopathies by iatrogenic or other mechanisms is difficult to predict or control.

The characteristics of these pathogens do not permit easy detection, clearance, or inactivation in routine biopharmaceutical production environments.

Introduction

The transmissible encephalopathies are neurodegenerative diseases of humans, domestic animals and possibly feral animals. Included in this group are Creutzfeldt-Jakob disease (CJD), Gerstmann-Straussler-Scheinker syndrome (GSS), and kuru of humans, sheep scrapie, transmissible mink encephalopathy (TME), chronic wasting disease of mule deer and elk (CWD), bovine spongiform encephalopathy (BSE), feline spongiform encephalopathy (FSE), and transmissible encephalopathies of captive old world bovidae (e.g. greater kudu, arabian oryx, gemsbok).

CJD, the most prominent of the human diseases, occurs with a world-wide incidence of fewer than 1:1,000,000 adults annually [1]. Approximately 10% of the CJD cases are documented as familial in nature, while the rest appear sporadically. GSS occurs at less than 1/10 the frequency of CJD, and kuru, which was restricted to a specific region of Papua New Guinea, has all but disappeared. CJD has been transmitted iatrogenically through corneal and dura mater transplants, through the use of «contaminated» neurosurgical instruments and electro-encephalographic probes, and through the administration of human growth hormone and gonadotropins derived in part from pituitaries of CJD-affected cadavers [1].

Scrapie is endemic in many countries with variable or undocumented incidences. The disease is transmitted both vertically and horizontally among sheep [2]. TME has been documented in ranch-reared mink in North America, northern Europe, and states of the former Soviet Union [4]. It has occurred infrequently, being the result of a food-borne infection, but with devastating results. CWD has been detected in captive and free-roaming animals in Wyoming and Colorado [5]. Its origin, mechanism(s) of transmission, and frequency of occurrence are unknown. BSE, FSE, and transmissible encephalopathies of captive old world bovidae are foreign animal diseases with respect to the United States, and have occurred primarily, though not exclusively, in the United Kingdom. They are thought to be transmitted through food-borne infection, similar to TME [6].

Detection

The common features of the transmissible encephalopathies are (i) an incubation period which lasts from months to years, (ii) a fatal outcome, (iii) no detectable immune response to a foreign pathogen in the affected individual during the preclinical or clinical periods of disease, (iv) an accumulation in central nervous system and other tissues of an abnormal form of a normal cellular protein variously referred to as PrP, protease-resistant protein, or prion protein, and (v) a low annual incidence, except for BSE. Clinical disease is detected when a sufficient number of neurons have died so that a motor deficit or behavioral change becomes apparent. Clinical diagnosis is by exclusion of a large number of differentials since there are no clinical signs which are pathognomonic for these diseases. Diagnostic confirmation is made through microscopic examination of CNS tissues for astrocytosis, pyknotic and vacuolated neurons, and vacuolated neurophil, bioassay of suspect tissues, or PrP immunohistochemistry or immunobiochemistry of CNS or lymphoid tissues. The long incubation period coupled with the low incidence of these diseases make them difficult to detect in human or domestic animal populations.

Clearance

The agents of transmissible encephalopathies are subviral in size. While the specific identity and composition of the transmissible entity remain a matter of controversy, it has been shown that filtration techniques do not provide a reliable method for the exclusion of infectivity until reaching nm pore sizes, in part due to the lipophilic nature of the pathogen [7]. If appropriate to the process, filtration may be used to reduce the pathogen burden, but should be followed by inactivation procedures tested through some type of validation experiment [8].

Inactivation

The literature describing experiments involving the inactivation of the agents of transmissible encephalopathies provides little consolation to process managers because of the variability of results from laboratory to laboratory. The differences in results have been due in part to the differences in laboratory strain of agent studied (e.g. mouse-passaged CJD versus hamster-adapted scrapie 263K versus mouse-adapted scrapie strains) and in part to the inherent variation between researchers' techniques and equipment. A common theme among reports of hamster-adapted scrapie inactivation studies is that very harsh treatments are required to eliminate all infectivity detectable through bio-assay. Usually these techniques are not amenable to production situations where the biological activity of the product must be maintained.

Examples of the results of a variety of inactivation treatments used on hamster-adapated scrapie strain 263K are presented in Table 1. While most of the treatments caused some decrease in detectable infectivity, the information cannot be assumed to be directly applicable to other systems for two reasons. First, hamster scrapie 263K is a laboratory artifact and is not necessarily representative of any of the agents of transmissible encephalopathies found in nature. Second, the mechanism(s) of inactivation can be assumed from the nature of the inactivant, but cannot be proven because a description of the molecular and biochemical nature of the infectious unit is not available at this time.

In addition, all inactivation studies of transmissible encephalopathies are based on bioassays, primarily in rodents. These bio-assays take up to two years to conduct in rodents, and longer in large animals [9], and still may not provide data that are appropriate to a given situation or question. An example of the problem in choice of recipient is illustrated by the observation that the TME agent from mink brain did not cause disease in mice even when inoculated intracerebrally into a wide variety of mouse strains [10], but the same agent caused disease in other species.

There is no way to know whether the agent of a given transmissible encephalopathy will cause disease in an experimental recipient except for historical precedent or trial and error. Also, the agent from one donor may display dissimilar dose-response properties in similar recipients, depending on the tissue source and preparation of tis-

Table 1: Inactivation of Hamster-adapted 263K.

Treatment	Temperature	Time	Inactivation	Reference
Dry heat	60°C 100°C 121°C	1 hr. 2 min. 1 min.	90% 97% 99.9999%	12 12 12
Autoclave with 1% SDS[a]	121°C 130°C	1 hr. 1.5 hr.	99.99999% 100%	13 13
Phenol	r.t.[b]	1 day	90%	14
Formic acid	r.t.	1 hr.	99.9999999%	15

[a] sodium dodecyl sulphate.
[b] r.t. = room temperature.

sue extracts [11]. It may be possible that organ-specific modification of these agents would allow the agent from brain tissue to cause disease in a given recipient, but the agent from spleen or some other tissue to cause no disease, leading to the erroneous conclusion that there was no agent in that tissue. It follows that a positive result in bio-assay may be significant depending on the question asked, but a negative result is meaningless in the absence of information that the bio-assay recipient is susceptible to the agent and tissue source in question.

Finally, it may be possible to show that a particular inactivation process eliminates all detectable infectivity for a given recipient species, but it cannot be shown directly that all the «infectious units» have been inactivated or that the bio-assay has specific relevance to humans or domestic animals. A negative result in a bio-assay for any transmissible encephalopathy should be interpreted with caution until an adequate definition of the infectious unit has been developed. These limitations must be borne in mind when designing and executing validation studies for products where there is any concern that an agent of a transmissible encephalopathy may be present.

Conclusions

None of the transmissible encephalopathies of domestic animals have been shown to be human health threats. The threat of BSE to human health remains questionable since the disease recently emerged and spread to a level which is uncharacteristic of most transmissible encephalopathies [3]. Historical precedent would suggest that BSE represents no threat to human health, but only time can prove that.

If the argument is made that the presence of any agent of a transmissible encephalopathy is unacceptable at any level in a biopharmaceutical product intended for humans or domestic animals, then the use of source and process materials that are known to be free of these agents is the best route for production. Once suspect materials enter the production chain, it may be impossible to prove that the «potential» infectivity has been eliminated through clearance or inactivation. From a scientific or medical viewpoint, the type of infectivity that might be present could be irrelevant to the application of the product in that the agent in question does not cause disease in the species in which the product will be used or that there is not enough of the agent to establish an infection. However, the cost of convincing regulatory authorities and the general public of these positions may, in the long run, make manufacture of the product less profitable than it would be to incur the expense of obtaining «agent-free» materials at the outset.

References

1 Brown P, Preece MA, Will RG: «Friendly fire» in medicine: hormones, homografts and Creutzfeldt-Jakob disease. Lancet 1992;340:24-27.

2 Dickinson AG: Scrapie in sheep and goats, in Kimberlin RH (ed): Slow Virus Diseases of Animals and Man. Amsterdam, North Holland Publishing Co., 1976, pp 209-241.

3 Wilesmith JW, Ryan JBM, Hueston WD, Hoinville LJ: Bovine spongiform encephalopathy: epidemiologic features 1985-1990. Vet Rec 1992,130:90-94.

4 Marsh RF, Hadlow WJ:. Transmissible mink encephalopathy. Rev Sci Tech Off Int Epiz 1992;11:539-550.

5 Williams ES, Young S: Neuropathology of chronic wasting disease of mule deer *Odocoileus hemionus* and elk *Cervus elaphus nelsoni*. Vet Pathol 1993;30:36-45.

6 Wells GAH, McGill IS: Recently described scrapie-like encephalopathies of animals: case definitions. Res Vet Sci 1992;53:1-10.

7 Diringer H, Hilmert H, Dietrich S, Werner E, Ehlers B: Towards purification of the scrapie agent. Eur J Biochem 1983;134:555-560.

8 Di Martino A, Safar J, Ceroni M, Gibbs CJ Jr: Purification of non-infectious ganglioside preparations from scrapie-infected brain tissue. Arch Virol 1992;124:111-121.

9 Clark WW, Hourrigan JL, Hadlow WJ: Encephalopathy in cattle experimentally infected with the scrapie agent. Am J Vet Res 1995;56:606-612.

10 Taylor DM, Dickinson AG, Fraser H, Marsh RF: Evidence that transmissible mink encephalopathy agent is biologically inactive in mice. Neuropathol Applied Neurobiol 1986; 12:207-215.

11 Robinson MM, Cheevers,WP, Burger D, Gorham JR: Organ-specific modification of the dose-response relationship of scrapie infectivity. J Infect Dis 1990;161:783-786.

12 Rohwer RG: Virus-like sensitivity of the scrapie agent to heat inactivation. Science 1984; 223:600-602.

13 Brown P, Rohwer RG, Green EM, Gajdusek DC: Effect of chemicals, heat, and histopathologic processing of high-infectivity hamster-adapted scrapie virus. J Infect Dis 1982;145:683-687.

14 Rohwer RG: Scrapie infectious agent is virus-like in size and susceptibility to inactivation. Nature 1984;308:658-662.

15 Brown P, Wolff A, Gajdusek DC: A simple and effective method for inactivating virus infectivity in formalin-fixed tissue samples from patients with Creutzfeldt-Jakob disease. Neurol 1990;40:887-890.

Dr. M.M. Robinson, Animal Production Unit, FAO/IAEA Agriculture and Biotechnology Laboratory, IAEA Seibersdorf Labs., A-2444 Seibersdorf.

Brown F, Lubiniecki AS: Viral Safety and Evaluation of Viral Clearance from Biopharmaceutical Products. Dev Biol Stand. Basel, Karger, 1996, vol 88, pp 243-245

Sourcing Pharmaceutical, Biological and Medical Components from Livestock in the United States

L.A. Detwiler

USDA, APHIS, Veterinary Services, Robbinsville, New Jersey, USA

Key words: Biopharmaceuticals U.S. livestock.

Abstract: The practice of using livestock as a source of pharmaceutical, biological and medicinal components is growing. Livestock donors may be kept in research-type settings where the majority of variables are controlled. Livestock or livestock by-products may originate from known farms where there is some input into management, or products may be obtained from random-source livestock at slaughter. In the case of random-source animals, little or no information may be available about the health status of the individual animal or the farm of origin.

There continues to exist a tremendous demand for the use of livestock as a source of pharmaceutical, biological and medicinal components. Depending on the nature and use of the final product, the animal source may or may not need to be closely monitored. The most controlled environment from which animals can be obtained would be a research-type setting where factors such as diet, water, housing, sanitation, and veterinary care are closely regulated. If necessary, the genetics of these animals can also be managed.

Another avenue which may be pursued for animal sourcing is the use of livestock from designated farms. A company may contract with a producer who maintains records of various husbandry practices and agrees to manage the herd or flock within certain specifications. The operations may vary from confinement to range or pasture. This approach provides the company with some degree of knowledge and input which may be less costly than the method described above.

To collect animal materials in high volume, many companies use slaughterhouse livestock. There are numerous unknowns involved in this route. Without undertaking extensive traceback efforts, one would not know the genetic background of the animals, the conditions under which they were raised or any information about herd or flock health.

Slaughter channels assemble livestock raised on pasture and in feedlots as well as livestock which are fed on balanced rations versus those which receive a bare maintenance diet. There are animals from herds or flocks with sound veterinary care and those originating from herds or flocks with disease problems.

The task of tracing animals to the farm of origin may be ominous in some cases. For example, a slaughter-plant in Tennessee which kills sows and boars slaughtered 12,000 hogs originating from 31 states in one month. A cull cow plant in Nebraska routinely receives cattle from every state west of the Mississippi with the exception of Arizona.

The ability to trace the movement of animals is, however, still a possibility. The United States Department of Agriculture's (USDA) Animal and Plant Health Inspection Service (APHIS) requires that identification such as backtags or official eartags be used on mature cattle moving in interstate commerce. These animals are bled at slaughter and the identification collected. The blood sample is tested for brucellosis. If the sample is found to be positive the source of the animal will be tracked in an attempt to locate the infected herd. This trace-back can also be done for other reasons such as drug residues or the evidence of tubercular lesions.

On a smaller scale, finding the farm of origin may not be too difficult. If a company must do this for a larger number of animals the efforts may be costly, time-consuming, and not always fruitful.

In the larger dairy states which have more than 1000 head of cows, locating the birth herd may not always be possible. For instance, in California, a heifer calf born in a large milking herd will most probably be moved to a contract calf raiser. The calf stays at this facility for two to three months. The calf then goes to a heifer raiser and may change hands several times (due to feed availability) until it is close to parturition. The heifer is then sold back into a milking herd until culled. It is estimated that only a quarter of the dairy herds will attempt to recycle their own calves. Hence, it is very probable that one could identify the most recent premises. The birth herd may be almost impossible to identify with any certainty.

Human error sometimes contributes to the unsuccessful completion of a trace-back. For example, when recording numbers, one may be transposed, or a dealer may record the wrong information, or a market may fail to maintain that particular information.

When obtaining material from animals at slaughter, inspections performed by the USDA's Food Safety Inspection Service (FSIS) would provide some degree of health protection. All animals slaughtered at federally inspected plants receive ante-mortem and post-mortem inspections. Animals which are in a moribund state will most probably be condemned at ante-mortem. Condemnations of carcasses at post-mortem inspections are usually the result of a systemic or septicemic condition or extensive pathological lymph node involvement.

As previously mentioned, cattle are required to be identified by a backtag or eartag. Retention of identification devices in cattle is optimal. Swine moving within interstate commerce are also required to be identified. To date, there is no identification device in swine which can always be retained. The hogs tend to eat the backtags off each other. In the United States it is not mandatory to identify or keep records on sheep and goats in commercial trade channels.

Several countries may require a company to certify that the materials used in the product were obtained from animals which were born in the U.S.A. This is a very difficult statement to make with complete certainty. If livestock is imported legally without restriction into the USA, the animals become US animals. There are no regulations which mandate that imported animals retain identification indicating they originate from another country. They may be transported, sold, and handled just like any other US animal. On occasion, it has been necessary to locate imported animals because of the threat of disease. For example, cattle imported from the United Kingdom in the 1980s have been traced due to the risk posed by Bovine Spongiform Encephalopathy (BSE). In these instances efforts must start at the import centres using their records. The breed associations and registries are often helpful in trying to find the animals. The greatest success is usually obtained when a minimal amount of time has elapsed between the time of introduction and tracing.

There are a number of routes a company may pursue when sourcing pharmaceutical, biological or medicinal ingredients from livestock. The choice may depend on the amount of information needed to ensure that the material is safe and effective for the intended use.

Dr. L.A. Detwiler, USDA, APHIS, Veterinary Services, 320 Corporate Blvd, Robbinsville, NJ 08691, USA.

Brown F, Lubiniecki AS: Viral Safety and Evaluation of Viral Clearance from Biopharmaceutical Products. Dev Biol Stand. Basel, Karger, 1996, vol 88, pp 247-256

Analysis of Risk to Biomedical Products Developed from Animal Sources

(With Special Emphasis on the Spongiform Encephalopathy Agents, Scrapie and BSE)

R.G. Rohwer

Medical Research Service, Veterans Affairs Medical Center, Baltimore, Maryland USA

Key words: Transmissible spongiform encephalopathies, scrapie, bovine spongiform encephalopathy, BSE, risk analysis, virus removal, virus inactivation, virus validation.

Abstract: Factors that must be considered in estimations of risk from exposure to adventitious contaminants of animal derived biologicals include: (i) the use of the product; (ii) the routes of administration and exposure to potential pathogens; (iii) the source of animal(s) and their history and maintenance; (iv) the tissue(s) used in the product and their likelihood of harbouring or being contaminated by an agent; (v) the methods by which the animal is slaughtered and the tissue(s) collected; (vi) the production process including steps that remove an agent either by inactivation or physical separation; (vii) the disposal of an agent that is sequestered but not inactivated; (viii) flow control, barriers, and between-batch cleaning; (ix) batch size; and (x) validations and quality assurance monitoring.

The transmissible spongiform encephalopathy (TSE) agents, scrapie and bovine spongiform encephalopathy (BSE) are among the most difficult to detect and remove from animal tissues. As such they constitute an especially stringent challenge for viral clearance.

INTRODUCTION

It seems likely that contamination of animal sources of biologicals by adventitious agents will always be with us. Fortunately, most animal pathogens appear to pose little or no threat to humans. Moreover, for many of those that do, refinements in detection methods are proving increasingly effective at tracking the level and fate of contaminations during the production of biologicals. Nevertheless, there will always remain an unquantifiable risk from pathogens that are rare and sporadic or are newly emergent, or are as yet unrecognized because they cause latent, subtle or infrequent diseases in humans which would be difficult to correlate with the primary exposure event. Thus, the ultimate goal of viral clearance should be the removal of all viruses, known or unknown, that may contaminate the raw materials used in production. Since the spectrum of virus properties is vast, generic strategies must be based upon the ability to clear those agents with the most extreme characteristics.

The spongiform encephalopathy agents are among the most cumbersome to detect, remove, and laborious to validate. Infectivity can only be detected after long incubation times in animals. The agents are notoriously difficult to disinfect. Quantal titration of inactivation or validation studies must be conducted under biohazard conditions and requires six to eighteen months of observation. It is not surprising that these agents are so imperfectly known. Even their viral nature is disputed [1]. The alternative prion hypothesis proposes that the agent is subviral in size and that the infectivity is borne by the amyloidotic form of the host prion protein. The infection is propagated by the recruitment and conversion of the native protein into amyloid by pre-existing amyloid [2].

Due to their extreme resistance to disinfection, the transmissible spongiform encephalopathy agents (TSE) are presumed to be an especially stringent challenge for viral clearance, and in general they are. However, even these agents have their vulnerabilities. They are more sensitive to lipophilic solvents and surfactants than some virus families. Also their adherent nature, while not unique, offers opportunities for clearance as well as exacerbating the hazard from cross-contamination. Because the spongiform agents are a contemporary problem of great concern, we will couch our discussion in terms of these agents.

Discussion

Use

Usage dictates the level of concern for animal-derived products. In general, parenterals are of greater concern than are foods, which in turn pose greater risks than do topical exposures or incidental contacts from non-biological uses. Higher risk is also more likely to be tolerated for essential/critical applications affecting small, high-need populations than for luxury or casual uses or for those exposing large segments of the population. Interestingly, we tolerate much higher levels of risk for traditional uses of animal products than could be sanctioned for new, untried, or experimental applications. In particular, the safety of many animal foods requires a level of consumer responsibility for their storage and preparation that would be unthinkable for products lacking similar historical precedence.

Route of exposure

Risk of infection can be strongly affected by the route of exposure to a given pathogen. In the case of the scrapie agent, an intracerebral inoculation is 10 times more potent than an intravenous inoculation, 10 to 100 times more potent than an interperitoneal inoculation, and 1000 to 10,000 times more potent than a subcutaneous exposure [3-5]. Gastro-intestinal exposure may be even less efficient [6, 7]. Nevertheless, consumption of rendered ruminant protein is almost certainly the cause of epidemic bovine spongiform encephalopathy (BSE) in cattle and is the likely source of feline spongiform encephalopathy and the spongiform disease in exotic zoo ruminants and *Felidae* that has occurred concurrently with the BSE epidemic [8].

Natural scrapie in sheep is also presumed to be transmitted by the GI route in spite of inefficiency and what is presumed to be a very low level of exposure [9]. Apparently, chronic exposure compensates for the other deficiencies. Of course, the efficiency of various routes of exposure is pathogen-specific and may even be agent-strain-specific, and so requires experimental or epidemiological knowledge of the pathogenesis to be used in a risk assessment.

Sourcing

When a manufacturing process lacks intrinsic clearance potential, sourcing may be the only solution. However, it is important to recognize the limitations of the sourcing strategy. Sourcing can be used only to avoid specifically identified contaminants and a source that is free of one contaminant may contain others.

Closed herds offer the potential for greatly limiting exposure to pathogens that are in general circulation. However, the term «closed» is unregulated and is often used in contexts that are irrelevant to pathogen control. Factors that are relevant to tissue quality would be: (i) documentation of the origin, genetic background and genetic maintenance of the herd; (ii) its health history, including viral diseases and serological testing for the presence of adventitious agents; (iii) its medication history including antibiotic use; (iv) housing and husbandry; (v) the level of containment and, ascertainment of containment; and (vi) the standard operating procedures for care, monitoring and personnel training.

If the objective is to avoid BSE, the most important consideration is the feeding history of the herd. The desired trait is a herd that has never been exposed to feed supplements containing ruminant-derived protein. Other considerations are lack of contact with sheep or other animals at risk for spongiform disease such as cats, deer and mink, and a breeding programme that avoids physical contact with outside cattle. While such herds exist, they are an unrealistic source for large quantities of tissue.

Since the United States still permits the feeding of ruminant-derived protein to ruminants, and since many foreign markets penalize the sourcing of bovine-derived biologicals from countries that permit this practice, many U.S. producers of exported biologicals are now sourcing bovine and ovine materials from Australia and New Zealand which are the only countries that are widely accepted to be free from sheep scrapie. The rationale for this is based on the unproven assumption that BSE is caused by exposure of cattle to sheep scrapie. If instead it turns out that BSE originates from some other source, for example a wild animal reservoir, exposure to sheep scrapie may prove irrelevant to the aetiology of BSE, thereby invalidating the sourcing rationale. Attempts to transmit scrapie to cattle have resulted in a neurological disease that is clinically and pathologically distinct from BSE.

There is no country currently claiming to be free from ruminant spongiform disease that has not been exposed to imported animals at risk. That such animals can carry BSE has now been demonstrated by outbreaks in Canada, Oman, Falkland Islands, Portugal, France, Denmark and Switzerland in imported stock. If such an animal were rendered during preclinical disease, the stage would be set for a recurrence of the BSE epidemic.

If the prion hypothesis is correct there could be no cattle population that is totally free from spongiform encephalopathy. It is suggested that prion disease occurs spontaneously at low frequency in any population susceptible to PrP amyloidosis [10, 11].

Such rare spontaneous conversions to the amyloidotic state might require years to progress to clinically recognizable disease. Thus, affected animals might never be recognized. Slaughter and rendering into high protein feed supplements would again set the stage for an eventual epidemic of BSE. Presumably this phenomenon could also occur in mammalian cell lines with the risk increasing with the level of propagation.

Whatever the origins of BSE, whether interspecies transfer of an exogenous pathogen, or intraspecies expansion of a spontaneously-generated intrinsic pathogen, as proposed by the prion hypothesis, the principal defence against a recurrence in other cattle populations is validated sterilization of foodstuffs and biologicals derived and administered intraspecifically, or bans on their use. It should be evident that any species that is similarly exposed to intraspecific recycling of foodstuffs and biologicals may also be at risk. In the case of biologicals, this includes man.

Tissues

Viruses often exhibit stringent tissue tropisms. For a given virus, tissues that are not infected should pose far less risk of contamination than the targeted tissues. However, this isolation is seldom complete. Blood, nerves and lymph provide global access to all tissues for viruses that use those avenues of infection. Viremias are frequent adjuncts to virus infection and result in systemic contamination. Finally, careless collection methods can result in the contamination of otherwise virus-free or virus-reduced tissue.

In the case of BSE, the British, WHO, and European Regulatory authorities have published lists of tissues ranked according to their presumed level of contamination by BSE. As expected, brain and other CNS tissues occupy *Category I, Highest infectivity, highest risk category*; ileum, lymph nodes, proximal colon, spleen, tonsil, dura mater, pineal gland, placenta, cerebrospinal fluid, pituitary, and adrenal occupy *Category II, Medium infectivity*; distal colon, nasal mucosa, sciatic nerve, bone marrow, liver, lung, pancreas, and thymus occupy *Category III, Low infectivity*; and blood clot, faeces, heart, kidney, mammary gland, milk, ovary, saliva, salivary gland, seminal vesicle, serum, skeletal muscle, testis, thyroid, uterus, foetal tissue, bile, bone, cartilage, connective tissue, hair, skin and urine are in *Category IV, No detectable infectivity* [12].

It is important to note that most of these assessments are based on titres from sheep and goats measured across a species barrier in mice with a great loss in sensitivity. Most involve inocula of 50 μl of a 10% tissue homogenate into four to eight mice for a total representation of the parent tissue of $\leq$20 microlitres. No detectable infectivity in Category IV means $\leq$100 LD50/ml. Moreover, many of these estimates are based upon only one or at most two determinations. In some cases rodent-to-rodent inoculations are at odds with some of these recommendations. Titration of bovine tissues in mice has revealed no infectivity outside the brain. However, it is likely that this is due to the insensitivity of the cross-species titrations rather than the complete absence of infectivity in peripheral organs.

Slaughter

Slaughter, while highly regulated, is directed towards the efficient collection of clean, safe food – not biologicals. Animal-derived foods are presumed to be contaminated by microbes and it is the storage and preparation of foods that ensures their safety. However, most of this contamination occurs during the slaughter. The viscera of a healthy animal are essentially sterile and yet this attribute is wasted in the typical slaughter environment which provides multiple opportunities for contamination. Animals from many sources are crowded together ensuring a high level of contact exposure. The slaughter itself exposes the desired tissue both to other organs from the same body and to the tissues of other animals.

In the United States the typical slaughter sequence is kill, bleed, de-limb, decapitate, skin, eviscerate, split, wash and chill [13]. Accidental slaughter of a BSE-infected animal would likely result in widespread contamination of the facility. The vast majority of spongiform infectivity is found in central nervous system tissue with lesser amounts in the lymphoreticular organs. Imperfect use of captive bolt killing frequently results in skull penetration, the release of CNS tissue, and contamination of the slaughterhouse from the kill floor forward with CNS tissue. The head is also removed early in the process with resultant contamination by the spinal cord. If it is further processed additional contamination would be inevitable. The evisceration step exposes the facility to the spleen and lymphatics. The carcase is split by bissecting the spinal column with a saw which likely results in the aerosolization of spinal cord tissue and widespread contamination.

Collection

The challenge of tissue collection is to overcome the obstacles to sterility in the slaughter environment. If the tissue requirements are modest and the product is valuable it may be feasible to collect the tissue under surgical conditions in a more controlled environment and dispose of the carcase as waste. Alternatively, if the manufacturer supplies the labour, it may be possible to retrieve a particular organ under more stringent conditions without unduly interrupting the slaughter. Unfortunately, these are not practical options for manufacturers with large mass requirements where lot sizes representing hundreds of thousands of animals are not uncommon.

Whatever the collection circumstances, much of the contamination of non-CNS tissues, if present, may be from exogenous sources, and superficial in nature. For Category IV, low risk tissues from cattle, simple surface washing of the tissue could be used to remove superficial CNS contamination. If one is dealing with a solid organ it may be possible to improve this wash by using NaOH as a disinfectant.

Inactivation

Virus removal can either be by inactivation or by separation and the manufacturing process may contain incidental or specific virus-removal steps. The most efficacious of these are inactivations which result in the destruction of virus infectivity. Because virus sensitivities to heat, pH, desiccation, solvents, surfactants, and chemical disinfectants vary widely, there are probably few processes that will not inactivate at least some viruses. The corollary is that there are also very few inactivants that are effective against all viruses and those that are, tend to be destructive of biologicals as well. Nevertheless, if the goal is to eliminate a particular virus, conditions can sometimes be found that result in its selective inactivation.

All inactivations proceed kinetically. Thus, if it is discovered that a step in the manufacturing process incidentally inactivates a virus of interest, it may be possible to achieve a much higher level of clearance with little effect on the product by simply extending the exposure at that step.

The spongiform encephalopathy agents are notoriously resistant to disinfection. However, where this has been studied, it appears that resistance is largely due to small subpopulations representing 0.1% or less of the infectivity that are refractory to killing. The vast majority of the infectivity is inactivated with kinetics similar to conventional viruses [1, 14, 15]. Sterility requires steam sterilization for at least one hour at 132°C, exposure to 1N NaOH for one hour, or incineration [16, 17]. Nevertheless, significant inactivations are achieved by phenols, lipophilic solvents, some detergents, strong chaotrophic agents, and in general, reagents that denature proteins and thereby viruses [18].

Separation

Viruses can also be removed from a product by filtration, chromatography, absorption, extraction, precipitation or other mechanical means. Such separations can often achieve high levels of viral clearance. However, they create downstream problems because they concentrate the infectivity and the viruses are still alive. In the case of filters and chromatography columns the infectivity may still be in the process stream. To avoid cross-contamination of the facility by sequestered virus, potentially contaminated filters and columns must, where possible, be sterilized in place before disassembly. Where virus has been removed from the process stream by precipitation or extraction, the virus-bearing phase must be disposed of as infectious waste.

Aggregation of virus results in a loss of titre which can be confused with inactivation or mechanical removal of infectivity when in fact virus numbers have remained constant. If unrecognized, this could lead to a very dangerous presumption of clearance where in fact none was achieved. The stability of aggregates often varies greatly with virus concentration, buffer conditions, presence of surfactants and many other factors. A later change in buffer conditions might disperse the aggregates and reintroduce the virus into the production stream.

Cross contamination

The risk of contaminating the facility is especially great when clearance is by removal rather than inactivation. In the case of the spongiform encephalopathies such contaminations would persist indefinitely unless specifically removed by disinfection. Well planned barriers and personnel and materials flow can reduce the likelihood of carrying infectivity past the removal steps. This is a practical daily consideration in the scrapie laboratory where high titre agent is used intentionally. Inadequate attention to this issue in the testing laboratory could compromise the results of validation studies.

Cleaning

Rigorous between-batch cleaning and disinfection of process equipment and the facility is essential to limit the consequences of any virus contamination that may have escaped notice. This could easily occur in the case of the spongiform agents and apparently did happen during the production of cadaveric human growth hormone [19, 20]. In most instances the disinfection options for equipment can be much harsher than those for the biologicals themselves. High pressure steam is optimal for cleaning and sterilization of vessels and lines. Work surfaces and floors can be disinfected with 1N NaOH. Maximum security is achieved when chemical and physical methods of decontamination are combined. Unfortunately, a commercial phenolic disinfectant, LpH, that was shown to be highly effective against the spongiform agents is no longer available in the original formulation. Other formulations should be validated before use with these agents.

Batch size

In general the more animals that are represented in a batch the greater the likelihood that one or more of them will harbour adventitious agents. If such contaminations have high titres and are highly dispersed a large fraction of product will be contaminated. Approximately 10,000 cadaveric human pituitaries were used to produce single lots of human growth hormone with catastrophic results [19, 20].

Small batch sizes are rarely compatible with current manufacturing approaches. However, when means can be found to implement such strategies, they go a long way towards limiting consumer exposure to rare sporadic contaminations of a product. However, the effectiveness of this approach depends on the adequacy of the decontamination cycle between lots.

Validation

As noted earlier, the ultimate goal of viral clearance is removal of all viruses, known and unknown. However, viruses vary greatly in their physical properties and sensitivities to inactivants so that no single virus can represent the entire spectrum. Rather, it is necessary to test several viruses representing extremes of behaviour that subsume the rest of virology. Where specific viruses are of concern they should, if possible, be tested directly.

In the case of BSE this is not practical. A validation using the BSE agent itself and titration in cattle would require eight to twelve years to complete. Alternatively, the process could be tested by titrating the BSE agent in mice, but in this case, the host-barrier effect would greatly reduce the sensitivity of detection, thereby reducing the clearance that could be demonstrated [21]. Moreover, the cross-species infection would itself have to be validated. A mouse-adapted BSE strain might be used but has not been well characterized [22]. Its importation and use in the United States would have to be approved by the USDA.

Spongiform agent removal has so far been validated exclusively using rodent-adapted laboratory strains isolated from sheep scrapie. This use presumes that the inactivation and separation properties of the rodent-adapted strains adequately represent the agents of concern. Use of rodent-adapted scrapie strains is also perceived as posing far less hazard to both agriculture and personnel than would the BSE agent or the human agents.

The mouse-adapted scrapie strains develop titres of $\lessapprox 10^9 ID_{50}/g$ of brain. Spiking at the level of 1% brain homogenate and inoculating 30 µl, five logs of clearance can be demonstrated in 12 months. The hamster adapted 263K strain has been favoured for research, due to its high titres and shorter incubation time. The hamster 263K strain develops agent titres of $\lessapprox 10^{11} ID_{50}/g$ of brain, and a 1% spike can demonstrate seven logs of clearance in six months.

A contamination can be either intrinsic or extrinsic. In an intrinsic contamination the agent is naturally present in the tissue of interest, either because it replicates or is sequestered there. Extrinsic contaminations result from exposure during collection or processing to extraneous sources of infectivity. Extrinsic contamination is easily modelled by spiking. Since central nervous system titres are orders of magnitude greater than those for any other tissue, CNS tissue is the most likely source of extrinsic contamination. Thus, high titre brain homogenate may constitute a perfect challenge with which to represent an extrinsic contamination.

In contrast, except in the rare instances in which brain or other CNS tissues are being processed, spiked, brain-derived infectivity may constitute a very imperfect representation of intrinsic contamination of infected tissues. First, it may be difficult to incorporate the spike into the tissue in a way that realistically recreates the in vivo association. Second, it is conceivable, though not established, that there may be differences in the properties of infectivity associated with non-CNS tissues that would affect its fractionation and inactivation properties. For example, the low titre infectivity associated with blood may represent infectivity that has been modified by exposure to the antigen elimination pathways. When brain homogenate is used as the challenge material, the validation is of a brain contamination of the process. High titre inoculum prepared from purified PrP amyloid or detergent-lipid-protein-complexes [2] are subject to the same caveats. If the tissue of interest contains only one or two logs of infectivity per gram, one cannot hope to demonstrate significant clearances using the infected tissue itself as the challenge. As a consequence there is usually no other choice than to spike with brain tissue. One hopes in that case that brain-associated infectivity represents a worst case challenge.

In an ideal manufacturing process, clearance is realized by more than one mechanism and at several different stages of production. It is generally, but not always, the case that each removal step is additive. While the objective is to demonstrate the safety of the overall procedure, one cannot by this means demonstrate more clearance than is present in the challenge inoculum. To adequately assess the maximal inactivation potential of the process, each step with removal potential must be challenged independently with high titre infectivity. Adding the step-wise clearances provides an estimate of the clearance potential of the entire process which, in many cases, is far in excess of the challenge titre, thereby providing a large margin of safety and attendant confidence in the product. The additivity of the step-wise clearances assumes that the surviving infectivity at each step is selected randomly and does not represent a refractory sub-population that will resist all steps. It is to eliminate this possibility that the entire process is tested for its ability to clear the agent.

It has been argued that dilution of infrequent, low-titre contaminations into huge tissue pools may disperse a virus contamination beyond the point of infectiousness [23, 24]. In the case of the spongiform agents, this argument is based on the observation that peripheral routes of inoculation are far less efficient than intracerebral inoculation. The presumption is that this reduction in efficiency is due to a requirement for multiple infections to establish disease. High dilution should greatly diminish the possibility of multiple infections of individuals. However, if the reduction in infection efficiency represents a reduction in the probability of a single productive infection in a background of abortive exposures, there is no a priori reason to expect that this probability would differ for a population compared to an individual. In this case the consequence of dilution would be to disperse and thereby increase the number of infected individuals. The iatrogenic transmissions of Creutzfeldt-Jakob disease by contaminated cadaveric human growth hormone warn against dilution as a means of removal [19, 20].

A distinction is sometimes attempted between the terms «clearance» and «sterility» with greater value attached to the latter. However, both terms must be operationally defined and to the same endpoint. Whether «clearance» or «sterility» is used, its significance is constrained by the titre of the challenge. It is always possible that a larger challenge will produce survivors.

Conclusions

Effective means for reducing the risk of TSE contamination to acceptable levels are available for even the most delicate biologicals. Unfortunately, they can seldom be imposed without considerable expense and inconvenience. However, once such measures are in place, the resulting product is often significantly safer with respect to all adventitious agents. Nevertheless, there are many potential pitfalls and even when a validation study indicates high levels of clearance for a process, the realization of that potential depends critically upon its implementation.

References

1 Rohwer RG: The scrapie agent: «A virus by any other name.» Curr Top Microbiol Immunol 1991;172:195-232.

2 Prusiner SB: Natural and experimental prion diseases of humans and animals. Curr Opin Neurobiol 1991;2:638-647.

3 Kimberlin RH, Walker CA: Pathogenesis of mouse scrapie: effect of route of inoculation on infectivity titers and dose-response curves. J Comp Pathol 1978;88:39-47.

4 Gibbs CJ Jr, Gajdusek DC, Amyx HL: Strain variation in the viruses of Creutzfeldt-Jakob disease and kuru, in Prusiner SB, Hadlow WJ (eds): Slow Transmissible Diseases of the Nervous System. NY, Academic Press, 1979, Vol 2, pp 87-110.

5 Outram G: The pathogenesis of scrapie in mice, in Kimberlin RH (ed): Slow Virus Diseases of Animals and Man. Amsterdam, North-Holland Publishing Co, 1976, pp 325-357.

6 Kimberlin RH, Walker CA: Pathogenesis of scrapie in mice after intragastric infection. Virus Res 1989;12:213-220.

7 Gibbs CJ Jr, Amyx HL, Bacote A, Masters CL, Gajdusek DC: Oral transmission of kuru, Creutzfeldt-Jakob disease, and scrapie to nonhuman primates. J Infect Dis 1980;142:205-208.

8 Bradley R, Wilesmith JW: Epidemiology and control of bovine spongiform encephalopathy (BSE). Br Med Bull 1993;49:932-959.

9 Hunter N: Natural transmission and genetic control of susceptibility of sheep to scrapie. Curr Top Microbiol Immunol 1991;172:165-180.

10 Cohen FE, Pan KM, Huang Z, Baldwin M, Fletterick RJ, Prusiner SB: Structural clues to prion replication. Science. 1994;264:530-531.

11 Gajdusek DC: The transmissible amyloidoses: genetical control of spontaneous generation of infectious amyloid proteins by nucleation of configurational change in host precursors: kuru-CJD-GSS-scrapie-BSE. Eur J Epidemiol. 1991;7:567-577.

12 World Health Organization Consultation on Public Health Issues Related to Animal and Human Spongiform Encephalopathies, Geneva, 12-14 November 1991. (WHO – Report: WHO/CDS/VPH/92.104), 1991.

13 Sanitation Handbook for Meat and Poultry Inspectors. United States Department of Agriculture, Food Safety and Inspection Service. US Government Printing Office: 1990;261-442:20210

14 Rohwer RG: Scrapie infectious agent is virus-like in size and susceptibility to inactivation. Nature 1984;308:658-662.

15 Rohwer RG: Virus like sensitivity of the scrapie agent to heat inactivation. Science 1984;223:600-602.

16 Rutala WA: APIC guideline for infection control practice. APIC guideline for selection and use of disinfectants. American Journal of Infection Control 1990;18:99-117.

17 Brown P, Rohwer RG, Gajdusek DC: Sodium hydroxide decontamination of Creutzfeldt-Jakob disease virus [letter]. N Engl J Med 1984;310:727.

18 Millson GC, Hunter GD, Kimberlin RH: The physico-chemical nature of the scrapie agent, in Kimberlin RH (ed): Slow Virus Diseases of Animals and Man. Amsterdam, North-Holland Publishing Co., 1976, pp 243-266.

19 Brown P, Gajdusek DC, Gibbs CJ Jr, Asher DH: Potential epidemic of Creutzfeldt-Jakob disease from human growth hormone therapy. N Engl J Med 1985;313:728-731.

20 Brown P: Human growth hormone therapy and Creutzfeldt-Jakob disease: a drama in three acts. Pediatrics 1988;81:85-92.

21 Kimberlin RH, Cole S, Walker CA: Temporary and permanent modifications to a single strain of mouse scrapie on transmission to rats and hamsters. J Gen Virol 1987;68:1875-1881.

22 Fraser H, McConnell I, Wells GA, Dawson M: Transmission of bovine spongiform encephalopathy to mice. Vet Rec 1988;123,472.

23 Schrieber R, Seybold U: Gelatine production, the six steps to maximum safety, in Brown F (ed): Transmissible Spongiform Encephalopathies. Dev Biol Stand. Basel, Karger, 1993, vol 80, pp 195-198.

24 Transcript of the 46th Meeting of the Blood Products Advisory Committee, US Department of Health and Human Services, Food & Drug Administration, Rockville MD, 12/15/1994.

Dr. R.G. Rohwer, Medical Research Service, Veterans Affairs Medical Center,
10 North Green Street, 3A-129, Baltimore, MD 21201, USA.

Brown F, Lubiniecki AS: Viral Safety and Evaluation of Viral Clearance from Biopharmaceutical Products. Dev Biol Stand. Basel, Karger, 1996, vol 88, pp 257-264

Transmissible Spongiform Encephalopathies (TSE): Minimizing the Risk of Transmission by Biological/Biopharmaceutical Products: an Industry Perspective*

R.W. Kozak[1], *C.F. Golker*[2], *P. Stadler*[2]

[1] Bayer Corporation, Berkeley CA, USA
[2] Bayer AG, Wuppertal, Germany

Key words: TSEs; transmission risk; biological products.

Abstract: Several guidelines and recommendations have been published on assessing the potential risk of a biological product being contaminated with an agent causing a Transmissible Spongiform Encephalopathy (TSE). Basic principales which can be used during the manufacturing of biological products to minimize the risk of transmission of TSE agents include the following: (i) obtaining animals, tissues or animal-derived raw materials from countries in which the relevant TSE agent is reported to be absent; (ii) screening animal-derived material for appropriate animal husbandry, feeding practices, health certification, tissue procurement and processing practices, and human-derived material for appropriate medical history information and, if relevant, plasma donor exclusion criteria; (iii) selection of tissue source of raw material with reference to potential risk for harbouring or amplifying TSE agents; (IV) human tropism of TSE agent of concern; (v) quantity of raw material used to manufacture a dose of product; (vi) number of daily doses required for indication; (vii) route of administration; (viii) indication and age of patient; and (ix) clearance capabilities by the product purification scheme.
A risk evaluation will be represented for Trasylol® (Aprotinin) which is manufactured from bovine lung tissue. Trasylol® is a protease inhibitor used in the control of blood loss during open heart surgery. Bovine Spongiform Encephalopathy (BSE) is a concern when a product is derived from a bovine source. Even though only bovine lung tissue from BSE-free countries is used to manufacture Trasylol®, it was decided, based on the primary tissue source selection, to evaluate the potential of the purification process to clear a TSE agent. Duplicate «spiking experiments» were performed using a mouse-adapted strain of scrapie as the model for BSE. Four steps in the purification process were evaluated. Titration of «spiked» loading material and output samples was performed at dilutions of 10^{-4} to 10^{-7} and undiluted to 10^{-7}, respectively. The sum of the purification steps evaluated provided a clearance factor in excess of 18 $\log_{10}$ for a TSE agent. An overall

* Part of this presentation was presented by P. Stadler before the DECHEMA panel «Sicherheit der Biotechnologie», on February 7, 1995.

safety factor of 30 is obtained using the notification document issued by the Federal Health Office (Germany), dated February 16, 1994, on «Safety Requirements for Medicinal Products Derived from Body Constituents of Cattle, Sheep, or Goats to Avoid Risk of Transmission of BSE or Scrapie». Based on a six-parameter risk assessment, this document considered a product of low-risk for transmitting TSE agents if obtaining or exceeding a score of 20. Principles for minimizing the risk of transmission of TSE agents in products derived from primary tissues are contrasted to strategies for biotechnology products.

Introduction

Transmissible Spongiform Encephalopathies (TSEs) are progressive neurodegenerative disease associated with infectious proteinaceous proteins (prions), and long incubation periods before onset of clinical symptoms. Currently, there are no known treatments and both animal and human manifestations are uniformly fatal. These agents are troublesome in that there are at present no available, rapid and sensitive, screening assays, even though many efforts continue to pursue development of in vitro assays, and the agent is extremely resistant to inactivating procedures. In addition, while there are many known TSE diseases that can be transmitted from animal-to-animal and human-to-human, the zoonotic nature of the animal TSE agents and their potential for infecting humans is still unknown. A concern for biopharmaceutical manufacturers is whether any of these agents can be contaminants of biological products either through raw materials used during manufacturing or from being in the starting cell subrate or biological source starting material from which the product is derived.

Bayer Corporation works with a variety of biological source starting materials (Table 1). These include organ and tissues, such as bovine lung used in the manufacture of Aprotinin, tradename Trasylol®; body fluids such as plasma, from which a wide variety of products is derived, or milk from transgenic animals and a variety of continuous cell lines used to manufacture and develop biotechnology products.

In performing TSE risk assessments on these different categories of biological source products, different conclusions and strategies may be obtained for minimiz-

Table 1: Strategies for manufacturing therapeutic proteins.

Biological source	**Production process**	**Product examples**
Organ/tissue	Isolation and purification	
i.e. lung		Aprotinin (Trasylol®)
Body fluids		
i.e. plasma		Immune Globulin (Gamimune® N)
i.e. milk		$Alpha_1$-Proteinase Inhibitor
Contiuous cell lines	Fermentation and purification	
i.e. BHK		rhuFVIII (Kogenate®)
i.e. CHO, HEK, Myeloma		ICAM
i.e. hybridoma		TNF MAb

ing the risk of TSE contamination of the final product. The focus of this paper is to provide an outline of how the risk was assessed and minimized for Trasylol®, a biopharmaceutical product derived from bovine lungs, and contrast this approach with strategies for minimizing the risk in biotechnology products derived from continuous cell lines.

There are a number of published guidelines and recommendations for risk-lowering measures for TSE agents [1-5]. Additional recommendations will also be available in transgenic points-to-consider documents as they become available and in position papers developed by the American Blood Resources Association (ABRA) and European Association of the Plasma Products Industry (EAPPI) on minimizing the risk of Creutzfeldt-Jakob Disease (CJD) in plasma-derived products. However, «Safety Requirements for Medicinal Products Derived from Body Constituents of Cattle, Sheep or Goats to Avoid Risk of Transmission of BSE or Scrapie» published in February 1994 by the German Federal Health Office [5] is probably the most detailed and complex risk evaluation document and is the guideline emphasized for the Trasylol® evaluation.

These various guidelines include the following basic principles for minimizing risk due to TSE agents (Table 2), many of which have been reviewed during this ses-

Table 2: Basic principales of risk assessment and reduction for TSE agents.

• Country of origin of source material
• Nature of starting material Organ, tissue, body fluid, cells or cell line
• Screening source material Animals: Husbandry Feeding procedures (ACM) Reliable health certificates Tissue: Donor age Procurement/processing methods Humans: Medical history Exclusion criteria Recipients of: a) pit-hGH (other than recombinant); b) pit-derived gonadotrophin; c) corneal, dura mater, or other tissue grafts; Family history of neurological disorders; Neurosurgery to donor
• Human tropism of specific TSE agent
• Quantity of raw material used to manufacture dose
• Number of daily doses required for indication
• Route of administration
• Age of donor and treatment patient population
• Reduction by purification scheme

sion. At least one of the guidelines mentioned above [3] recommends that continuous cell lines that can amplify TSE agents should not be used for manufacturing. The literature would indicate that lines of neuronal lineage are the predominant source of successfully infected lines by TSE agents, while rapidly dividing cell lines analogous to most of those used to manufacture biotechnology products have not been susceptible to TSE infection. How this can be used in the risk assessment of biotechnology products is dicussed later.

Trasylol® (Aprotinin) is a biopharmaceutical product derived from bovine lung. It is a serine protease inhibitor and was introduced by Bayer in 1959 for the treatment of pancreatitis, hyperfibrinolytic haemorrhages and more recently for the control of blood loss during open-heart surgery. Some properties of Trasylol® used for purification are shown in Table 3.

Table 3: Trasylol® (Aprotinin).

Biological source:	Bovine lung
Biological activity:	Serine protease inhibitor
Indications:	Pancreatitis Hyperfibrinolytic haemorrhages Control blood loss during OHS
Structure:	Single chain polypeptide 58 aa; 6512 MW; 3 disulphide bridges
Isoelectric point:	10.5
Solubility:	70% MeOH or EtOH
Stability	Over wide pH and temperature range

Since Trasylol® is derived from bovine lung, it was decided from the start that the purification scheme would be evaluated for removal of a TSE agent. During 1990/91 discussions with regulatory authorities, it was established that scrapie would be the relevant model to use for establishing Bovine Spongiform Encephalopathy (BSE) clearance. BSE is the TSE agent of concern when working with cattle. Thus, the TSE reduction evaluation study was performed with the mouse-adapted scrapie strain (ME7) and outlined below.

Results

Four purification steps were selected for evaluation and included: (a) extraction from lung, where differences in solubility were expected between Trasylol® and scrapie; (b) adsorption/desorption with ion exchange resin I, expecting the high basicity of Trasylol® to enhance the binding of product to the column preferentially; (c) an additional ion exchange column with a different resin, expecting differences in paritioning; and (d) a membrane filtration step specifically added to remove TSE agents. The production process was scaled-down and validated to mimic large-scale production. The product was «spiked» with scrapie agent before each purification step. The extraction step was spiked with supernatant from 10% mouse-brain homogenate. All other steps were spiked with scrapie preparations containing the detergent lysolecithin to prevent scrapie aggregation and provide a «worst case» removal evaluation of the step. End-point titration had to be performed by the in vivo mouse bio-assay since in vitro assays were not available. Stocks and spiked product «input» samples were titrated in C57BI/6N mice by ic injection of samples at dilutions of 10^{-4} to 10^{-8}. Titration of the starting stocks was performed twice using 10 mice per dilution. Confirmation that the product did not reduce the stock titre was performed by titration of the input samples, using five mice per dilution. «Output» samples following the purification step were titrated at dilutions from undiluted to 10^{-7}, using 10 mice per dilution. Histopathology was done on all animals either following clinical symptoms (appearing post-inoculation day 157) or at the end of the 18-month observation period. This study was performed in collaboration with Microbiological Associates Inc. and required in excess of 1600 mice and three years from planning to final reports*.

The results of the scrapie clearance study for the four Trasylol® purification steps is shown in Table 4. The input spike titres are shown along with data from two separate experiments. The removal factor represents the average removal of the two separate studies. Overall, there was greater than 18 logs of clearance by the four different purification steps. These results were then entered into an overall risk assessment of Trasylol® based on the TSE guideline issued in February 1994 by the German Federal Health Office (Table 5). This risk assessment is based on six parameters.

(i) The first takes into account the country of origin of the bovine lungs. Lungs have been sourced from the UK and have been obtained from countries free from native cases of BSE which allows a class factor score of 6, or a one-in-a-million chance of a sporadic case of BSE occurring. In addition, all the tissues came from inspected abattoirs and animals inspected by veterinarians.

Table 4: Scrapie clearance for four Trasylol® manufacturing steps.

Sample	**Log_{10}**
Spike Methanol extraction Removal factor	7.6 ≤ 3.2; ≤ 3.2 ≥ 4.4
Spike Ion exchange resin I Removal factor	8.4 2.8; 3.6 5.2
Spike Ion exchange resin II Removal factor	7.6 4.0; 3.2 4.0
Spike Membrane filtration Removal factor	8.0 ≤ 3.3; ≤ 3.4 4.7
Summary of overall clearance	≥ 18.3

* Publication in preparation.

Table 5: Overall safety of Trasylol® based on TSE guidelines.

Parameter	Class factor	Description
1. Origin Country, husbandry	6	No cases of BSE
2. Starting material Organ, tissue, fluids, cell	5	Group III: Lung
3. Reduction	18.3	Four steps evaluated
4. Quantity of raw material to make daily dose	-1	$\leq$ 10 kg and > 1 kg
5. Number of daily doses	2	$\leq$ 10 doses/yr
6. Route of administration	1	Into blood
Sum of risk factors	> 31	$\geq$ 20 BSE/Safe

(ii) The second relates to the potential risk a tissue is thought to possess for transmitting a TSE agent if the host was infected, based on native scrapie sheep and goast tissue titration studies assessed in the mouse bioassay. Lung is categorized as a group III risk, or minimum infectivity risk, and thus receives a high score of 5.

(iii) The third relates to clearance by the purification scheme which was demonstrated to be in excess of 18 logs.

(iv) The fourth is the quantity of lung required to make a single dose. Since Aprotinin is present in very low concentrations in the lung, a low score is obtained (-1) because of the large quantities of starting material needed.

(v) The fifth is the number of daily doses administered per year, which for Trasylol® is essentially one, given at the time of surgery. A class factor of two is the best score obtainable in this category.

(vi) The sixth is the route of administration. Trasylol® is administered directly into the blood and received a score of 1.

Overall, however, the Trasylol® product received a very high score of in excess of 30, where the guideline target is 20 or greater for products thought to be safe from transmitting TSE agents. This high score was achieved by both incorporating a membrane filtration step in the purification process and by performing a scrapie clearance study of the purification scheme.

Biotechnology products derived from continuous cell lines can also score well in this type of risk assessment without necessarily having to perform evaluation on the clearance capabilities of the purification scheme. This can be achieved in part by eliminating animal and human-derived raw material during manufacturing, purification, and in excipients. While often in the history of a cell line, there is a gap in the knowledge of the sources of all raw materials to which the cell line could have been exposed, different possible approaches to alleviating the concerns regarding the past exposure raw materials of unknown origin are listed in Table 6 for the purposes of discussion.

Table 6: Biotech products derived from continuous cell lines (CLL).

How do we alleviate concerns regarding past exposure to raw material of unknown origin?
• Test for TSE presence by in vivo bio-assay using animal of similar species?; (for non-human CLL)
• Test for susceptibility to infection by TSE agents? – Are PrP genes present? – Can cell line be infected; does infection persist?
• Literature references – BHK and CHO cell lines reported not to be susceptible to murine or hamster adapted scrapie – Cell lines other than neuronal or related lineages have not been successfully infected

In vivo animal studies may be useful for evaluating reduction of high titre TSE samples as used in the Trasylol® study, but may prove less useful for screening samples for low levels of TSE agents. Negative results in these assays may not be interpretable unless information is generated regarding assay sensitivity and duration for detecting low levels of these agents.

Some of the parental lines, Chinese Hamster Ovary (CHO) and Baby Hamster Kidney (BHK), used to manufacture biotechnology products, have already been shown not to be susceptible to relevant TSE agents, such as mouse and hamster adapted scrapies [6,7]. Basic assumptions that should be made include: (i) the lack of susceptibility to a TSE agent of the same species as the cell line, which indicates that the cell line would be highly unlikely to be infected with TSE agents of other species; (ii) genetic manipulation of the cell line does not alter the susceptibility to TSE agents; and (iii) rapidly growing cell lines cannot amplify TSE agents.

Conclusion

Different biological source starting materials used to manufacture products may require or use different strategies for minimizing the risk of transmitting a TSE agent. The strategy used to minimize the risk for the biopharmaceutical product Trasylol®, derived from bovine lung, was presented. By using many risk reduction measures, and in addition incorporating a specific filtration step to the purification scheme and evaluating the Trasylol® purification process for removal of a TSE agent, a very favourable risk assessment was achieved in accordance with the available international guidelines. In contrast, other products and in particular biotechnology products may not carry the same risk factors and achieve acceptable risk evaluations through other strategies.

References

1 Letters to Biological Manufacturers: FDA May 3, 1991; and Dec. 17, 1993.

2 Recommendations for the Reduction of Risk of Infection Through Zoonotic and Other Animal Pathogens in the Production of Pharmaceuticals: Fed. Journal of 15 Aug 1991.

3 Guidelines for Minimizing the Risk of Transmitting Agents Causing Spongiform Encephalopathy via Medicinal Products: Ec-Document N° III/3298/91-EN.

4 Evaluation of the Theoretical Risk for a Product Being Contaminated by Non-conventional Infectious Agents (NCIA): Italian Ministry of Health, Rome, 20 Sept 1993.

5 Safety Requirements for Medicinal Products Derived from Body Constituents of Cattle, Sheep, or Goats to Avoid Risk of Transmission of BSE or Scrapie: Fed. Journal of 16 Feb 1994.

6 Race R: The scrapie agent in vitro. Current topics in Microbiol and Immun 1991;172:181-193.

7 Butler D, Scott MR, Bockman JM, Borchelt DR, in Taraboulos A, Hsiao KK, Kingsbury DT, Prusiner SB (eds): Scrapie-Infected Murine Neuroblastoma Cells Produce Protease-Resistant Prion Proteins. J Virology 1988;62:1558-1564.

Dr. R.W. Kozak, Bayer Corporation, 800 Dwight Way, P.O.B. 1986, Berkeley, CA 94710, USA.

Brown F, Lubiniecki AS: Viral Safety and Evaluation of Viral Clearance from Biopharmaceutical Products. Dev Biol Stand. Basel, Karger, 1996, vol 88, pp 265-268

Minimization of Viral Contamination in Human Pharmaceuticals Produced in the Milk of Transgenic Goats

C.A. Ziomek

Genzyme Transgenics Corporation (GTC),
Framingham, MA USA

Key words: Transgenics, transgenic goats, animal-derived proteins, milk.

Abstract: The minimization of viral contamination in therapeutic proteins produced in transgenic goats' milk can be achieved by a combinatorial approach. It begins with reduction in the risk in the starting material followed by appropriate clearance/inactivation steps in the purification process. To minimize risk in the starting material, Genzyme Transgenics Corporation (GTC)'s closed goat herds are subjected to routine serological surveillance for known viral diseases, especially those transmitted through milk. Although scrapie is defined as a slow-acting virus of sheep and goats, its incidence in goats in the US is rare (only four cases) and all four were in goats co-mingled with scrapie-infected sheep. All GTC's domestic goats were selected for previous non-exposure to sheep, cows or scrapie. In addition, milk, which is the starting material for transgenic protein production, is categorized as non-infectious for prions. Standard operating procedures are in place at GTC Farm sites to minimize human, animal or vehicular vectoring of viral diseases and the transgenic production animals are milked according to high standard Good Agricultural Practices (GAP). The transgenic protein (ATIII) purification process contains steps that should provide a high level of viral reduction. Validation of viral and prion removal will also be undertaken.

Transgenic Pharming for the production of human therapeutic proteins in the milk of transgenic dairy animals is becoming a more widely accepted production system, particularly for complex or high dosage pharmaceutical proteins. For materials required in very large amounts for therapy, transgenic protein production may be one of the few viable production scenarios. It also offers the further advantage that it should considerably reduce the cost of producing the starting material as compared to existing culture methods.

Although goats, sheep, pigs and cows have all been used as barnyard bioreactors, dairy goats have proven to be an excellent target animal. Due to their short gestation (five months) and maturation (six months) periods and their moderate milk yields (600-800 litres/lactation), goats expressing a transgene in excess of 1 gm/litre in their milk can produce large amounts of a human therapeutic protein in a relatively short time. Scaling up the protein manufacturing process is inexpensive and involves simply breeding more transgenic goats of a specified lineage and adding these transgenic animals to the milking string for that therapeutic protein. Following extraction of the transgenic protein from the goat's milk using a proprietary high yield process, the purification of the transgenic protein can be performed using standard protein purification technology.

The production of a transgenic goat begins with the engineering of an appropriate expression construct, which contains a mammary specific promoter, such as the goat beta casein promoter, linked to the gene or cDNA encoding the desired target protein. The expression construct is micro-injected into the pronucleus of a one-cell goat embryo, which is then transferred to a suitable recipient doe. Approximately 7-10% of the resultant offspring carry the transgene. GTC currently has transgenic goats expressing over 4 gm/litre of three different proteins namely, longer-acting tissue plasminogen activator, antithrombin III (ATIII), and an anti-cancer human monoclonal antibody. For the older lines, stability of expression between siblings, from lactation to lactation, and from generation to generation has been demonstrated (unpublished data).

The production of medicinal products from animal sources and of similar recombinant products from mammalian cell culture have been continuing for many years. Therefore, the production of medicinal products from transgenic animals should pose no new regulatory challenges in the area of minimization of risk of transmission of animal disease pathogens. The European Community has recently issued its guidelines for «Use of transgenic animals in the manufacture of biological medicinal products for human use» [1] in which it has stated that «The considerations that apply are therefore a blend of those relevant to recombinant DNA derived materials and materials from less defined sources». The FDA has similarly written its own set of Points to Consider that are due to be released in draft form in mid-1995.

The reduction in risk of viral contamination in a therapeutic protein produced in transgenic goats' milk can be achieved by a combinatorial approach beginning with minimization of the risk in the starting material followed by appropriate clearance/inactivation steps in the purification process. Viral contamination can be assessed at three levels: the goat, the milk and the final product. The minimization of risk in the starting material can be achieved by starting with a relatively disease-free goat herd maintained under high standards of animal husbandry. GTC maintains four closed goat herds of approximately 850 head total (including 500 goats imported from New Zealand) at four separate sites. No other type of livestock occupies these sites. Rigorous selection criteria were used in the acquisition of both the domestic and imported goats. Only dairy goats of the Alpine, Saanen and Toggenburg breeds were considered. The environment in which the animals were raised was assessed, eliminating domestic farms that also raised sheep or cows. No domestic animals that had co-mingled with sheep or cows at any time were selected to minimize any perceived risk of scrapie. (New Zealand is accepted internationally as scrapie-free.) The medical histories of the animals were evaluated and they were given a physical examination by a qualified veterinarian. They were also subjected to disease testing for tuberculosis, brucellosis, Q-fever (zoonotic), Johne's disease and caprine arthritis encephalitis (CAE).

Once animals were selected for inclusion in one of the herds, they then became subject to the Good Agricultural Practices (GAP) in place at GTC farm sites. The GTC Farm sites were chosen for lack of occupancy by any animals or lack of occupancy by sheep or cows for at least 10-20 years. The GAP include: high standards of animal husbandry, high quality barn facilities with double fencing around the perimeter of the animal occupancy sites, adherence to Standard Operating Procedures (over 100 in number), exclusion of animal-derived protein or fats from the diet, on-site veterinary care by a full-time staff veterinarian in our clinic facility, rigorous periodic animal health monitoring (especially for diseases transmitted through milk), disease prevention programmes, and state-of-the-art milking practices. Standard operating procedures are in place to minimize human, animal or vehicular vectoring of viral diseases on to the production site.

Although scrapie (spongiform encephalopathy) is defined as a slow acting agent of sheep and goats, its incidence in goats in the US is rare (only four cases) and all four were in goats co-mingled with scrapie-infected sheep [2]. As mentioned earlier, all GTC's domestic goats were selected for previous non-exposure to sheep, cows or scrapie. In goats and sheep there is transmission of scrapie from dam to offspring in contemporary kidding/lambing situations. In goats, this transmission of scrapie from dam to offspring can be prevented by removing the offspring at birth [3]. In GTC's domestic and New Zealand herds, the offspring are separated from the dams at birth and reared in separate facilities. Finally, milk, which is the starting material for transgenic protein production, is categorized as non-infectious for the scrapie prions [4]. This equates to a less than theoretical risk of scrapie transmission from proteins produced by this method. To gain formal recognition of the scrapie-free status of our goat herds, all GTC herds are enrolled in the five-year Massachusetts Voluntary Scrapie Certification Program. The domestic goats advanced to Class B status in January 1995, signifying completion of year 1 in the programme, and the New Zealand goats were awarded Certified Scrapie-free status upon their arrival in the US in July 1994.

The first therapeutic product derived from transgenic goats that will enter human clinical trials in early 1996 is transgenic human ATIII. The ATIII transgenic animals are milked according to high standard Good Agricultural Practices and their milk meets the requirements of the Pasteurized Milk Ordinances and Interstate Milk Shippers Guidelines. Once transgenic animals enter the production herd, the level of health monitoring increases and any animals that show signs of disease are removed from production. It is at the point of commencement of collection of milk for clinical grade product purification that milk testing begins. The ATIII downstream purification process itself contains several steps that are likely to yield a high level of viral reduction or inactivation. Validation of viral and prion removal/inactivation will be undertaken at appropriate times in the ATIII Development Program currently underway. The non-exotic, zoonotic and non-zoonotic viruses that infect goats in North America are presented in Table 1. Validation for viral removal/inactivation will include actual or model viruses to demonstrate removal of at least two enveloped viruses (one RNA, one DNA) and two non-enveloped viruses (one RNA, one DNA). Validation for scrapie removal can be performed using the standard 18 month mouse scrapie assay. Even without scrapie validation, transgenic ATIII scores a safety factor of 25 on the German Pharmaceutical Industry Spongiform Encephalopathy score sheet [5]. A score of 20 or higher is currently required for product licensing in Germany.

In conclusion, the minimization of risk of viral contamination of human pharmaceuticals produced in transgenic goats can be achieved with the combinatorial approach that has been effectively used for other animal-derived products. No new regulatory challenges are anticipated and transgenically produced medicinal products will be safe from viruses and suitable for use in humans.

Table 1: Relevant goat viruses in North America.

	Family	**Nucleic acid**	**Shape**	**Size (nm)**
Enveloped				
Rabies (z)	Rhabdoviridae	RNA(ss)	bullet	75 × 180
Coronavirus	Coronoviridae	RNA(ss)	unsymmetric	70-120
Caprine Arthritis Encephalitis	Retroyiridae	RNA(ss)	spherical	80-110
Caprine Herpes Virus	Herpesviridae	DNA(ds)	spherical	120-200
Pseudorabies	Herpesviridae	DNA(ds)	icosahedral	180-200
Non-enveloped				
Bluetongue	Reoviridae	RNA(ds)	spherical	60-80
Rotavirus	Reoviridae	RNA(ds)	spherical	60-80
Adenovirus	Adenoviridae	DNA(ds)	icosahedral	70-90
Capripox (z)	Poxviridae	DNA(ds)	complex	200-300
Contageous ecthyma (z)	Poxviridae	DNA(ds)	complex	200-350

z = zoonotic; ss = single stranded; ds = double stranded.

Acknowledgements

The author thanks Russell Herndon, Drs. William Gavin, Edward Cole and Michael Hayes for their input and review of this manuscript.

References

1 European Commission: Use of transgenic animals in the manufacture of biological medicinal products for human use, 1994.

2 U.S. Department of Agriculture: Scrapie flock certification and animal identification procedures. Federal Register 1992;57:33625-33633.

3 Hourrigan J; Klingsporn A; Clark WW; Camp M de: Epidemiology of scrapie in the United States, in: Slow Transmissible Diseases of the Nervous System. Academic Press, 1979, vol 1, pp 331-356.

4 Commission of the European Communities: Guidelines for minimizing the risk of transmission of agents causing spongiform encephalopathies via medicinal products, 1991.

5 Federal Ministry of Health: Guidelines on safety measures in connection with medicinal products containing body materials obtained from cattle, sheep, or goats for minimizing the risk of transmission of BSE and scrapie. Federal Bulletin, 1994, N° 40.

Dr. C.A. Ziomek, Genzyme Transgenics Corporation, One Mountain Road, Framingham, MA 01701-9322, USA.

Brown F, Lubiniecki AS: Viral Safety and Evaluation of Viral Clearance from Biopharmaceutical Products. Dev Biol Stand. Basel, Karger, 1996, vol 88, pp 269-276

Licensing of Protein Products from the Milk of Transgenic Animals. Validation for Pathogen Removal – A Strategy

G. Wright, A. Colman, D. Cottom, M. Williams

PPL Therapeutics, Roslin, Edinburgh, Scotland

Key words: Pathogen removal; transgenic animal milk.

Introduction

PPL Therapeutics is a biotechnology company which produces therapeutically useful proteins using transgenic technology. As well as assuring that these proteins are beneficial to patients and are produced at low cost, it is essential that they present the minimum level of risk to patient health. One advantage of the technology is that it is free from human pathogens. However there is a potential risk from animal pathogens which are zoonotic for humans. This paper presents some of the approaches being adopted at PPL to ensure that its products are free from such pathogens.

The technology employed by PPL for production of therapeutics involves the introduction of a human gene into animals such as sheep, cows or pigs (normally by micro-injection of DNA into fertilised eggs). These animals, when mature, express the human protein from the transgene. Controlling sequences on the transgene cause the proteins to be expressed only in the mammary gland of the animal which allows the protein to be harvested by conventional milking.

Following initial work by the AFRC Institute in Edinburgh, the first of these transgenes was injected by PPL into sheep eggs in 1989. This coded for the human protein Alpha-1-Antitrypsin (AAT) and PPL now has a small production flock of about 90 sheep, all of which consistently produce 14 g/l of AAT in their milk. We intend to enter clinical trials with this protein in 1996 to assess its usefulness in treating patients suffering from cystic fibrosis. The high levels of free neutrophil elastase in the lungs of these patients leads to extensive tissue damage and premature death. AAT is the natural inhibitor of neutrophil elastase.

Not only is this a new use for AAT but the technology used to produce it is also new. PPL has therefore been expending considerable effort to develop an effective strategy to ensure that products are safe as well as effective.

The strategy to ensure freedom of products from pathogenic organisms adopted by PPL is multi-faceted but can be broadly categorised as follows: (i) highest standards of animal husbandry; (ii) identification of viruses of concern; (iii) assessment of potential levels of pathogens in raw product (milk); (iv) routine viral screens; (v) validation of the purification process for viral removal/inactivation.

In the following sections some of these approaches are discussed in more detail, in addition to some new experimental work commissioned by PPL to address them.

Strategy to Exclude Pathogens from Production Flocks

The exclusion of pathogens or potential pathogens from the milk of transgenic production flock animals relies on the following: (i) defined health status in nucleus and production flocks kept under quarantine conditions; (ii) monitoring of nucleus flock to maintain health status definition; (iii) monitoring of producer flock to a further defined health status; (iv) collection of milk only from healthy animals; (v) hygienic collection of milk;

The following sub-sections discuss these points in more detail.

1. Origin and quarantine

Nucleus flocks, used to generate transgenic ewes as production animals and production flocks, will be kept *either* in a country recognised by the regulatory authorities as free of scrapie (ovine spongiform encephalopathy) e.g. New Zealand or Australia, *or*, if not, in a recognised scrapie-free country, maintained as a quarantined, closed flock consisting of sheep imported from a recognised scrapie-free country and their offspring. Any genetic material not originating from a scrapie-free country could only enter the flock following passage through a «barrier» generation produced via either washed embryos or semen (dependent on health status of donor animals).

Such flocks will be maintained under defined quarantine conditions including: (i) double perimeter fencing -one of which should be high enough to exclude deer; (ii) restriction on entrance of feed, personnel, vehicles, equipment and other materials; (iii) requirements for personnel to charge clothing.

The following will also be taken into consideration when siting the quarantine facility, particularly where this is outside New Zealand or Australia: (i) no adjacent stock farms; (ii) secure natural boundaries.

Within the quarantine facility, production flock ewes which are being milked will be kept in conventional agricultural sheep housing or kept at grass. Such animals will be kept in separate pens or fields from animals not being milked. Agricultural sheep housing is typically made of a steel frame with block walling to one to two metres high, a metal perforated cladding above and metal roof with a ventilated ridge. The floor is concrete laid to drains. Straw will be used as bedding.

These buildings are not environmentally controlled. Lights are provided and switched on or off as necessary by staff. Natural or forced air ventilation is provided as required. There is no temperature control. Records of the location of all animals on the quarantine facility at all times are kept i.e. in which field or building.

2. Adventitious agents and latent/endogenous flora

These will vary depending on the country in which the flocks are kept. In the UK – using New Zealand-derived animals held in quarantine – the following organisms must be considered:

2.1. *Zoonotic organisms*

Viruses	Contagious Pustular Dermatitis (Orf) Tick-Borne Encephalitis Virus (Louping Ill)
Bacteria	Anthrax Brucella spp. Campylobacter fetus Chlamydia psittaci Clostridia spp.* Corynebacterium pseudotuberculosis (Caseous Lymphadenitis) Erysipelothrix rhusiopathiae Leptospira spp. Listeria monocytogenes Mycobacterium bovis (Tuberculosis) Pasteurella spp.* Salmonella spp. Yersinia paratuberculosis
Parasites	Cryptosporidia Toxoplasma gondii
Others	Coxiella Burnetti

2.2. *Organisms pathogenic for sheep*

Viruses	Border Disease Pulmonary Adenomatosis (Jaagsiekte) Maedi-Visna Parainfluenza Virus 3* Rotavirus
Bacteria	E. coli* Mycobacterium paratuberculosis (Johnes) Staphylococci spp.* Streptococci spp.*

Others	Mycoplasma spp. Scrapie

2.3 *Organisms producing inapparent infections in sheep*

Viruses	Adenovirus* Bovine Leukaemia Virus Ovine Herpes-2 (Bovine Malignant Catarrhal Fever)* Reo Virus 1-3*

* Latent/endogenous

3. Health status and monitoring

The nucleus and production flock animals will be maintained as specific pathogen-free flocks. The health status of nucleus flock animals will be equivalent to that of production flock animals kept on the same facility. Flocks will be maintained free from: (i) scrapie; (ii) Maedi-Visna; (iii) louping-ILL (tick-borne encephalitis); (iv) Border disease.

Criteria for entry to the nucleus flock will be that they originate from a country free of these diseases (e.g. New Zealand is declared free of the first three) or will be monitored free of them (e.g. blood tested negative for Border disease).Where there are government, or equivalent, sheep health schemes to show freedom from diseases of concern, flocks will be registered under such schemes.

All sheep will be cared for by suitably trained and experienced personnel. All sheep will be visually inspected for ill-health twice each day. Any animals showing signs of ill-health will be clinically examined by a veterinary surgeon, treated as necessary and subsequently confirmed free of disease and drugs. Individual records will be kept for each animal, specifying parental background, date of birth, vaccination, anthelmintic and other routine veterinary procedures and health status. Records of any treatments, including antibiotics, given to any animal will be kept. Periodic serological testing will be carried out, as will routine post mortem screens.

Appropriate laboratory investigations will be performed in cases of: (i) abortion; (ii) diarrhoea: (iii) mastitis; (iv) vaginal discharge; (v) suspected zoonosis; (vi) diseases following an epidemic course; (vii) death.

4. Feeding

Production flocks will be maintained on a mix of standard and supplementary feeds (including as appropriate natural grazing on conventionally prepared grasslands). All feeds will be free of other mammalian-derived proteins. All supplementary feeds will be free of agents such as antibiotics and hormones. Diet will be adjusted in line with availability of grass, breeding status and age of the animals. In general, nutritional practices will not differ significantly from those used in high quality dairy flocks. Feed intake will be monitored during lactation in line with conventional dairy practice, animals will be put out to graze at appropriate opportunities. This has a favourable impact on animal health and milk yield.

5. Milking

Following birth, lambs will remain with the mother for a few days. The offspring will then be removed and normal machine milking commenced. Collections of milk from the first 10 days will be discarded. After that material will be used for processing. Colostrum is produced only for the first 3-4 days of lactation.

Milk from any dairy flock is subject to variation, both between animals and in the same animal from lactation to lactation and even within one lactation. This is completely normal and is due to a variety of factors, genotypic, phenotypic and environmental. It has been shown that this normal variation will not have any effect on the quality of the product. Any variation in milk entering the process which could affect the processing characteristics will be minimised by a combination of blending and having a defined AAT yield criteria for individual production animals. The use of bulk milk as opposed to milk from individual ewes will minimise variations. Milking will be carried out to conventional dairy standards, that is to say on clinically healthy animals, by healthy operators (subject to similar health controls in pharmaceutical manufacturing operatives) and in clean conditions. Milking equipment will be standard automatic milking equipment similar to that employed in the dairy industry. It will be sanitised after every milking using standard dairy procedures.

Any production animals failing the visual inspection before milking and showing signs of ill-health (other than minor skin, eye or foot lesions which require either no treatment or only topical treatment known not to enter the milk) will be physically removed from the milking flock and clinically examined by a veterinary surgeon. Immediately before milking the udders and milk of each animal will be subject to a brief visual examination and palpation (as is standard in the dairy industry). Any animals with udder or milk abnormalities will be rejected from that collection and physically removed from the flock and given a veterinary examination. Animals removed from the milking flock will be kept separate from the flock until released back to the flock as healthy by a veterinary surgeon.

Only essential medication will be administered to production flock animals. During lactation, milk from a production flock ewe receiving treatment (other than topical treatments known not to be absorbed or excreted in the milk) will not be used for processing. Ewes producing milk unsuitable for processing will be removed from the milking flock.

Animals which have finished a course of treatment may only be returned to the milking flock on veterinary certification to say they are healthy and out of the stated withdrawal time for the medicine(s) received. Routine treatments e.g. anthelmintic, antiparasitic treatments will only be given to milking animals as directed by a Standard Operating Procedure.

Volumes of milk from each animal will be recorded on a daily or weekly basis and records of milk yields and lactation periods maintained. Animals will be kept in the production flock for a defined maximum period in a given lactation. If daily yield drops below a certain level they will be removed from the flock before the maximum period. Their overall life in the production flock will also be affected by yield per lactation and general health. It is not currently proposed that artificial induction or maintenance of lactation be used, nor is this believed to be necessary. The milk collected will be of a high quality (not less than that seen in the highest quality dairy flocks); however it will not necessarily be completely sterile upon collection (and will contain a significant cellular load from the animal).

POTENTIAL LEVELS OF VIRUS IN MILK

1. Selection of virus

To address the issue of how much virus could be present in milk, PPL set up a study in conjunction with the Moredun Insitute of Animal Health (Edinburgh, UK) and Quality-1-Biotech (Glasgow, UK). In this study sheep were infected intentionally with different virus types and the level of virus in their milk was measured. The viruses chosen were:

1.1 Ovine Herpes Virus-2 (OHV-2)

This is ubiquitous in sheep where the infection is inapparent. Little is known of its transmission and it is difficult to isolate, although it is often associated with white blood cells. It is believed not to be infectious in humans.

1.2 Orf

This is a double-stranded DNA (paravaccinia) virus which causes painful crusty lesions on the lips of lambs and on the lips, nostrils and udder of ewes. The scabs from lesions contain infectious virus and the disease spreads generally by direct contact with infected lesions. It is a very common infection in sheep and can be passed to humans where similar lesions occur.

1.3 Adenovirus

Although infection from the many serotypes of this virus does not normally result in ill-health of the sheep it has been associated with respiratory disease. Transmission is probably aerosol. Infection is again common and most sheep have antibodies against this virus. It is not thought to be transferable to humans.

2. Selection of ewes

A total of 107 pregnant ewes of various breeds were screened between five and six weeks before parturition for antibody titres to orf and adenovirus. Nine ewes were selected on the basis of having low antibody titres to orf and/or adenovirus and all were shown to have a positive antibody titre to ovine herpes virus-2. The sheep were split into three groups as follows:

Group 1	Herpes positive
	Low Adenovirus titre
	Low Orf titre
Group 2	Low Orf titre
Group 3	Low Adenovirus titre

3. OHV-2

Group 1 acted as a control group for the orf and adenovirus study but was also used to assess whether the latent herpes infection could result in virus being found in milk. Since no in vitro assay is currently available for OHV-2, detection was solely by PCR for genomic viral sequences. This was not quantitative.

Samples of blood taken from all three sheep in Group 1 were positive by PCR indicating that all were latently infected with herpes virus. PCR carried out on whole milk from one sheep gave positive results but in another they were negative. The third sheep tested positive in only one of the two PCR replicates. A cellular fraction from milk was produced by centrifuging milk and positive samples were obtained by this technique for all three sheep.

4. Orf

Lesions developed on sheep 6-14 days after they were infected and antibodies against orf virus had increased in the plasma of all three sheep by day 13 post infection. Electron microscopy of scabs taken from the sheep later confirmed the presence of orf virus.

Whole milk, cellular fractions of milk and samples of milk taken from sheep by cannula all failed to show infectivity for foetal lamb kidney (FLK) cells. This indicated a level in milk of less than one infective virus particle per ml. Many of the cellular fractions prepared from milk (days 5-13) however, were positive by PCR and one whole milk sample taken from a sheep on day 13 was also positive by PCR.

5. Adenovirus

Sheep infected with adenovirus showed no evidence of ill-health throughout the study (42 days). However, anti-adenovirus antibodies did increase, and infectious particles were found in the blood of two of the three sheep between days 12 and 14 post infection. The same infectivity assay failed to detect adenovirus (FLK cells) in any of the milk samples taken.

PCR assays for ovine adenovirus DNA sequences were able to detect virus in about 50% of the milk samples. These positive results by PCR were apparently random and showed no signs of following the progress of infection.

6. Study conclusion

This study indicates that in cases of orf or adenovirus infection of sheep the concentration of infective virus in milk is likely to remain very low. No infectivity assay was available for OHV-2. However, several milk samples were positive by PCR assay for this virus, as were samples from orf – and adenovirus – infected sheep which were assayed by PCR. The significance of the positive PCR results warrants further investigation.

Process Considerations

1. Raw product screening

The above goes some way to indicate the maximum possible level of virus which might be potentially present in milk but only a few selected viruses can be investigated in this way. A more general approach to assuring products are free from virus is required. PCR assays are highly virus-specific and although perhaps useful for detection of a specific virus of concern, they do not offer the required general virus screen. The significance of a positive result is also questionable since it does not indicate whether what is being detected is a live virus or simply free DNA. All the above PCR assays were also non-quantitative. All the samples of milk taken in the above work were negative by cell-based infectivity assays and yet many were positive by PCR. This was not the result of PCR being able to detect lower levels or virus since titrations have shown that the infectivity assays used in the study were in fact more sensitive than the PCR assays.

Cell-based assays, particularly if carried out on the unprocessed milk when virus concentration would be at its highest, offer a wider screen and a greater chance of detecting an unexpected problem.

2. Process validation

Although bacteria and fungi can be removed from biological products by use of sterilising filters, no single step can be used for similar removal of virus. However, possibly the greatest contribution to ensuring viral safety is the demonstration that the purification process, when taken in its entirety, will remove and/or kill more virus than is likely to be present in the starting material. This is done by purposely spiking virus into a scaled-down version of the process and measuring the capability of that process to remove the added virus. PPL intends to carry out such work shortly on the purification process for AAT. A broad range of virus types will be used as models to demonstrate that the process is capable of removing/killing any virus, regardless of its morphology or resistance to harsh physical treatment. Additionally, specific viruses of concern will be used in the study.

A similar study has already been carried out in conjunction with Inveresk Research International (Edinburgh, UK) to assess the capability of the AAT purification process to remove scrapie particles. Scrapie infectivity survives normal autoclave cycles and other extremely harsh conditions and therefore eradication of scrapie, if it was present, will be based on removal rather than inactivation.

Conclusion

We believe that no one approach can be used as a strategy for ensuring that products are free from virus. The strategy being adopted at PPL is to prevent infections from occurring in its production flocks by using the best standards of animal husbandry, obtaining animals of the highest health status and controlling and maintaining the flock. We have also initiated studies to investigate the virus levels that could potentially occur in milk. No infective virus was detected in any of the milk samples tested. Viral assays carried out on milk will allow rejection of milk unsuitable for further processing. Lastly, the process used to purify the product will be validated to demonstrate that even if virus were present in milk, it would not be present in the final product.

Dr. G. Wright, PPL Therapeutics, Roslin, Edinburgh EH25 9PP, Scotland, UK.

Brown F, Lubiniecki AS: Viral Safety and Evaluation of Viral Clearance from Biopharmaceutical Products. Dev Biol Stand. Basel, Karger, 1996, vol 88, pp 277-281

Process Considerations for Minimizing Virus Transmissions from Primary Porcine Tissues

P.J. Gil

Grace Biomedical, Lexington MA, USA

Key words: Virus transmission; swine tissue.

Introduction

W.R. Grace has been conducting research toward the development of artificial organs over the past ten years. A new business, Grace Biomedical, located in Lexington, MA has been formed to develop and commercialize innovative products which serve as therapies for selected acute and chronically diseased patients.

Bioartificial Organs

The company's two products are a Bioartificial Liver (BAL) and a Bioartificial Pancreas (BAP). These products are hybrid artificial organs which contain both synthetic and biological components. Since these medical devices comprise viable and functional primary porcine tissue, they are regulated as «combination products» by FDA's Center for Biologics (CBER).

Bioartificial Liver (BAL)

The Bioartificial Liver is an extracorporeal device used to support patients in acute liver failure by temporarily providing a broad spectrum of functions. It consists of a bundle of specially prepared biocompatible hollow fibre membranes enclosed within a cylindrical housing. Isolated and purified normal porcine hepatocytes attached to selected microcarriers are placed around the hollow fibres within the housing.

Plasma, continuously separated from the blood of patients, is perfused through the lumen of the fibres supplying nutrients and oxygen to the hepatocytes. The pore size of the hollow fibres allows free exchange of most proteins between the plasma and the hepatocytes. However, the hepatocytes and other cells are restricted from passing across the fibre wall. As the plasma passes through the BAL, the sequestered hepatocytes provide key synthetic, metabolic and detoxification liver functions. The plasma, once treated by the BAL, is returned to the patient.

Bioartificial Pancreas (BAP)

The Bioartificial Pancreas is an implantable device intended to provide insulin in response to changes in the blood glucose levels to improve glucose control in diabetics. The device consists of a membrane, porcine pancreatic islets and a housing. The biocompatible membrane is designed to provide free exchange of small components between the blood and the islets (e.g. nutrients, glucose, insulin and oxygen), but prevent the passage of larger white cells and immunoglobulins. The device is anastomosed as an arterio-venous shunt into the circulatory system of the patient. The porcine islets in the device detect changes in blood glucose levels and produce insulin for appropriate glucose control. The device has been designed so that the islets can readily be replenished without the need for complex surgery.

Farm Inspection

The farm inspection is the critical first step in minimizing the risk of transmission of infectious agents in animal-derived tissue. While it is important to identify and define the animal breeds (origins and genetic backgrounds) which are appropriate for performance, farm biosecurity is an equally important criteria in evaluation of a source farm. Important animal donor considerations include: (i) disease history; (ii) clinical evidence; (iii) slaughter inspections; (iv) vaccination programme; (v) rodent and bird control practices; (vi) shower in/out; (vii) dedicated, freshly laundered attire; (viii) footbaths; (ix) isolation/fencing; (x) feed programme; (xi) storage/loading chutes.

Also, statistics such as the number of litters/female/year and the number of pigs weaned/female/year as well as the mortality during birth-to-weaning, nursery, and growing-finishing in the breeding herd are important measures of the production profile, herd management and biosecurity of the farm.

Serological Titres

In addition to the source farm programme description, serological titres from a representative sample of current breeding stock bled within the past 60 days needs to be obtained. An AAVLD accredited veterinary diagnostic laboratory such as Iowa State Diagnostic Veterinary Laboratory should be contracted to perform the testing. As with any serological screen, it is not uncommon to have some inherent titre variability due to the nature of the serological assay itself.

As an example, a clean biosecure herd in the serological titres should be negative for brucella, pseudorabies, PRRS, TGE, vesicular stomatitis (both Indiana and New Jersey serotypes) and BVD.

Even a clean, isolated and biosecure herd may show low titres for leptospirosis, parvovirus, encephalomyocarditis, mycoplasma hyopneumoniae, eperythrozoon and swine influenza virus. If the serological results exceed these levels, the herd animals should be re-tested and the source veterinarian should be contacted to determine whether the titre is the result of exposure/vaccination or whether there is evidence of clinical signs of infection.

Source Farm Qualification

Once the source farm inspection and serological titres are found to be acceptable, the source farm is considered qualified. Should the serological titres exceed established specifications, the cause needs to be investigated. This usually involves both the Grace and source farm veterinarians. Re-testing may be warranted. If the high titre is the result of recent exposure or there is clinical evidence of disease, the herd is disqualified as a source. If the high titre is a result of a disease outbreak or the reason is unknown, a sentinel quarantine is performed to determine whether animals are contagious as a result of clinical evidence or rising titres in the sentinel animals.

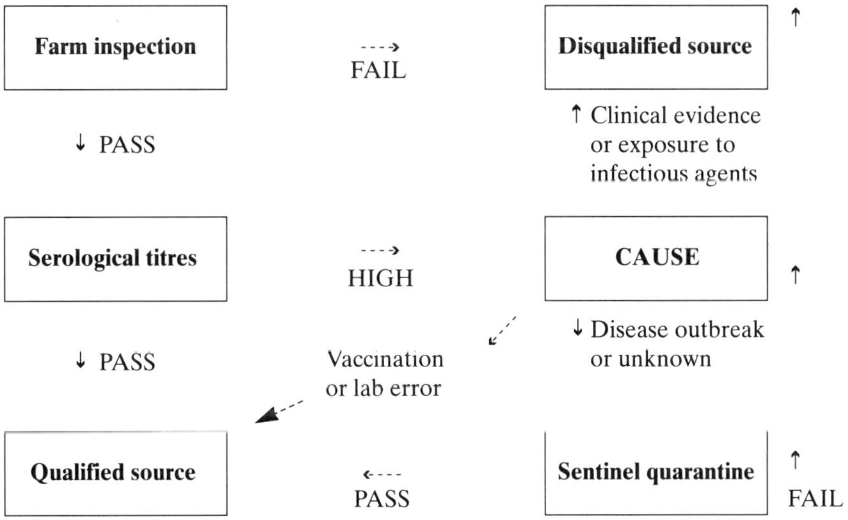

Sentinel Quarantine

Following a disease outbreak, the source herd must be re-qualified. The infectious agents must be identified, the means of outbreak must be determined, the measures for decontamination and control must be defined and the measures for reconfirming herd health at the source must be defined and documented. To ascertain whether a herd is shedding virus or whether a disease state is contagious, a sentinel quarantine is performed. The sentinel quarantine is established at a remote site for a minimum period of 30 days. A minimum of five source swine are co-mingled with three sentinel swine cohorts from a clean, qualified source and are observed daily. At the end of the 30-day period, the sentinel animals are bled with subsequent serological, mycoplasma and viral co-culture testing on a lymph node and suspected tissue as well as necropsy. The necropsy will include micro- and macroscopic examinations with particular interest to tissues affected by suspected infectious agents.

Biomanufacturing Process

A routine biomanufacturing process includes QC sampling and quality assurance checks at all critical steps: (i) swine receipt and quarantine; (ii) organ procurement; (iii) organ processing/cell isolation; (iv) cryopreservation; (v) product release; (vi) shipment to clinical site.

Processing Control

Implementation of training and Good Manufacturing Practices (GMPs) need to be incorporated at all intermediate process steps to minimize the potential for contamination by infectious agents. A comprehensive sampling plan at each key intermediate step with timely testing provides valuable feedback on process control: (i) defined processing solutions (buffers, medium, serum); (ii) aseptic practices; (iii) antibiotics; (iv) storage.

Process QC

The process quality control includes a number of animal screening and process intermediate assays: (i) bioburden; (ii) sterility; (iii) endotoxin; (iv) blood chemistries, diffs & serology; (v) necropsy; (vi) cell viability and function; (vii) viral cocultivation; (viii) mycoplasma culture.

Timely analysis and trending of the data provide important feedback to assess the process and implement appropriate corrective actions where necessary. Archive samples are maintained for reference.

Cell Lines & Susceptibilities

Virus co-cultivation assays are performed on both the lymph node samples procured at the time of organ harvest and also on the final cell product. As a minimum, the FDA «Points to Consider in the Characterization of Cell Lines used to Produce Biologicals» recommends that one primary monkey cell line, one porcine cell line and one human diploid cell line be used. These cell lines are sensitive to virus types: *The primary monkey kidney cells, RMK* are derived by enzyme treatment and are used for the isolation and replication of a variety of human and other primate viruses. *The porcine cell line, PK-15* was initiated from the kidney of an adult pig. *The human diploid lung cell line, MRC-5* was derived from the normal lung tissue of a 14 week old male foetus and is sensitive to a wide range of viruses. Other cell lines may be used to enhance the range of sensitivity.

Growth Validation Studies

Positive controls or growth validation studies need to be performed to ensure the cell lines used promote viral replication.

A sample positive control scheme follows:

Cell line	Porcine parvovirus 7 strain NADL-2 ATCC VR-742	Parainfluenza type 3 strain C-243 ATCC 93-V	Coxsackie-virus type A9 clinical isolate	Vesicular stomatitis strain Indiana ATCC VR-158	Swine influenza A swine/Iowa 15/30
RMK	NT	+	NT	+	+
PK15	+	+/-	0	+	0
MRC-5	0	0	+	+	NT

Key: (+) = positive for presence of virus
(+/-) = equivocal
(0) = no virus detected
(NT) = not tested

In addition, haemagglutination and haemabsorption controls need to be performed with each test.

Conclusion

A comprehensive programme of controls in four major areas is the right approach to ensure quality of the artificial organ products produced by Grace Biomedical: (i) **source animal qualification and control** – including farm biosecurity, herd management practices, herd history, slaughter checks, serological testing and sentinel quarantine; (ii) **raw material control** – particularly serum, collagenase and buffer components; (iii) **GMP$_S$ with appropriate QC sampling during bioprocessing** – including sterility, bioburden and endotoxin on intermediates before and after use; (iv) **adventitious agents testing** – including viral and mycoplasma cocultivation on animal tissue and final product.

Dr. P.J. Gil, Grace Biomedical, 128 Spring Street, One Ledgemont Center, Lexington, MA 02173, U.S.A.

Brown F, Lubiniecki AS: Viral Safety and Evaluation of Viral Clearance from Biopharmaceutical Products. Dev Biol Stand. Basel, Karger, 1996, vol 88, pp 283-290

Adventitious Agents from Animal-Derived Raw Materials – a Method of Risk Assessment

L.G. Foster

HyClone Laboratories, Inc., Logan, UT USA

Key words: Adventitious animal agents; risk assessment.

Abstract: The use of animal-derived raw materials in biopharmaceutical manufacture has a long history. On a risk/benefit basis the use of these raw materials has certainly been beneficial. With proper precautions animals will continue to be the source of raw material for many years to come. As scientific sophistication increases, continuing examination of animal-derived raw material sourcing is important. Even as biopharmaceutical production has made significant progress so has the quality and safety of available animal-derived supplements. It is now possible to perform a reasonable assessment of risk as it relates to adventitious agents. An easy-to-use, preliminary risk-assessment exercise will be discussed. It will allow users of animal-derived raw materials to make more informed decisions regarding the need and type of pre-use processing before the final risk/benefit analysis. Areas to be assessed include geographical origin, age and species of animal, method of harvesting, and anatomical origin of the material.

Much has been said about the assessment of risk from the use of animal-derived raw materials in pharmaceutical and diagnostic products. Many parameters are important in that assessment. Clients have asked on numerous occasions for assistance in assessing specific risks. As a result of these requests a worksheet has been developed by HyClone to be used as a preliminary assessment guide. The purpose of the worksheet is to focus attention on the most critical, common issues affecting risk. It allows the respondent flexibility in assigning biases to each issue based on individual circumstances. The sample worksheet included here has already had biases established by pre-assigning numbers to the «response range» column. The higher numbers indicate increased risk. However, these numbers can be reviewed carefully and changed to reflect the judgement and circumstances of the client better. Selection of a number from zero to five corresponds to low risk to higher risk. Extreme risk is denoted by selecting the number ten.

The worksheet is easy to use. First, each issue is reviewed in the «consideration» column to determine whether the bias assigned in the «response range» column agrees with individual circumstances. Next, each issue is considered as it relates to the specific animal-derived raw material being assessed and a score assigned. A separate assessment is made for each raw material. At the conclusion of the exercise, the worksheet is reviewed in an attempt to reduce the total score. This score reduction may be done, for example, by changing geographical origin or donor species origin, adding pre-processing steps, or selecting other supply methods/logistics which result in a lower total.

The following is a brief discussion for each of the considerations listed on the worksheet to assist you in making informed decisions about the biases.

Geographical Origin

Geographical origin plays an important role in risk because of the animal disease spectrum by location. Here is an opportunity to consider origin as it relates to the product. Traditionally, the world is geographically divided into at least three major disease groups for bovine species, the most common source of animal-derived raw materials. These divisions are: (i) «BSE» countries; (ii) foot and mouth/rinderpest disease countries; and (iii) others. The «other» category may also be sub-divided using the United States Department of Agriculture (USDA) position on the distribution of exotic strains of bluetongue virus.

The next issue is verification of origin through certification and vendor traceability audits. If geographical origin is important, great care should be taken to assure authentic certification. An unfortunate effect of supply by geographical location based on animal disease is price differential. This differential may induce fabrication of documentation in pursuit of profit. Therefore, careful audit of traceability is essential.

We recommend selection of suppliers to provide the fewest companies/individuals involved in collection and supply of raw materials. This would simplify traceability.

Animal Donor

The donor animal itself impacts risk. The developmental stage is the first consideration. Generally, foetal material is less risky than material from newborn, adolescent, or adult animals. This is largely because a healthy placenta is a remarkable organ that shields the developing foetus from many adventitious agents. Newborn animals, on the other hand, have a high risk of disease because of gut permeability and relatively underdeveloped immune systems which necessitates passive immunity (colostrum). Adolescents have a generally lower incidence of disease because they have successfully survived the newborn stage.

Species similarity promotes disease similarities. Therefore, it is generally better to use raw materials from animals that are a different species than the recipient of the final product.

Depending on the diseases of concern, anatomical origin of raw material is important. Generally, adventitious agents harboured in the central nervous system present a higher risk than those present elsewhere, such as blood or milk. This is particularly true for encephalopathies. Therefore, ancestry deserves review in light of slow virus diseases.

Finally, risk is more assessable if the donor species has been characterized for disease potential and if those diseases are monitored and reported by organized surveillance programmes.

Animal Husbandry

Intensive feeding of animals is associated with crowded conditions and is thus more conducive to disease transmission among animals than is open-range grazing. Additionally, feed associated with intensive feeding has a small but increased chance of containing rendered ruminant animal-derived protein supplements which have been implicated in the epidemiology of bovine spongiform encephalopathy (BSE). Healthcare is dependent on qualitative judgement of the veterinary care available in the country of origin.

Harvesting and Post-Harvest Handling

Harvest procedures should be taken into consideration during risk assessment. In the case of serum, closed-donor herds are subjected to continuing health monitoring and management. By contrast, animals presented for slaughter have a less-well-characterized health status. Generally harvest methods must be designed to maintain quality of the original material since it can be impacted negatively both during harvest procedures and post-harvest handling. The benchmark is rapid, aseptic procedures with proper equipment and supplies accomplished by trained, task-dedicated workers.

Final Preparation of Material

Processes should be selected that are validated to remove, reduce, or eliminate adventitious agents and avoid concentrating them. Centrifugation, ultra-centrifugation, filtration through highly retentive filters, and gamma irradiation are among the most commonly chosen. Finally, reliable testing for relevant agents germane to your finished products minimizes risk.

Supplement Use

Lower risk is associated with the use of animal-derived raw materials as cell culture media supplements in low concentrations, particularly if the cells are not susceptible to relevant viruses. Additionally, inclusion of the animal-derived supplements in product-processing before purification of the final product is desirable.

Recipient Risk

A recipient preliminary-risk-assessment worksheet is included. Presented in the same format to provide flexibility, it is an attempt to focus discussions on the component of risk associated with the final recipient of the pharmaceutical product. This assessment is, of course, unnecessary if the final use is an extra-corporeal diagnostic product.

CONCLUSION

Animals are extremely useful sources of raw material in the manufacture of biopharmaceutical products. On a risk/benefit basis the use of these raw materials has certainly been beneficial. With proper precautions animals will continue to be the source of raw materials for many years to come. As scientific sophistication increases, continuing examination of animal-derived raw material sourcing is important. Even as biopharmaceutical production has made significant progress so has the quality and safety of the available raw material supply. Use of the worksheet presented will direct consideration of risks. It will be a useful tool.

Table 1: The aim of this exercise is to aid in planning procurement and processing to ensure the total risk is as small as possible. Upon completion, the assessment should be reviewed carefully to determine changes in procurement and handling practices which could lower the total score even further and thus the risk.

Preliminary assessment of risk instructions for use.

Recipient Risk Total	+	Raw Material Risk Total	=	TOTAL RISK

RISK LEVELS

Minimal	*Moderate*	*High*
< 40	40-100	> 100

Minimal risk
Additional pre-use processing is probably not necessary.

Moderate risk
If a careful review of the assessment does not lead to a lower total, additional pre-use processing should be seriously considered.

High risk
If a careful review of the assessment does not lead to a lower total, additional pre-use processing is required.

Table 2.

Recipient preliminary risk assessment*		
Consideration	**Response range**	**Score**
I. Recipient		
1. Species	Different from donor = 1 Same as donor = 5	
2. State of health	Healthy = 1 Compromised = 4	
3. Age	Adolescent, adult = 1 Very young, very old = 4	
II. Administration		
1. Dose	Single, small = 1 Large, continuous = 4	
2. Method	Transdermal = 1 Oral = 2 Injection = 4 Intraneural = 10	
Total		

* Unnecessary if used for extra-corporeal diagnostics.

Table 3.

Preliminary raw material risk assessment*		
Consideration	**Response range**	**Score**
I. Geographic origin		
*1. Obtained from animals originating from countries from which bovine serum imported into the US is –	Permitted without safety testing = 1 (US, Canada, New Zealand, etc.) Permitted with safety testing = 2 (Australia, Central America, etc.) Not permitted (non- «BSE» countries) =5 **«BSE» countries = 10	
2. Documentation of traceability –	Complete, authentic certificate of origin YES = 1 NO = 4	
	Number of companies involved in supply chain from collection to end user NONE = 0 ONE = 1 TWO = 3 THREE or more = 5	

Preliminary raw material risk assessment*		
Consideration	**Response range**	**Score**
II. Animal donor		
1. Age	Fetus = 1 Adolescent = 2 Adult = 4 Newborn = 5	
2. Species	Different from recipient = 1 Same as recipient = 5	
	Domesticated, common = 1 Exotic, uncommon = 10	
3. Anatomical origin of material	Blood, milk = 1 Organs = 2 CNS, CNS – associated glands = 4	
4. Ancestry	No known ancestry linked to «BSE» countries = 1 Known ancestry linked to «BSE» countries = 5	
5. Status as virus reservoir	Viral infections well known = 1 Viral infections not well known = 4	
	Viral infections reported and monitored = 1 Viral infections not reported = 4	
III. Animal Husbandry		
1. Feeding method	Open = 1 Intense = 4	
2. Feed type	Natural = 1 (grass, hay, grain) Formula-fed without rendered supplements = 1 Rendered supplements = 3	
3. Health care	Veterinary care is generally good and available = 1 Veterinary care is inconsistent and not easily available = 4	
IV. Harvesting		
1. Location	Closed herd donor = 1 Abattoir = 3	
2. Method	Aseptic (venipuncture, cardiac puncture) = 1 Open (exposed to uncontrolled environment) = 4	
3. Personnel	Ante-/Post-mortem veterinary examination = 1 No Ante-/Post-mortem veterinary examination = 3	
	Collection personnel task dedicated, trained = 1 Collection personnel multi-tasked, untrained = 3	

Preliminary raw material risk assessment*		
Consideration	**Response range**	**Score**
V. Post-harvest handling		
1. Time	Rapidly achieve suitable storage state = 1 Slow in achieving suitable storage state = 3	
2. Method	Aseptic handling = 1 Non-aseptic handling = 3	
VI. Final preparation of materials		
1. Adventitious agent reduction	Several validated reduction processes (filtration, centrifugation, etc.) = 1 At least one validated reduction process = 2 No validated reduction process = 4	
2. Adventitious agent inactivation	At least one validated inactivation process = 0 No validated inactivation process = 4	
3. Adventitious agent concentration process	No concentrating processes = 0 At least one concentrating process = 4	
4. Testing for relevant adventitious agents (viruses which are known to, or likely to contaminate raw material)	Tests available and results reported = 1 Tests difficult or not available = 4	
VII. Supplement use		
1. Cell culture	Cells of different species = 1 Cells of same species or known to be susceptible = 5	
	Low concentration of material (≤ 1% serum) = 1 High concentration of material (≥ 10% serum) = 5	
2. Time of addition to product	Prior to purification procedures designed to reduce nucleic acids = 0 Directly to finished product = 5	
Total		

* Based on requirement by the USDA for safety testing of bovine serum.

** «BSE» countries = United Kingdom (including Great Britain, Northern Ireland, and the Falklands) Republic of Ireland, France, Oman, Switzerland and Portugal.

References

1 1991. Report of a WHO Consultation on Public Health Issues Related to Animal and Human Spongiform Encephalopathies. World Health Organization WHO/CDS/VPH/92.104.

2 Buchanan CR, Preece MA, Milner RDG: Mortality, Neoplasia, and Creutzfeldt-Jakob Disease in patients treated with human pituitary growth hormone in the United Kingdom. Br Med J 1991;302:824-828.

3 Committee for Proprietary Medicinal Products: ad hoc Working Party on Biotechnology/Pharmacy and Working Party on Safety of Medicines. EEC Regulatory Document. Not for Guidance. Guidelines for Minimizing the Risk of Transmitting Agents Causing Spongiform Encephalopathy via Medicinal Products. Biologicals 1992;20:155-158.

4 Hodgson J: Fetal Bovine Serum Revisited. Bio/Technology 1993;11:49-53.

5 Hodgson J: Checking Sources: The Serum Supply Secret. Bio/Technology 1991;9:1320-1324.

6 Hanson G, Wilkinson R, Black J, Gamma Radiation and Virus Inactivation: New Findings and Old Theories. Art to Science 1993;12:2:1.

7 Hodgson J:. To Treat or Not to Treat: That is the Question for Serum. Bio/Technology 1995; 13:333-334.

8 HyClone's Position on Smuggled FBS. Art to Science 1988;7:4.

9 Newborn Calf Serum. Art to Science 1989;8:3.

10 Announcement of the Safety Requirements for Pharmaceuticals Made From Body Components of the Cow, Sheep, or Goat to Avoids the Risk of Transmission of BSE or Scrapie. German Federal Health Office. Federal Gazette N° 40 February 26, 1994 p. 1851.

11 Substances Prohibited From Use in Animal Food or Feed; Specified Offal From Sheep and Goats Prohibited in Ruminant Feed; Scrapie. August 29, 1994. Federal Register vol 59 n° 166 p. 44584.

12 Qualitative Analysis of BSE Risk Factors in the United States. 1991 USDA:APHIS:VS.

13 Quantitative Risk Assessment of BSE in the United States. 1991. USDA:APHIS:VS.

14 Wilkinson R: Bovine Spongiform Encephalopathy. Art to Science 1993;12:1.

15 EEC Regulatory Document: Note for Guidance: Validation of Virus Removal and Inactivation Procedures. Biologicals 1991;19:247-251.

Dr. L.G. Foster, HyClone Laboratories, Inc., 1725 South Hyclone Road, Logan,UT 84321, USA.

Brown F, Lubiniecki AS: Viral Safety and Evaluation of Viral Clearance from Biopharmaceutical Products. Dev Biol Stand. Basel, Karger, 1996, vol 88, pp 291-294

SUMMARY TO BREAKOUT SESSION F

Chairmen: *K. Hellman, FDA/CDRH, USA*
C. Vincent, FDA/CDER, USA
J. Honstead, FDA/CVM, USA
R. Rohwer, Medical Research Service, USA

Participants: *F. DeLustro, Collagen Corporation, USA*
L. Detwiler, APHIS/USDA, USA
M. Egan, Diacrin Inc., USA
L. Foster, Myclone Laboratories, USA
P. Gill, Grace Biomedical, USA
R. Kozak, Bayer Corporation, USA
M. Robinson, ARS/ADRU/USDA, USA
G. Wright, PPL Therapeutics, UK
C. Ziomek, Genzyme Transgenic Corporation, USA

INTRODUCTION

The transmission of adventitious agents, including viruses and agents of the transmissible spongiform encephalopathies (TSEs), is a concern for all animal-derived raw materials and production systems used in the manufacture of medical products. As a result, the U.S. Food and Drug Administration (FDA) recommends that appropriate material sourcing documentation, screening tests and process validation be performed to ensure that the product is free of biological contaminants.

Dialogue between the public and private sector of the international community is important for the development of guiding principles for the use of these materials and systems in product manufacture. The Session Organizing Committe, which included representatives from the FDA's Center for Devices and Radiological Health (CDRH), Center for Drug Evaluation and Research (CDER), and Center for Veterinary Medicine (CVM), identified the following issues as critical elements for this development:

- definition and charaterization of the animal source and its maintenance;
- methods of material procurement and storage;

- material processing and process monitoring procedures;
- evaluation of viral clearance with identification of model protocols, and the rationale for model selection.

Goal of Session

The primary aim of the session was to exchange current information and present management approaches from the perspective of the government regulatory communities and representative production industries, addressing the critical elements identified above with the aim of developing common principles for minimizing/eliminating virus transmission through medical products.

Participants

The participants represented the following U.S. Federal agencies and biomedical production industries:

U.S. Federal Agencies:
U.S. Department of Agriculture (USDA); U.S. Food and Drug Administration (FDA); Veterans' Administration (VA)

Biomedical Production Industries:

Biopharmaceuticals:
Bayer Corporation

Production Systems: Animals as bioreactors (transgenic systems)
Genzime Transgenics Corporation
PPL Therapeutics

Tissue-Engineered Products: Xenotransplantation; Cell Encapsulation
Diacrin, Incorporated
Grace Biomedical

Processed Materials:
Collagen Corporation

Product Components:
Hyclone Laboratories

Presentations and Discussion

The presentations focussed on three main themes:

1. an update of the regulatory considerations for the manufacture of medical products from animal sources;

2. discussion of certain guiding principles for the development of medical products from animal sources;
3. process considerations for minimizing/eliminating virus transmission by medical products from the perspective of the production industries.

The concern posed by the potential of TSE agents in regulated products prompted the FDA to issue a December 1993 Letter to Manufacturers of Biomedical Products for Administration to Humans which proposed recommendations for identification of animal-derived raw materials, documentation of raw material sourcing, and maintenance of traceable records at the manufacturing site. Additional letters in this area have also been issued by the FDA.

The Agricultural Research Service (ARS) and the Animal and Plant Health Inspection Service (APHIS) of the U.S.D.A. have been monitoring the problems associated with viruses and the TSEs in animals very closely. Since agents such as the TSEs are difficult to detect, the identification of animals used as the source of medical products assumes great importance. Tracking and monitoring procedures as well as information on the transportation and importation of animals are required for appropriate identification.

The end-use of the product, and the route(s) of administration and exposure to potential pathogens should dictate the level of concern in the estimates of risk from exposure to adventitious agents of animal-derived products. Other factors for consideration include the

- source of animal(s), their history and maintenance;
- tissue(s) used and their likelihood of harbouring or being contaminated by the agent;
- method by which the animal is slaughtered and the tissue(s) collected;
- production process steps that remove the agent either by inactivation of physical separation;
- disposal of agent sequestered but no inactivated;
- batch size;
- validation and quality assurance monitoring.

The process considerations for minimizing virus transmission by biomedical products presented by the reprensentative production industries incorporated many of the factors identified above and are indicated in the following highlights of the session.

It is generally recognized that the TSE agents are the most difficult adventitious agents to detect, thus constituting a stringent challenge for the medical product industry. Bioassay, using the mouse/scrapie or hamster/scrapie model, is the only available detection method. Since there are great time and cost factors involved, research on other detection assays comparable to bio-assay is critical.

The most effective inactivation processes for the TSEs are not always compatible with maintaining the integrity of the biomedical product of interest. Therefore, adequately controlled clearance studies must be performed, i.e. titration of «spiked» loading materials and output samples. The only approach for non-destructive removal of the TSE agents is through proper sourcing. The control of tissue/cell collection procedures is also critical.

Common elements in the risk assessment approaches developed and used by the different industries included: stringent material sourcing; quantification of the risk in starting materials; validation of removal of agent(s); screening of animals and products; final product testing. Biosecurity approaches and testing of the initial animal source as well as in-process controls and the use of sensitive cell lines validated as susceptible to an array of viruses are also important.

An easy-to-use, preliminary risk assessment exercise was proposed to enable users of animal-derived raw materials to make more informed decisions on the need and type of pre-use processing before the final risk/benefit analysis.

Areas for Consideration

Areas for further consideration by the international scientific community evolved from the session. Among others is the need to develop a common language/definitions regarding animal-derived raw materials and systems. For exemple, in animal sourcing, what are the differences between a closed herd and one that is isolated of contained? Is containment feasible?

In discussions of different risk assessment strategies, it is apparent that the development of common guiding principles and generic risk assessment strategies are needed for minimizing/eliminating virus transmission through biomedical products where animal-derived materials and systems are used as the inital sources. This effort should be devloped between the public and private sectors.

Much of the discussion centred on the TSE agents because of the special problems that they pose for the safety of biomedical products and the lack of information and approaches regarding their origin, detection, and inactivation. This session, while just « scratching the surface », nevertheless provided the opportunity for initiation of the dialogue towards resolution of some of these issues.

Dr. K. Hellman, FDA/CDRH, 5600 Fishers Lane, HFZ-113, TRL 40E, Rockville, MD 20852, USA.

Brown F, Lubiniecki AS: Viral Safety and Evaluation of Viral Clearance from Biopharmaceutical Products. Dev Biol Stand. Basel, Karger, 1996, vol 88, p 295

BREAKOUT SESSION G

Retrovirus Clearance for Early Clinical Trials

Chairpersons: K.E. Stein, FDA/CBER, USA
P. Martin, Fred Hutchinson Cancer Center, USA

Brown F, Lubiniecki AS: Viral Safety and Evaluation of Viral Clearance from Biopharmaceutical Products. Dev Biol Stand. Basel, Karger, 1996, vol 88, pp 297-304

Validation of Retroviral Detection for Rodent Cell-Derived Products and Gene Therapy Applications

J.V. Hughes, K. Messner, M. Burnham, D. Patel, E.M. White

Quality Biotech Inc., Camden, NJ 08104, USA

Key words: PG4 S^+L^-, *Mus Dunni*, retrovirus, MuLV, IFA detection, validation, RCR.

Abstract: The availability of sensitive assays for detecting infectious murine retroviruses has become critical for the development and acceptance of a number of biopharmaceuticals, including monoclonal antibody-derived products and gene therapy vectors. Comparative studies demonstrated that the PG4 S^+L^- retrovirus infectivity test routinely yields higher titres than the mink cell test for xenotropic, amphotrophic and MCF murine retroviruses. A validation study for the PG4 S^+L^- assay demonstrated very good linearity (r^2 of 0.95 to 0.99), reproducibility within a study (±0.35 log_{10} units), and precision between tests (± 0.45 log_{10} units). Interference (or selectivity) in the presence of a non-specific antibody was insignificant (less than 0.2 log_{10} units). Sensitivity levels established from measurements as virus titres approach zero demonstrated a threshold value of 2-3 focus forming units (FFU)/ml. Two methods for increasing assay sensitivity were used including: (i) increased product samplings combined with a Poisson distribution analysis, and (ii) a 14-day co-cultivation with *Mus dunni* cells. Each of these methods was shown to increase sensitivity by at least one log_{10} unit. Murine retroviruses may also be detected by a less sensitive immunofluorescence assay (IFA) using specific monoclonal antibodies; this assay is essential for detecting certain recombinant ecotropic MuLVs. In summary, murine retroviral detection ranked by sensitivity is mink S^+L^- < IFA with monoclonal antibodies < PG4 S^+L^- < *Mus dunni* co-cultivation followed by PG4 S^+L^-.

Introduction

The detection of murine retroviruses is very critical for at least three different areas of biopharmaceutical product including: (i) testing of rodent-derived cell lines, in-process materials and final products [1, 2]; (ii) viral clearance and inactivation studies associated with the purification of these products [2]; and (iii) monitoring gene therapy vector banks or lots for replication-competent retroviruses (RCRs). Retrovirus particles are readily detected by electron micrographic techniques in rodent cell lines, including murine hybridomas and hamster cell lines such as Chinese Hamster Ovary (CHO) and Baby Hamster Kidney (BHK). The number of infectious viruses detected is somewhat lower than the number determined by electron microscopy and may be due either to a lowered sensitivity for infectivity assays, or to the defective nature of the retroviral particles that are detected. The testing of rodent cell lines and at least some of the first production lots (to characterize potential viral load) typically requires determination of retroviral presence by particle counts (with electron microscopy) and tests designed to demonstrate infectious virus.

Highly sensitive assays for retroviruses are obviously important in the early stages of a project, particularly for the testing of a Master Cell Bank and the Manufacturer's Working Cell Bank. Subsequently, validation studies for viral elimination need to be performed as outlined in the Points to Consider of 1993 [2] and CPMP Guidelines [3]. As some of the virus removal steps are very robust, with complete removal or inactivation expected, it is important for viral validation studies that the most sensitive assay be in place to demonstrate complete removal. As most murine and rodent cell lines have retroviral particles, their efficient removal or inactivation is probably the most critical stage for validating a purification process. Removal of 10 to 15 $\log_{10}$ units (or greater) of retrovirus will often have to be demonstrated to satisfy requirements for production [2, 3]. Clearly, sensitivity of these assays is of great importance, as it can result in greater $\log_{10}$ reduction values.

In some special cases, proposals have been made by investigators to use selected monoclonal antibodies in clinical trials, even though only limited data are available from the cell bank characterization and for which little or no viral validation studies are available. These antibodies are proposed for use only in life-threatening situations, typically for cancer treatments. Sensitive viral assays that can be performed on the final antibody preparation are important to demonstrate that the materials, which are often purified by generic antibody production steps, are free of any contaminating infectious retrovirus.

Lastly, in the generation of replication-defective retroviral vectors for gene therapy, a test for replication-competent retroviruses (RCRs) which may have been produced through recombination during the viral amplification process is essential for safe vector preparation. This testing should include an examination of the Master Cell Bank (MCB), the Manufacturer's Working Cell Bank (MWCB), the production lots and the transduced target cells if ex vivo technology is used [4]. This testing process also requires a very sensitive assay that can detect infectious murine leukemia viruses (MuLVs).

Evaluation of a variety of infectivity assays was undertaken to determine the most reasonable and sensitive testing regimen for each of these situations; namely production of monoclonals with and without all the detailed characterization, viral validation studies, and gene therapy vector production.

Retrovirus assays

Two distinct cell substrates have been used in effective infectivity assays for a variety of murine leukaemia viruses (MuLVs). One system uses mink cells ($MiCL_1$) which contain a defective murine sarcoma virus genome (sarcoma positive, leukaemia negative, or S^+L^-). Superinfection of this S^+L^- cell line with xenotropic, amphotropic or Mink Cell Focus-Inducing (MCF) MuLVs, allows rescue of the murine sarcoma virus, resulting in transformation and the formation of foci [5]. The alternative system uses a PG4 S^+L^- line which originated from a feline embryonic cell line infected with the Moloney isolate of murine sarcoma virus; this line has proven to be a very sensitive line for most MuLVs [6]. Regardless of the detection cell used, the assay begins by establishing cells in six-well tissue culture plates. The following day each virus is diluted with medium for several dilutions (2- to 10- fold) and inoculated on to the subconfluent monolayer using 0.5 ml/well of each dilution of the virus. The assay is read when the cells reach confluency and are scored for FFU as FFU/ml of the original inoculum.

The immunofluorescence assay (IFA) with the 83-A25 monoclonal antibody [7] is performed on *Mus dunni* cells grown on coverslips and infected with various MuLVs. The cells are washed gently with PBS and then fixed in a methanol acetone mix (50:50). The antibody is added for 30 to 60 minutes, followed by washing and reaction with a secondary antibody (FITC labelled anti-rat immunoglobulin). The cells are then observed with a fluorescence microscope.

Comparison of mink cell and PG4 S⁺L⁻ assays

Different strains of MuLV were tested in both the mink cell and PG4 S⁺L⁻ cell systems to compare the relative sensitivities of these assays (Table 1). When the identical stocks of virus were tested on the two lines, a higher titre was always observed on the PG4 S⁺L⁻ cells. However, the level of enhancement on the PG4 cells was quite variable. In particular, the xenotropic virus ranged from 2- to 55-fold enhancement. The highest yield for the different MuLVs was consistently demonstrated for the MCF viruses, particularly for the HIX strain. As the PG4 S⁺L⁻ system is significantly more sensitive than the mink cell assay, validation data were obtained on the PG4 S⁺L⁻ assay system.

Table 1: Comparison of virus titres obtained in mink cell and PG4 S⁺L⁻ assays.

Retrovirus type	Mink S⁺L⁻	PG4 S⁺L⁻	Ratio PG4/Mink
Xenotropic	**FFU/mL**	**FFU/mL**	
X-MuLV (A)	2.6×10^2	9.0×10^3	35
X MuLV (B)	3.4×10^3	7.4×10^3	2
X-MuLV (C)	3.8×10^3	2.1×10^5	55
MCF			
CB208 (A)	2.4×10^2	1.3×10^4	54
CB208 (B)	7.0×10^1	4.0×10^3	57
HIX (A)	2.5×10^2	7.0×10^4	208
HIX (B)	8.0×10^1	5.0×10^4	625
Amphotropic			
A-MuLV (A)	3.3×10^3	2.6×10^5	79

(A), (B), (C) refer to different stocks of the same virus type.

Validation of the PG4 S^+L^- assay

In the validation study of the PG4 S^+L^- assay system, a number of parameters, as described in the USP [8], were selected for detailed examination. These parameters included Precision (or Reproducibility), Limit of Detection (Sensitivity), Selectivity (with interfering factors present), Range and Linearity. In each assay that was used in the validation study, several dilutions of the virus stock were made such that the virus titre would fall within a countable range, typically from 1 to 100 FFU. Reproducibility within a single assay was evaluated by determining the grand mean and standard error from four separate dilution series; the resulting data are shown in Figure 1 and summarized in Table 2. The reproducibility within a given study was shown to have a value of 0.35 $\log_{10}$ units, which is low for a viral infectivity assay. To assess the level of variability between assays, these were performed 11 times with the same stock of virus over a 5 month period. The reproducibility between these assays was 0.48 $\log_{10}$ units (Table 2).

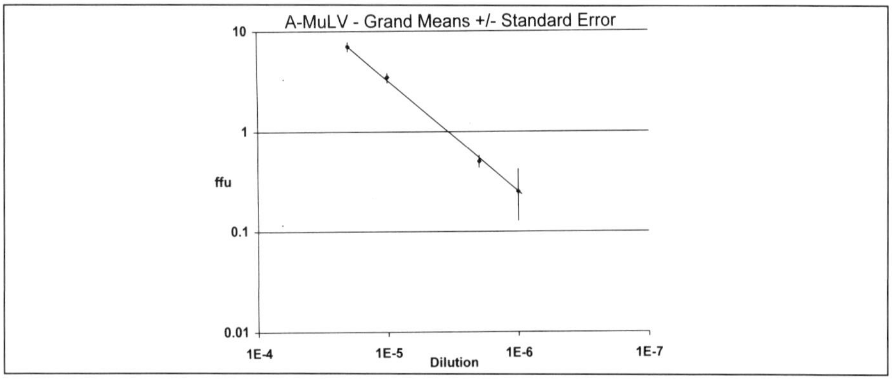

Fig. 1: Four separate dilutions of a stock of amphotropic MuLV were made and tested in the PG4 S^+L^- assay, using three wells per dilution. At the conclusion of the assay, the FFU were determined for the dilutions in figure (less dilute samples contained too many foci to count accurately). The grand mean and standard errors were calculated from these dilutions.

Table 2: Summary of validation study of PG4 S^+L^- assay.

Validation parameter	Measurement
Reproducibility • Within study • Between studies	 0.35 $\log_{10}$ 0.48 $\log_{10}$
Linearity	r^2 of 0.94 to 0.99
Sensitivity • With no additions • With added antibody	 1-2 FFU/ml 3 FFU/ml
Selectivity - with added antibody	0.2 $\log_{10}$
***Mus dunni* co-culture enhancement**	1.0 $\log_{10}$

Linearity measurements indicated that the FFU over a fairly wide range of dilutions resulted in excellent correlation. The r^2 values calculated from the tests used in the large-scale validations ranged from 0.94 to 0.99. Naturally, as one approaches the range where there are too many foci to count accurately (at approximately 100 FFU), it is expected that some of the foci will be produced from two or more viruses, resulting in a non-linear response.

Sensitivity, or the limit of detection as one approaches zero, was first determined empirically from the data. For this determination, four separate dilutions (with three wells per dilution) were made within each experiment. At the highest dilution that contained infectious virus, i.e. dilutions which did not contain a zero, calculations were made of the grand mean and 95% confidence level values. These 95% upper-bound confidence levels were defined as the sensitivity value closest to zero for this assay. While this represents a highly conservative estimate, it allows us to calculate the sensitivity level directly from the observed experimental data. Sensitivity measured in this manner was determined at 1 and at 2 FFU/ml in two independent tests (Table 2).

To measure selectivity (viral infectious units in the presence of other potential interfering proteins or products), a non-specific antibody was added at 10 µg/ml to the concentrated virus stock before dilution and titration. When the infectivity assay was performed in parallel with the untreated virus stock, only a small difference in titre (a decrease of 0.2 $\log_{10}$ units) was measured. The sensitivity measurement in the presence of this added antibody was also only minimally affected. In two tests the sensitivity level was 3 FFU as compared to 1 to 2 FFU without the added protein.

Methods for enhancing assay sensitivity

Two methods have been used to enhance the sensitivity of this assay.

1. *Statistical enhancement.* This method uses a statistical analysis of the probability of detection or non-detection of an infectious unit as a function of sampling number. The equation below can be used to describe the probability of non-detection of virus, which is expected to follow a Poisson distribution [9]:

$$P = [\, e^{-S/N} \,]^{KZ}$$

P = Probability of non-detection
S = Theoretical Sensitivity in IU/mL
N = Dilution Factor
K = Number of Wells or Samplings
Z = Volume (in ml) tested per well

If the probability is set equal to 0.05 (only 5% non-detection level), N equal to 1 (for no sample dilution), sampling (K) to three wells, and Z to 0.5 ml per well, and solving for S

$$P = [\, e^{-S/N} \,]^{KZ}$$

$$\ln P = KZ(-S/N)$$

$$\ln (0.05) = 3(0.5)(-S/1)$$

$$S = 2.0$$

This theoretical sensitivity of 2.0 FFU/ml using a three-well format closely matches the value we determined from our experimental data (1 to 2 FFU/ml in Table 2). If the number of samplings of the test material is increased to eight wells, then the sensitivity drops to 0.75 FFU/ml. As might be expected with increased samplings, the probability of detecting the presence of a virus can be greatly enhanced, reaching an increase in excess of one log if up to 32 wells are used (Table 3).

2. *Co-cultivation or amplification enhancement.* The second method developed for enhancing assay sensitivity for retrovirus detection is to perform a co-cultivation or amplification on highly sensitive cells such as *Mus dunni*. The primary advantage of such a co-cultivation is that low levels of virus and slowly growing viruses can be amplified, resulting in increased detection in the PG4 S^+L^- assay. Additionally, any interfering factors in the initial sample will be diluted out in the subsequent co-cultivations. The *Mus dunni* cell is well characterized and will support the replication of most classes of MuLVs including the ecotropic, amphotropic, xenotropic and mink cell focus-forming, or polytropic, viruses [10]. Because there is little homology between MuLV and endogenous *Mus dunni* DNA sequences, there is a reduced possibility of aberrant results from recombination between input virus and endogenous viral sequences.

In two separate experiments the co-cultivation of a known amount of an amphotropic MuLV on *Mus dunni* cells enhanced the amount of retrovirus detected by almost one $\log_{10}$ unit (Table 4). Although two passages of the *Mus dunni* cultures are typically performed, the second test shows that no further enhancement was gained by one or two additional passages. This enhanced detection through co-cultivation has been demonstrated with a number of different murine retroviruses and is routinely employed for monitoring retroviruses in a variety of biopharmaceutical products. In particular, for gene therapy applications using replication-incompetent vectors, the co-cultivation/amplification on *Mus dunni* cells followed by a sensitive detection assay has been mandated by the FDA and other agencies to ensure the absence of replication-competent retroviruses [4].

Detection of MuLV by immunofluorescent assay

There are a number of monoclonal antibodies against the MuLVs that can be used to detect viral replication [7]. Typically, sensitive cells such as *Mus dunni* are infected and passed to ensure replication of low levels of virus. At the conclusion of

Table 3: Improved sensitivity with increased sampling.

Number of samplings (Wells)	Sensitivity: FFU/ml
3	2.0
8	0.75
16	0.37
32	0.19

Table 4: Enhanced detection of MuLV by *Mus dunni* co-cultivation.

Passage level	Initial titre* FFU/ml	Results from co-cultivation # Positive flask/Total # flasks			End titre** $TCID_{50}$
Test 1		10^{-9}	Dilution 10^{-8}	10^{-7}	
Post passage 2	7.5	1/6	5/6	6/6	8.5
Test 2					
1st growth	7.65	0/6	6/6	6/6	8.5
Post passage 1	7.65	0/6	6/6	6/6	8.5
Post passage 2	7.65	0/6	6/6	6/6	8.5

* Initial titre was taken when the stock virus was initially titrated on PG4 S^+L^- cells.

** The End Titre was calculated using the Reed-Muench formula for determining $TCID_{50}$ from the number of positive flasks per total flasks inoculated following passaging on *Mus dunni* cells.

the cultivation, the cells are fixed in acetone: methanol (1:1) and the viral antigens are detected by incubation with specific monoclonal antibodies followed by an FITC-labelled second antibody. One can readily measure replication in *Mus dunni* cells for ecotropic, xenotropic, amphotropic and the MCF MuLVs. While some antibodies can be used to detect almost all the MuLVs, specific antibodies that react only with unique classes of MuLVs can also be used. The sensitivity of this assay is no greater than some of the other assays described above. Because some recombinant ecotropic MuLVs will not be detected by other routine infectivity assays that are used for ecotropic viruses, the IFA test, must be used for detection of these viruses.

Discussion and Conclusions

Validation studies on the various methods for detection of MuLVs suggest the following ranking from least to most sensitive: Mink Cell S^+L^- assay < IFA techniques, with specific monoclonals on infected *Mus dunni* cells, < PG4 S^+L^- assay < Co-cultivation or Amplification on the *Mus dunni* cells followed by PG4 S^+L^- assay. Two methods increase assay sensitivity by approximately one $\log_{10}$ unit: amplification on sensitive cell lines and increased samplings of the test material. A combination of the methods revealed by these validation tests allows a highly sensitive rational approach to be devised for the detection of various MuLVs. Some of these methods are obviously also suitable for application to other retroviruses and to viruses in general.

References

1 Points to Consider in the Manufacture and Testing of Monoclonal Antibody Products for Human Use FDA, CBER, 1994.

2 Points to Consider in the Characterization of Cell Lines Used to Produce Biologics FDA, CBER, 1993.

3 Revised CPMP Guideline on Virus Validation Studies: The Design, Contribution and Interpretation of Studies Validating the Inactivation and Removal of Viruses, CPMP, 1995.

4 Kuta A: Presentation to the Vaccine Advisory Committee by the FDA on Proposed Testing for Replication Competent Retroviruses, 1993.

5 Bassin RH, Tuttle N, Fischinger PJ: Rapid cell culture assay technique for murine leukemia viruses. Nature (London) 1971;229:564-566.

6 Bassin RH, Ruscetti S, Ali I, Haapala DK, Rein A: Normal DBA/2 mouse cells synthesize a glycoprotein which interferes with MCF virus infection. Virology 1982;123:139-151.

7 Chesebro B, Britt W, Evans L, Wehrly K, Nishio J, Cloyd, M: Characterization of monoclonal antibodies reactive with murine leukemia viruses: use in analysis of strains of Friend MCF and Friend Ecotropic Murine Leukemia Virus. Virology 1983;127:134-148.

8 Validation of Compendial Methods. (1994) USP XXIII, Section 1225: 1982-1984.

9 Walpole RE, Myers RH: Probability and Statistics for Engineers and Scientists, Fifth Edn., New York, Macmillan Publishing Company, 1993.

10 Lander MR, Chattopadhyay SK: A Mus dunni cell line that lacks sequences closely related to endogenous Murine Leukemia Viruses and can be infected by Ecotropic, Amphotropic, Xenotropic and Mink Cell Focus-Forming Viruses. J Virol 1984;52:695-698.

Dr. J.V. Hughes, Quality Biotech, Inc., 1667 Davis Street, Camden, NJ 08104, USA.

Brown F, Lubiniecki AS: Viral Safety and Evaluation of Viral Clearance from Biopharmaceutical Products. Dev Biol Stand. Basel, Karger, 1996, vol 88, p 305

SUMMARY TO BREAKOUT SESSION G

Chairmen: *K. Stein, FDA/CBER, USA*
P. Martin, Fred Hutchinson Cancer Center, USA

Participants: *N. Bander, Cornell University, USA*
S. Eck, University of Pennsylvania, USA
J. Hughes, Quality Biotech, Inc., USA
E. Morgan, Microbiological Associates, Inc., USA
H. Schlesinger, Tektagen, Inc., USA
C. Wilson, FDA/CBER/OTRR, USA
G. Weiner, University of Iowa, USA

The discussion in this session focussed on retrovirus testing of murine hybridomas or products produced in murine hybridoma cells, for early phase I studies in patients with immediately life-threatening diseases, that is those who have a median life expectancy of less than one year. The discussion covered the following issues: the assays used to test for replication, competent retroviruses in final container material, how much sample to test, the patient population for whom this would be appropriate, the risk of murine retroviruses for humans, and the informed consent. There was agreement that the preferred assay involved amplification on Mus dunni cells followed by an infectivity assay in PG4 cells. Participants generally agreed that three times the total human dose was appropriate, although not always possible and that a larger sampling volume would add to the sensitivity of the assay. It was also generally agreed that where the purification involved at least two, orthogonal, robust steps, final container testing might not be necessary. In the absence of data on the purification steps, the final container would need to be tested. It was suggested that it would be appropriate to use such antibodies in patients who had undergone TBI and others ablative or immunosuppressive regimens. It was emphasized that any monoclonal antibody given to patients should be tested for bacteria, fungi, mycoplasma, endotoxin and adventitious viruses. There was general agreement that adventitious viruses in monoclonal antibodies were of greater concern and posed greater risk to the patient than endogenous murine retroviruses. There was concern that adventitious viruses and appropriate testing were of increased importance when the monoclonal antibodies were produced in ascites rather than tissue culture cells. It was agreed that the informed consent should disclose all known risks and clearly state the possibility of unknown risks.

Dr. K. Stein, FDA/CBER, 8800 Rockville Pike, HFM 555, Bethesda, MD 20892, USA.

Brown F, Lubiniecki AS: Viral Safety and Evaluation of Viral Clearance from Biopharmaceutical Products. Dev Biol Stand. Basel, Karger, 1996, vol 88, p 307

BREAKOUT SESSION H

Process Evaluation

Chairpersons: J. Fogh, Novo Nordisk, Denmark
A. Moreira, University of Maryland, USA

Brown F, Lubiniecki AS: Viral Safety and Evaluation of Viral Clearance from Biopharmaceutical Products. Dev Biol Stand. Basel, Karger, 1996, vol 88, pp 309-315

Process Evaluation for Biopharmaceuticals: What is Appropriate in Process Evaluation?

A.S. Lubiniecki, P.R. McAllister, T.M. Smith, P.J. Shadle

SmithKline Beecham Pharmaceuticals, King of Prussia, PA USA

One of the central features of any biopharmaceutical process design is to reduce the risk of putative viral contamination to acceptably safe levels. This is typically achieved by using a multifaceted approach that includes (i) minimization of external sources of risk; (ii) characterization of cell lines; (iii) design of purification processes to remove and/or inactivate risk factors of concern; (iv) evaluation of the effectiveness of those procedures; and (v) prevention of forward flow of contaminants through engineering and procedural controls. As part of this effort, it is important to define how much clearance of virus is adequate to meet the safety goals of the product. Paula Shadle presented our approach to the statistical aspects of study design, data analysis, and interpretation of the clearance values [1]. We will build upon her presentation and try to present some of our thoughts on safety goals and the management of viral concerns through process evaluation.

To review briefly, one answer to the question «How much risk is acceptable?» is «When excess clearance (EC) is large enough». EC equals clearance measured minus risk measured. For example, if 19 logs of total clearance are observed, and in the worst case assuming complete co-purification of viral particles with the active ingredient, a dose might contain 9 logs of particles, then EC = 19 logs - 9 logs = 10 logs. Thus, the average theoretical probability of a particle-contamined unit is 10^{-10}, ignoring the statistical considerations presented earlier (and to which we will return later).

This analysis is straightforward and, in some form, is practised by many organizations. However, the above question and its answer beget another question, namely, «How much excess clearance is large enough?» There is no clear universal answer to this question, but there are lessons learned in other fields which yield information worthy of examination.

For example, in the field of sterilization, a critical issue related to our question is «What level of risk of unspecified microbial contamination is acceptable for parenterally administered products». There are two answers to this question, whose implications are widely known. For terminally sterilized products, the standard is that the average theoretical probability of contamination is 10^{-6} per unit, or less. This is an arbitrary [2] but universally accepted standard, and in the model which Shadle presented, EC could be calculated to be 6. It has been used to provide products which are safe and trouble-free from the perspective of microbial contamination when the process is maintained in the validated state. Furthermore, attainment of this standard in a validated process is the foundation for parametric release of terminally sterilized products. No sterility test is performed for lot release; verification that the validated conditions were met during processing is adequate to ensure freedom from microbial contamination.

Another example from parenteral manufacturing is the use of aseptic processing, usually for products whose ingredients are not thermostable. Here the standard is that the probability of contamination is 10^{-3} or less, corresponding to an average value for EC of 3 or more. This is widely accepted as a limit of convential aseptic technology. Despite the validation of processes to this standard, and compliance with Good Manufacturing Practices, occasional sterility test failures occur for products manufactured by these methods. Furthemore, occasional product recalls are performed for sterility problems in the field; virtually all such recalls for field sterility problems involve aseptically processed rather than terminally sterilized products.

Several conclusions can be drawn from these related examples. First, larger values of EC correlate with a greater likelihood that the product will be free of the safety concern in terms of both laboratory testing and actual field use. Second, for unspecified microbial contamination, EC should at least attain an average value of 3, and preferably 6, in the absence of statistical evaluation. It should also be made clear that the analogy of comparing microbes to virus particles overlooks two important points. One is that microbes may replicate in dosage forms, but viruses cannot. A second point is that this argument addresses only viable particle concentration, and ignores the potential pathogenicity of the contaminant. In other words, the microbial sterility assurance limits treat every colony-forming microbe as capable of causing disease in humans. This is a worst-case assumption for viruses, since the nature of the relationship between particles and disease is not straightforward. While experimental studies for many viruses have shown that a single particle is capable of generating an infection of a cell in vitro, infection of normal healthy adults usually requires more than one in vitro infectious unit. A number of biological barriers to viral diseases are known. For most viruses, there are more viral particles than infectious units in a given virus suspension. In short, one particle does not equal one case of disease. The relationship between particles and disease is a component of the pathogenicity of a given virus, and should be a component of the risk assessment for that concern.

As examples of this concept, history shows us that millions of children received live attenuated viral vaccines which contained bateriophage in the 1950s and 1960s [3], hundreds of millions of humans have received avian retroviruses which have contaminated live attenuated yellow fever vaccines since the 1940s [4], and tens of millions of children received live SV40 in their inactivated poliovirus vaccine during the 1950s [5]. When examined by electron microscopy, fresh water sources around the world typically contain 10^5 to 10^8 virus particles per ml [6]. Clearly there are major biological barriers that prevent most virus particles from causing disease in most people most of the time. These barriers may differ depending on the route of exposure. Again, a virus particle does not equate to a case of disease.

That differences exist in risk among different patient populations is also clear from the historical record. Immunocompromised persons who received smallpox vaccine frequently suffered from a generalized vaccinia infection when the immune system failed to contain the virus disease to the site of inoculation. Another clear demonstration of the concept that host factors may interact with biological contaminants to affect the likelihood or severity of disease was recently witnessed in a preclinical study for a gene therapy preparation. Irradiated immunosuppressed monkeys were given preparations containing about 10^9 rDNA-modified retrovirus particles which also contained 10^3 to 10^4 replication-competent retrovirus (RCRV) particles. Of eight monkeys exposed, three died of malignant thymomas from which the RCRV genomes could be recovered [7]. Previous studies had failed to detect any pathological effects of exposure of normal macaques to xenotropic retroviruses [8]. This experiment also illustrates why it has been prudent to evaluate the ability of recovery processes for biopharmaceutical products to remove or inacivate RVLPs.

Another set of lessons comes from efforts to prepare previous generations of biological products. For example, from 1980-1986, plasma-derived hepatitis B surface antigen vaccine was distributed to millions of recipients without infectious sequelae. The source material for this product was plasma from chronic carriers of hepatitis B, whose plasma contained infectious hepatitis B virus and may well have been cryptically contaminated with other infectious agents. What is equally important to grasp is that the use of a validated manufacturing process with virus clearance steps (pepsin at low PH, urea, and formalin) was capable of providing a safe product, even from an overtly unsafe starting material [9]. A similar case can be made for the use of appropriate inactivation procedures to prepare other inactivated viral (influenza, polio, rabies) and bacterial (diphtheria, pertussis, tetanus) vaccines. These have been shown to be free of risks attributable to the live infectious agent by decades of safe use in humans. Another example is the safe use of plasma-derived human albumin after heat tratment (typically 60°C for 10 hours or equivalent) for several decades. Thus, there is ample evidence that robust validated inactivation processes work effectively to ensure freedom from risk of contamination by infectious agents. In some cases, the infectious agents were known to be present, but in others, their presence was not known specifically or only suspected. Greater confidence in validation studies exists when data are available for each specific risk factor, than when the treatment is assumed to apply to all possible concerns. Despite this generality, two examples (human serum albumin and plasma-derived hepatitis vaccine) exist where the field experience with the products over extensive periods clearly demonstrates that suitably robust inactivation procedures are effective in ensuring freedom from viral concerns.

An additional set of lessons from other fields is found in the use of conventional biologicals where no opportunity for inactivation of infectious agents exists, or where incomplete knowledge led to subsequent problems. Here a question can be asked, «What level of adverse reactions/sequelae are acceptable for products used to treat or prevent serious diseases?» Table 1 shows observed levels of sequelae for live attenuated vaccines and other biological products. The products can be grouped into two categories, based on the discovery that the product led to sequelae in some recipients. In one group, which includes brain-derived rabies vaccines, cadaver-derived growth hormone, and Urabe mumps vaccine post-licensing, the probability of sequelae ranged from $10^{-3.3}$ to $10^{-4.0}$, and the product was withdrawn from distribution *when a safer alternative existed* [10, 11]. In the second group, which includes oral live attenuated poliovirus vaccine, live attenuated measles vaccine, and Urabe mumps vaccin at licensing, the probability of sequelae ranged from $10^{-4.8}$ to $10^{-6.3}$, *no safer alternative existed,* and the product remained in distribution [12-

Table 1: Incidence of some transmissible illnesses in humans receiving biologicals.

Product	Sequela	Log_{10} (incidence)[a]	Removed (ref)
Yellow fever vaccine[b]	Neuroparalysis	3.7–4.0	Yes
Urabe mumps vaccine	Aseptic meningitis	3.3–4.0	Yes [10]
Growth hormone[c]	Creutzfeld-Jakob disease	3.3–3.5	Yes [11]
Measles vaccine	SSPE	5.7–6.0	No [12]
Urabe mumps vaccine	Aseptic meningitis	4.8–5.6	No [13]
Oral polio vaccine	Paralytic polio	6.0–6.3	No [14]

[a] Number of humans using product needed to observe one adverse event.
[b] Brain-derived.
[c] Pituitary-derived.

14]. For clarification, the Urabe mumps story is that at the time of licensing, the incidence of aseptic meningitis was believed to be $10^{-4.8}$, very similar to that of the Jeryl Lynn mumps vaccine. However, after about a decade of distribution, the incidence of aseptic meningitis for Urabe mumps vaccine was determined to be significantly higher than originally believed, and the vaccine was withdrawn from worldwide distribution. The availability as well as the superior safety record of the Jeryl Lynn strain were clearly factors in this decision.

In a larger sense, these examples reinforce the concept that all products and services possess potential side effects which are undesirable and unintentional, but unavoidable or unforeseeable. The probability that the average American automobile driver will die in an auto accident is about $10^{-3.7}$ per year. From the perspective of a lifetime of driving lasting at least 50 years, this implies that the lifetime risk of dying in an auto accident is about $10^{-2.0}$, yet the pobability of dying on a given auto trip is probably very small (perhaps $10^{-6.0}$ per trip if one makes 200 trips per year). Almost no normal healthy adult decides that it is appropriate to stop driving their automobile permanently simply because of the risks inherent in driving. However, this may represent an important option for someone with a significant physical impairment, or for whom driving has no benefits. The probability that an American will die in a commercial airline accident (major air carriers) is $10^{-6.7}$ per passenger mile or $10^{-5.8}$ per departure. This risk does not deter many millions of frequent airplane passengers. Life itself has inherent risk – the average American has a risk of death by stroke of $10^{-3.3}$ per year. Thus, the realities of life argue that absolute freedom from risk is not scientifically possible, and that the prudent course is to evaluate the risks and benefits of a given option together in concerted fashion for a given use.

Let us now consider how this relates to biopharmaceuticals, and especially in our case, to recombinant DNA-derived proteins expressed in CHO cells. What types of concern exist? How much of the putative risk factor exists? Answers to these questions usually come from cell line characterization studies, process evaluation studies and product characterization efforts. For CHO cells, a typical answer is that modest amounts of biologically inert retrovirus-like particles (RVLPs) exist. These particles probably do not co-purify with the product; however this is usually assumed as the worst case. For CHO cells, typically there is non measurable infectitvity in these RVLPs. When virus assays give negative results, one is usally suspicious that the host cell used for the assay may have been incapable of supporting virus replication. However, Anderson et al [15] have shown that there is a molecular explanation for the lack of infectivity for CHO RVLP in that all open reading frames for the three genes essential for retrovirus replication contain at least one stop codon. This fact, coupled with uniformly negative results from many laboratories for dozens of CHO rDNA construction, argue that CHO RVLPs are truly defective viruses.

How much excess clearance is enough for RVLPs? If one temporarily ignores statistics, the lessons of the past argue that where average EC =6, then the product is very safe, and where average EC=3, this may represent greater potential risk which may or may not be acceptable, depending on the specifics. Further, regulatory approval may conceivably be withdrawn where $-\log_{10} P < 4.0$ *when perceptively safer alternatives exist.* Considering the statistics of risk and clearance, estimation prompts consideration of larger levels of virus clearance to compensate for the variability inherent in experimental estimation of these parameters. Studies presented by Shadle showed that an excess of clearance (F_s) of 8.2 logs is needed to attain 95% confidence.

One should also consider the effect of statistics on the economics of evaluation studies. Table 2 compares the cost of conducting the clearance evaluation studies for RVLP only, including the contributions of internal scientific labour and contract laboratory fees for virus titration, with and without statistical evaluation. There is a clear difference in the cost of the two experimental designs. The total amount of clearance observed is similar (> 21 logs vs > 23.1 logs), but the information gained regarding excess clearance is qualitatively different. Without replication, a numerical value for average EC can be calculated without any idea of its accuracy or precision. With replication, an average value for EC can be calculated as well as a 95% confidence limit or lower bound about this mean value of EC. In practical terms, without statistics, there is an unknown level of confidence in the average value of EC (but subjectively, it is presumed to be good), and with statistics, it is known that 95% of the time, the value of EC will exceed a specified lower bound value (in this case, at least 4.6 logs). It is important at this juncture to ask more practical questions: are these two products (with and without replication) the same? If a differ-

Table 2: Attributes of evaluation studies.

Item	One replicate[a]	Three replicates[b]
External cost	$ 28.0	$ 83.0
Internal cost	$ 3.0	$ 12.0
Clearance	> 21 logs	> 23.1 logs
Excess clearance	12.8 logs	–
Lower bound of excess clearance	–	> 4.6 logs
Confidence	? (GOOD)	95%

[a] No knowledge of variation.
[b] Variation accounted for.

ence between the products is perceived, is the difference worth the difference in cost? At least for the one example presented here, it is proposed that the answer to these questions are yes and no, respectively. Although the two products are likely to be the same, it seems to us that the table is mostly concerned with the *decision* that will be made about two *processes*. The question is whether the quality of the decision is changed by having information about variation.

It is also worth mentioning that human data now exist for several products prepared from mammalian cell lines with RVLP. Since 1987, over a million people have received tissue plasminogen activator, and over a half million have received erythropoietin repeatedly since 1989. Since 1992, recombinant Factor 8 has been administered repeatedly to tens of thousands of haemophiliacs. No evidence of viral infection attributable to these products has been reported. We believe that this is not accidental, but rather reflects our ability to characterize cell banks, certify raw materials, and design and evaluate processes to provide levels of safety assurance not previously possible.

We would like to summarize all these thoughts into a proposal for assessing RVLPs:

1. Processes with multiple inactivation and removal steps are adequate to clear RVLP to EC = 3 to 6 which exhibit:

   (i) no evidence of infectivity in vitro; (ii) no open reading frames in key vitral gene functions; (iii) no evidence of viral infection in large numbers of product recipients.
2. Elaborate experiments for full statistical treatment of clearance of RVLPs are not necessary in view of the molecular defects of RVLPs.
3. Simple duplication of clearance results and monitoring of process parameters are adequate to assure that a process reliably provides a product which is free from the putative risks of RVLPs.
4. One study is sometimes enough to define adequately the status of a given problem or concern. For example, defining the spectrum of viral replication in rDNA-modified CHO cells was provided by a single large study of parental CHO cells and CHO cells modified to secrete human tissue plasminogen activator [16]. Regulatory bodies seem to accept that this study is valid of all rDNA-modified CHO cells in terms of the basic findings and conclusions, and to accept that additional studies would not provide substantially different results. Similarly, one adequate study of the statistics of clearance of RVLPs is probably sufficient for an entire industry.

References

1 McAllister PR, Shadle PJ, Smith TM, Scott RG, Lubiniecki AS: Use of a statistical strategy to evaluate sources of variability in viral safety experiments for a recombinant Biopharmaceutical, in Brown F, Lubiniecki AS (eds): Viral Safety and Evaluation of Viral Clearance from Biopharmaceutical Products. Dev Biol Stand. Basel, Karger, 1996, vol 88, pp 111-121.

2 Sharp J: What do we mean by «sterility»? PDA J Pharmaceutic Sci Technol 1995;49:90-92.

3 Petricciani JP, Chu FC, Johnson JB: Bacteriophages in live virus vaccines. Proc Soc Exp Biol Med 1973;144:789-792.

4 Draper CC: A yellow fever vaccine free from avian leukosis viruses. J Hyg Lond 1967;65:505-513.

5 Sweet BN, Hilleman MR: The vacuolating virus, SV40. Proc Soc Exp Biol Med 1960;108:420-427.

6 Bergh O, Borsheim KY, Bratbak G, Heldal M: High abundance of viruses found in aquatic environments. Nature 1989;340:467-468.

7 Donahue RE, Kessler SW, Bodine D, McDonagh K, Dunbar C, Goodman S, Agricola B, Byrne E, Raffeld M, Moen R, Bacher J, Zsebo KM, Nienhuis AW: Helper virus induced T cell lymphoma in nonhuman primates after retroviral mediated gene transfer. J Exp Med 1992;176: 1125-1135.

8 Cornetta K, Moen RC, Culver K, Morgan RA, McLachin JR, Sturm S, Selegue J, London W, Blaese RM, Anderson WF: Amphotropic murine leukemia virus is not an acute pathogen for primates. Hum Gen Ther 1990;1:15-30.

9 Hilleman MR, McAleer, Bunyak EB, Mclean AA: The preparation and safety of hepatitis B vaccine. J Infect (Suppl) 1983;1:3-8.

10 Miller E, Goldacre M, Pugh S, Colville A, Farrington P, Flower A, Nash J, MacFarlane L, Tettmar R: Risk of aseptic meningitis after measles, mumps, and rubella vaccine in UK children. Lancet 1993;341:979-983.

11 Fradkin JE, Schonberger LB, Mills JI, Gunn WJ, Piper JM, Wysowski DK, Thomson R, Durako S, Brown P: Cruetzfeld-Jakob disease in pituitary growth hormone recipients in the United States. J Amer Med Assn 1991;265:880-884.

12 Center for Disease Control. Morbid Mortal Weekly Rep 1977, 26: 309.

13 Champagne S, Thomas E, Furesz J. A case of mumps meningitis: a post immunization complication. Can Dis Weely Rep 1987;13:155-157.

14 Assad F, Cockburn WC: The relation between acute persisting spinal paralysis and poliomyelitis vaccine - result of a ten year enquiry. Bull WHO 1982;60:231-242.

15 Anderson KP, Lie YS, Low MAL, Williams SR, Wurm FM, Dinowitz M: Defective endogenous retrovirus-like sequencesand particles of Chinese hamster ovary cells, in Brown F, Horaud F (eds): Virological Aspects of the Safety of Biological Products. Dev Biol Stand Basel, Karger, 1991, vol 75, pp 123-132.

16 Wiebe ME, Becker F, Lazar R, May L, Casto B, Semense M, Fautz C, Garnick R, Miller C, Masover G, Bergmann D, Lubiniecki AS: A multifaceted approach to that recombinant tPA is free of adventitious virus, in Spier RE (ed): Advances in animal cell biology for bioprocesses. London, Butterworths & Co, Ltd, 1989, pp 68-71.

Dr. A.S. Lubiniecki, Depts of Biopharmaceutical Development, SmithKline Beecham Pharmaceuticals, 709 Swedeland Road, UE-0428, King of Prussia, PA 19406, USA.

Brown F, Lubiniecki AS: Viral Safety and Evaluation of Viral Clearance from Biopharmaceutical Products. Dev Biol Stand. Basel, Karger, 1996, vol 88, pp 317-318

Experimental Experience with Virus Validation Studies

H. Willkommen

Paul-Ehrlich-Institut, Langen, Germany

Key words: Virus validation.

The risk of virus transmission by administration of plasma derivatives cannot be completely excluded. Occasional transmissions of viruses have been observed with different products in the past. Analysis of these events showed that the reason for virus contamination could be found in an over-evaluation of the capacity of the manufacturing process for virus removal or inactivation. Also a change in the prerequisite of manufacture can influence the safety of products. The conditions of manufacture are changed, for instance if the level of antibodies in plasma pools has been reduced by the introduction of a screening programme or a change in the epidemiological situation.

The last would also be observed in the case of HCV transmission by intravenous immunoglobulins. Although sporadic cases of HCV transmission were observed earlier, about 200 infected patients showed symptoms after administration of an intravenous immunoglobulin last year. The manufacturing process for immunoglobulins based on ethanol fractionation without an additional virus inactivation step was not effective enough in removing or inactivating viruses. The elimination of antibodies against HCV from the plasma pool by the screening programme could have contributed to the reduced removal of HCV by the process.

We investigated the Cohn ethanol fractionation procedure with model viruses (togavirus, pestivirus, orthomyxovirus). The results confirmed that virus was not inactivated during the process. A reduction of virus load occurred only by partitioning the virus into the precipitates and removal of the precipitate (Cohn-fraction III) from the process. The partition into the precipitate of model viruses for HCV such as pestiviruses was not as effective as shown for other viruses. To guarantee the virus safety of immunoglobulins, the manufacturing process for immunoglobulins should, therefore, contain a distinct virus inactivation step.

In general, we conclude from virus transmissions by blood derivatives that very small changes in the prerequisites or conditions of the manufacturing process can influence the efficacy of the process to inactivate or remove viruses. They have, therefore, to be considered very carefully.

Dr. H. Willkommen, Paul-Ehrlich-Institut, Paul-Ehrlich-Str. 51-59, D-63225 Langen, Germany.

Brown F, Lubiniecki AS: Viral Safety and Evaluation of Viral Clearance from Biopharmaceutical Products. Dev Biol Stand. Basel, Karger, 1996, vol 88, pp 319-326

Virus Removal Studies Using Nanofiltration Membranes

J. O'Grady[1], A. Losikoff[2], J. Poiley[2], D. Fickett[1], C. Oliver[1]

[1] Department of Process Biochemistry, MedImmune, Inc. Gaithersburg, MD, USA
[2] Biological Research Faculty and Facility, Inc. Ijamsville, MD, USA

Key words: Virus removal.

Abstract: Nanofiltration of immunoglobulin products was investigated as a virus removal step using Planova-controlled pore size membranes. For these studies, purified preparations of an IgG and an IgM were tested with 15 nm and 35 nm Planovaœ™ nanofiltration membranes, respectively. The results of spiking studies with four model viruses indicated that both the Planova 15 and 35 membranes removed 6-7 $\log_{10}$ plaque or focus forming units of murine xenotropic retrovirus, simian virus 40 and pseudorabies virus. Although two tandem Planova 35 membranes were tested for clearance of poliovirus, only the Planova 15 membrane removed this virus. Nanofiltration experiments were carried out with protein concentrations up to 12 mg/ml for the IgG and 1-2 mg/ml for the IgM, and protein recovery was excellent.

INTRODUCTION

Nanofiltration of immunoglobulin products was evaluated as a virus removal step following chromatographic purification and before final sterile filtration. For these studies, Planova™ membranes (Asahi Chemical Industry Co. Ldt., Tokyo, Japan) were tested with four model viruses, namely, poliovirus (Type 1), murine xenotropic retrovirus, simian virus 40 and pseudorabies virus (Table 1). Viruses were chosen on the following criteria: the viruses represented a range of families with different characteristics such as size, presence or absence of an envelope, different genomic structure and because of their relevance as models for those likely to be present as endogenous or adventitious contaminants in a murine hybridoma cell culture production process. This was done to address the possible range of potential contaminants and to challenge adequately the size-exclusion capability of the Planova membrane filters. Both IgG (~ 150 kDa) and IgM (~1,000 kDa) preparations were spiked with each high titred stock virus and filtered through either a 15 or 35 nm membrane. The respective prefiltrate fractions were analysed using 60 nm and in most cases 100 nm dishes to test sample volumes up to 2 ml for enhanced assay sensitivy. The report presented here describes the experiments and results of these studies.

Table 1: Viruses used in nanofiltration studies.

Virus	Family	Size	Genome	Envelope
Pseudorabies	Herpesviridae	120-250 nm	dsDNA	Yes
Murine xenotropic Type C retrovirus	Retroviridae	80-120 nm	ssRNA	Yes
Simian virus 40	Papovaviridae	45-55 nm	dsDNA	No
Poliovirus (Type 1)	Picornaviridae	22-30 nm	ssRNA	No

Materials and Methods

Proteins

Purified IgG and IgM were prepared according to standard chromatographic methods. Approximately 250 ml of each protein solution in a standard phosphate buffered saline was tested in each spiking experiment. The protein concentrations ranged from 5-12 mg/ml for the IgG preparation and 1-2 mg/ml for the IgM preparation. Protein concentration was determinated by absorbance at 280 nm using an extinction of 1.5 $(mg/ml)^{-1}$ cm^{-1} for the IgG and 1.68 $(mg/ml)^{-1}$ cm^{-1} for the IgM.

Planova membrane cartridges

For each of the spiking studies, 0.03 m^2 Planova membrane cartridges were obtained from Asahi Chemical Industry CO., LDT., Tokyo, Japan. The cartridges were operated in a dead-end mode at a pressure of 10-15 psi.

Viruses

The target viruses were laboratory-adapted strains as follows: pseudorabies, Aujeszky strain, ATCC VR-135 (American Type Culture Collection, Rockville, MD) passaged in porcine kidney cells (PK-13, American BioResearch, Inc., Seymour, TN); murine xenotropic Type-C retrovirus (American BioResearch, Inc., Seymour, TN) passaged in Mink S+L- cells (American BioResearch, Inc., Seymour, TN); SV_{40}, PA-57 strain, ATCC VR-239 (American Type Culture Collection, Rockville, MD) passaged in CV-1 cells (American BioResearch, Inc., Seymour, TN); and poliovirus 1, Chat strain (American BioResearch, Inc., Seymour, TN) passaged in Vero cells, ATCC CCL 81 (American Type Culture Collection, Rockville, MD). Laboratory strains were chosen for these studies because they can be grown to very high infectivity titres and because sensitive, standardized plaque and focus assay methods are available for their titration.

Virus spiking procedure

With the exception of pseudorabies virus, the spiking viruses were concentrated by ultracentrifugation (100,000 × *g* for 90 minutes) and the pellet resuspended in serum-free stabilization buffer to reduce the effect of extraneous proteins from the culture media on the filtration procedure. For filtration with 15 nm Planova filters, four 250 ml aliquots of the IgG were spiked with 1.9 ml of pseudorabies virus 0.8 ml of poliovirus, 0.3 ml of SV40 and 0.8 ml of murine xenotropic retrovirus, respectively, and 2.0 ml of each spiked IgG pre-filtrate was taken for assay. The remainder was then filtered through a 0.3 m^2 Planova 15 filter.

For filtration with 35 nm Planova filters, four 250 ml aliquots of the IgM were spiked with 1.0 ml of pseudorabies virus, 0.9 ml of poliovirus, 0.8 ml of SV40 and 1.0 ml of murine xenotropic retrovirus, respectively, and 2.0 ml of each spiked IgM pre-filtrate was taken for virus assay. The remainder was then filtered through a 0.3 m^2 Planova 35 filter.

Plaque assay procedure

Appropriate indicator cells (PK-13 cells for pseudorabies virus, CV-1 cells for SV40 and Vero cells for poliovirus) were plated in 60 nm and 100 mm Petri dishes and labelled with virus dilution and an appropriate code to identify assay control and test samples. When the cultures were 80-100% confluent the cell sheets were washed with balanced salt solution (BSS) and infected with 0.2 ml/60 mm dish or 2.0 ml/100 mm dish or appropriate concentration of test article, positive control (virus) or negative control (medium). Sensitivity controls for 100 mm Petri dishes of 1-5 and 30-50 pfu were prepared by adding appropriate dilutions of positive control virus in 2.0 ml/100 mm Petri dish. After a 30 to 90 minute incubation in a 36°C humidified CO_2 incubator the inocula were removed and the dishes overlayed with 3 ml/60 mm dish or 15 ml/100 mm dish of agarose overlay medium which was allowed to solidify. The dishes were then incubated at 36°C in a humidified CO_2 incubator until plaques formed in the positive control cultures. The cultures were then overlayed with agarose containing neutral red and the assay read when the neutral red penetrated to the cell layer and the plaques were fully formed.

To assess the effects of the test materials on the growth of indicator cells and expression of virus in the plaque assay they were spiked with a 1:10 dilution of the viruses previously adjusted to provide 30-50 pfu/dish in the selected dosing volume. The plaque assay was then carried out as described above. Cultures were observed daily for signs of any effect on indicator cell growth and plaques counted to see if there was any reduction in plaque formation.

Focus assay procedure

Mink S+L– cells were plated into 60 mm and 100 mm dishes and incubated for 18 to 24 hours at 36°C in a humidified CO_2 incubator. Following incubation the medium was removed and the cells re-fed with complete medium containing 1-3 μg/ml polybrene and the cultures again incubated at 36°C in a humidified CO_2 incubator for one to eight hours. The cultures were then inoculated with 0.2 ml/60 mm dish or 2.0 mm/100 mm dish of test material positive or negative control, incubated at 36°C in a humidified CO_2 incubator for 18 to 24 hours and then re-fed with complete medium without polybrene. The dishes were re-fed again on the fourth day following inoculation and as necessary after words. Cultures were examined every day following the day 4 re-fed until foci were fully formed in the positive controls, at which time the assay was read. Sensitivity controls for the focus assay in 100 mm Petri dishes of 1-5 and 30-50 focus forming units (ffu) were prepared by using appropriate dilutions of the positive control virus in 2.0 ml/100 mm Petri dishes.

To assess the effects of the test materials on growth of indicator cells and virus expression in the focus assay they were spiked with 1:10 dilution of the virus adjusted appropriately to give 30-50 ffu/dish in the selected dosing volume. The focus assay was then carried out as described above. Cultures were observed daily for signs of any effect on indicator cell growth and foci were counted to see if there was any reduction in focus formation.

Sample preparation

All samples were assayed in 60 mm dishes, unless otherwise indicated. Pseudorabies virus samples from 15 and 35 nm filtration were tested as follows: the assay positive control and spiking virus were tested diluted $1{:}1\times10^6$-10^8. The pre-filtrate was tested diluted $1{:}1\times10^4$-10^6. The filtrate was tested undiluted and diluted 10 fold. The filtrate was also tested undiluted in 100 mm dishes.

Poliovirus samples from 15 nm and 35 nm filtration were tested as follows: the assay positive control and spiking virus were tested diluted $1{:}1\times10^6$-10^8. The pre-filtrate was tested diluted $1{:}1\times10^4$-10^6. The filtrate from 15 nm filtration was tested undiluted and diluted $1{:}1\times10^1$ and also tested undiluted in 100 mm dishes. The filtrate from 35 nm filtration was tested undiluted and diluted $1{:}1\times10^1$-10^6.

Simian Virus40 samples form 15 nm and 35 nm filtration were tested as follow: the assay positive control and spiking virus were tested diluted $1{:}1\times10^6$-10^8. The pre-filtrate from 15 nm filtration was tested diluted $1{:}1\times10^5$-10^7. The pre-filtrate from 35 nm filtration was tested diluted $1{:}1\times10^4$- 10^6. The filtrate from 15 and 35 nm filtration were tested undiluted and diluted $1{:}1\times10^1$ and also tested undiluted in 100 mm dishes.

Murine xenotropic retrovirus samples from 15 nm and 35 nm filtration were tested as follows: the assay positive control was tested diluted $1{:}5\times10^2$ and $1{:}1\times10^3$. The spiking virus was tested diluted 1:1×10^5, 1:1×10^6 and 1:2×10^6. The pre-filtrate was tested diluted 1:1×10^2, 1:1×10^3, 1:2×10^3 and 1:1×10^4. The filtrate was tested undiluted and diluted 1:1×10^1 and also tested undiluted in 100 mm dishes.

Plaque focus assay results were considered acceptable if the titre of the assay positive control virus was within one log_{10} of the previously established titre of that virus pool and if no plaques or foci were observed in the negative control.

Results and Discussion

To test the permeability of the membrane to the IgG, a purified preparation was filtered through the 15 nm Planova membrane. With protein concentrations of 5-12 mg/ml, the recovery of the IgG in the filtrate was 92-97% (data not shown). For all spiking studies, nanofiltration cartridges were operated in a dead-end mode at constant pressure. All the spiking studies were carried out with high titred virus stocks resulting in a virus load for all viruses in the range of 8-9.5 log_{10} total focus forming or plaque units in a volume of approximately 250 ml. The results showed that in all experiments with the Planova 15 membranes, no virus was detected in the filtrate (Table 2). The total log_{10} reduction for murine xenotropic retrovirus was ≥ 6.43 and ≥ 7.0 each of the other three viruses. Moreover, it is important to note that assay methods using the larger test volumes in 100 mm dishes permitted more accurate determination of endpoint titres and enhanced assay sensitivity compared to similar assays carried out in either smaller dishes or microtitre wells. This method, coupled with the use of high titred virus stocks, contributed to achieving removal of 6-7 log_{10} plaque or focus forming units for each virus tested. Even when the IgG preparation was spiked with the very large pseudorabies virus (120-250 nm), no effect on protein recovery was observed, although the flux rate decreased due to the added serum protein in the virus preparation. Control studies indicated that the sample buffer did not show any toxicity in the virus test systems and the IgG had no effect on the growth or expression of the viruses.

Similar experiments were carried out with an IgM preparation. Because this immunoglobulin is much larger than IgG and has a calculated hydrodynamic diameter which is approximate to that of very small viruses, it failed to penetrate the Planova 15 membranes but penetrated the Planova 35. In addition to testing the Planova 35 with the three larger viruses, studies were designed to test the possibility that two tandem Planova 35 membranes might permit sufficient discrimination between the IgM and the smallest virus tested (poliovirus, 30 nm) to remove the virus effectively. Spiking studies showed that although simian virus 40 (45-55 nm), murine xenotropic retrovirus (80-120 nm) and pseudorabies virus (120-250) were effectively removed by a single Planova 35 membrane with no detectable virus in the filtrate, little or no poliovirus (≤ 0.75 Log_{10} PFU) was removed by the two tandem membranes (Table 3). Again, in buffer control studies, no toxicity was observed with any virus test system and the protein did not reduce the plaquing or focus forming efficiency of the test viruses.

Recent advances in the technology of manufacturing controlled pore size membranes in the nanometer range (typically 15, 35 and 75 nm) have made it feasible to consider nanofiltration as a virus removal step in the purification processes for biological macromolecules. The Planova membranes are composed of a unique polymerized form of cuprammonium regenerated cellulose with a membrane thickness consisting of 100-200 layers [1, 2, 3]. The membranes are compatible with biological macromolecules, they are hydrophilic and exhibit low protein binding properties.

Table 2: Viral removal using Planova 15 membrane filtration with IgG product.

Poliovirus 1				
Sample	**PFU/ml (a)**	**Volume (ml)**	**Total PFU**	**Total log_{10} PFU**
Spike control	4.87×10^9	NA(b)	NA	NA
Prefiltrate	7.95×10^6	249	1.98×10^9	9.30
Filtrate	≤ 0.17(c)	246	≤ 4.18	≤ 1.62
Total log_{10} reduction				≥ 7.68(d)
Murine xenotropic retrovirus				
Sample	**FFU/ml**	**Volume (ml)**	**Total FFU**	**Total log_{10} FFU**
Spike control	1.4×10^8	NA	NA	NA
Prefiltrate	4.61×10^5	249	1.15×10^8	8.046
Filtrate	≤ 0.17	249	$\leq 4.23 \times 10^1$	≤ 1.63
Total log_{10} reduction				≥ 6.43(d)
Simian virus 40				
Sample	**PFU/ml (a)**	**Volume (ml)**	**Total PFU**	**Total log_{10} PFU**
Spike control	1.33×10^9	NA	NA	NA
Prefiltrate	1.67×10^6	248	4.15×10^8	8.62
Filtrate	≤ 0.17	247	$\leq 4.2 \times 10^1$	≤ 1.62
Total log_{10} reduction				≥ 7.0(d)
Pseudorabies virus				
Sample	**PFU/ml**	**Volume (ml)**	**Total PFU**	**Total log_{10} PFU**
Spike control	2.1×10^9	NA	NA	NA
Prefiltrate	1.29×10^7	250	3.22×10^9	9.51
Filtrate	≤ 0.17	245	$\leq 4.17 \times 10^1$	≤ 1.62
Total log_{10} reduction				≥ 7.89(d)

(a) average of all countable dilutions;
(b) not applicable;
(c) limit of assay detectability using 2 ml dosed/100mm dish with no observable plaques;
(d) log reduction = (log_{10} PFU or FFU pre-filtrate) – (log_{10} PFU or FFU filtrate).

Table 3: Viral removal using Planova 35 membrane filtration with an IgM product.

Poliovirus 1				
Sample	**PFU/ml (a)**	**Volume (ml)**	**Total PFU**	**Total $\log_{10}$ PFU**
Spike control	1.98×10^9	NA(b)	NA	NA
Prefiltrate	7.89×10^6	249	1.96×10^9	9.29
Filtrate	1.40×10^6	250	3.5×10^8	8.54
Total $\log_{10}$ reduction				0.75(d)
Murine xenotropic retrovirus				
Sample	**PFU/ml**	**Volume (ml)**	**Total FFU**	**Total $\log_{10}$ FFU**
Spike control	1.29×10^8	NA	NA	NA
Prefiltrate	4.37×10^5	249	1.09×10^8	8.04
Filtrate	≤ 0.17(c)	245	$\leq 4.17 \times 10^1$	≤ 1.62
Total $\log_{10}$ reduction				≥ 6.42(d)
Simian virus 40				
Sample	**PFU/ml**	**Volume (ml)**	**Total PFU**	**Total $\log_{10}$ PFU**
Spike control	1.25×10^9	NA	NA	NA
Prefiltrate	4.38×10^6	249	1.09×10^9	9.04
Filtrate	≤ 0.17	252	$\leq 4.28 \times 10^1$	≤ 1.63
Total $\log_{10}$ reduction				≥ 7.41(d)
Pseudorabies virus				
Sample	**PFU/ml**	**Volume (ml)**	**Total PFU**	**Total $\log_{10}$ PFU**
Spike control	2.24×10^9	NA	NA	NA
Prefiltrate	1.06×10^7	249	2.64×10^9	9.42
Filtrate	≤ 0.17	245	$\leq 4.17 \times 10^1$	≤ 1.62
Total $\log_{10}$ reduction				≥ 7.80(d)

(a) average of all countable dilutions;
(b) not applicable;
(c) limit of assay detectability using 2 ml dosed/100mm dish with no observable plaques;
(d) log reduction = ($\log_{10}$ PFU or FFU pre-filtrate) – ($\log_{10}$ PFU or FFU filtrate).

They are manufactured by a proprietary process in a hollow fibre format assembled into a polycarbonate housing ranging in surface area from 0.001 to 1.0 m^2. Considerable information is available on the structure, properties and applications of the membrane [1-6]. Because copper is used in the manufacture of the Planova membranes, analysis of the washes and product was made to ensure that copper levels were below detectable limits (data not shown). Although the hollow fibre membrane cartridges can be operated in either a tangential flow or dead-end mode, the experiments described here used the membranes in a dead-end mode at a constant pressure. The advantages of using a dead-end mode for some proteins include enhanced protein recovery, reduced shear for shear-sensitive proteins and ease of operation.

The results of these studies showed that the Planova 15 and 35 membranes effectively removed simian virus 40, murine xenotropic retrovirus and pseudorabies virus but only the Planova 15 membranes removed poliovirus. Although the IgM was expected to be somewhat deformable compared to the rigid structure of the poliovirus capsid, it readily fouled the Planova 15 membranes. Two Planova 35 membranes in tandem failed to remove poliovirus. These results are consistent with the size exclusion properties and the pore size estimate of 15 nm for the Planova 15 membranes which exclude poliovirus (30 nm) and 35 nm for the membranes which exhibit complete permeability for poliovirus. The pore size of these membranes has also been estimated using calibrated gold particles [7] and can be validated by this method. The possibility exists, however, that an intermediate size, 20, 25 or 30 nm membrane might effectively remove parvoviruses or picornaviruses from products with a molecular weight or hydrodynamic diameter similar to that of IgM molecules. In the spiking studies, protein concentrations tested ranged from 10-12 mg/ml for the IgG and 1-2 mg/ml for the IgM. Protein recovery was excellent for both the 15 nm and the 35 nm membranes in experiments using 0.03-1.0 m^2 surface area filters. The lowest recovery of 84% was obtained in a test run with an 0.03 m^2 surface area filter that was slightly undersized for the total protein loaded. In all other experiments, the recovery was greater than 90%. Limited experience with scale-up has indicated that the larger cartridges gave performances consistent with the smaller test cartridges. Appropriate sizing of the membranes for a manufacturing process would depend on variables such as volume, time, temperature, flux, protein concentration and viscosity of the solution.

References

1 Planova™: A validatable process filter for effective removal of viruses. Asahi Chemical Industry Co., Ltd., Tokyo, Japan, 1994.

2 Sekiguchi S, Itoh K, Kobayashi M, Ikeds H, Tsurumi T, Ishikawa G, Manabe SI, Satani M, Yamashiki T: Possibility of Hepatitis B virus (HBV) removal from human plasma using regenerated cellulose hollow fiber (BMM). Membrane 1989;14:253-261.

3 Tsurumi T, Osawa N, Hitaka H, Hirasaki T, Yamamaguchi K, Manabe SI, Yamashizi T: Structure of cuprammonium regenerated hollow fiber (BMM hollow fiber) for virus removal. Polymer J 1990;22:751-758.

4 Ishikawa G, Hirasaki T, Manabe SI, Uematsu S, Yamamoto: Novel determination method of size of virus in solution using cuprammonium regenerated cellulose membrane (BMM) Membrane 1991;16:101-111.

5 Tomokiyo K, Shigenobu A, Ogata Y, Yoshikawa H, Tanikawa H, Funatsu H, Ohashi T: Studies on virus elimination and inactivation effect of highly purified F-VIII concentrate. Clinical Report 1991;25(1):271-275.

6 Manabe SI, Tsurumi T: Mechanism of human immunodefficiency virus (HIV) removal by regenerated cellulose hollow fiber (BMM). Membrane 1989;14(1):77-83.

7 Tsurumi T, Osawa N, Hirasaki T, Yamaguchi K, Manabe SI, Yamashiki T: Mechanism of removing monodisperse gold particles from a suspension using cuprammonium regenerated cellulose hollow fiber (BMM hollow fiber). Polymer J 1990;22:304-311.

Dr. J. O'Grady, Dept. of Process Biochemistry, MedImmune, Inc., 35 West Watkins Mill Road, Gaithersburg, MD 20878, USA.

Brown F, Lubiniecki AS: Viral Safety and Evaluation of Viral Clearance from Biopharmaceutical Products. Dev Biol Stand. Basel, Karger, 1996, vol 88, pp 327-328

SUMMARY TO BREAKOUT SESSION H

Chairmen: *J. Fogh, Novo Nordisk, Denmark*
A. Moreira, Universty of Maryland, USA

Participants: *R. Burian*
A. Lubiniecki, SmithKline Beecham Pharmaceuticals, USA
P. Minor, NIBSC, UK
C. Oliver, MedImmune, Inc., USA
D. Ramies, Bayer Corporation, USA
A. Tominaga, Seikagaku Corporation, Japan
H. Willkommen, Paul-Ehrlich-Institute, Germany

The Process Evaluation Session covered the following topics: Riks Management, European Viral Guidelines, various methods for viral inactivation and scale-up considerations.

The question arose in risk management of how much excess clearance is needed to provide statistical confidence for the viral studies? Are the present viral validation data sufficient or is a statistical probability-based study package necessary? Is the product after this more costly indeed safer and are the costs justified? Guidance from the regulatory agencies on this issue is valuable to the biopharmaceutical industry.

The new common European Guidelines will emphasize «real data», not generic, on actual processes. The design of the study will be very important and should include good modelling of the process, identification of the mechanism of viral removal/inactivation and include the crucial parameters for the various steps. The aim for the validation process is to show that viruses will be removed if they are known to be present, and to test the ability of the process to remove any virus that might be present. Justification for the clearance levels should be developed.

Filtration of viruses seems to be an effective tool. Performance is dependent on the composition of the streams being processed and the tests must be performed on actual processes.

In scaling-up, one should always be aware of the process, any process changes and of the process steps that give an improved margin of clearance.

During the discussions, questions were raised regarding the necessity for retesting already tested or certified raw materials for validated viruses in the product manufacturing process. The opinion was that validated raw materials need not be retested. Another question raised was the need for revalidation when a duplicate facility is brought into use. The response was that if a process can be documented to be «fully equivalent», then no additional studies for virus are necessary. This would be applicable for the same company or a contract manufacturer. It is of course dependent on the robustness of the process, and therefore case-by-case decisions might be needed.

The audience was inquisitive on the issue of partition versus inactivation and how to handle these results in viral validation studies. This is an area where additional discussions and information might be valuable.

Dr. J. Fogh, Novo Nordisk, Niels Steensensvej 1, DK-2820 Gentofte, Denmark.

Brown F, Lubiniecki AS: Viral Safety and Evaluation of Viral Clearance from Biopharmaceutical Products. Dev Biol Stand. Basel, Karger, 1996, vol 88, p 329

Summary and Conclusion to the Meeting

Brown F, Lubiniecki AS: Viral Safety and Evaluation of Viral Clearance from Biopharmaceutical Products. Dev Biol Stand. Basel, Karger, 1996, vol 88, pp 331-332

Future Development of Harmonized Guidelines

T. Hayakawa

National Institute of Health Sciences, Tokyo, Japan

The goal of the meeting was to gather data from real life experiences so that realistic risk assessments for viral safety of biological products could be made in the context of historical experience, product information and the intended clinical setting. The history of production of conventional biological products contains documented episodes of viral contamination. However, in no case was the contamination due to a lack of knowledge or carelessness in the manufacturing procedure. Products met all criteria applicable at the time. To minimize chances of outbreaks, systems for detecting contaminating viruses must keep pace with the advance of science. This can, however, never be an absolute. Although scientifc progress, in conjunction with infrequent contamination events, has led to new approaches, emphases and subsequent guidance, these were often not enough to prevent new contamination events.

The clinical risk for a biopharmaceutical product is practically nil. Preventive approaches, when rationally applied to anticipated risks, can be very effective. The risks should reflect the source of reagents used in manufacture. For cell banks, testing paradigms are not necessarily obvious. Often the most susceptible cell lines for in vitro testing bear no species relationship to the chosen cell substrate. As judged by experience with viral vaccines, the in vitro and in vivo tests for detection of adventitious viruses are complementary; therefore, both types of test are desirable. The best policy is to attempt to exclude all viral contaminants from human products since it would be unwise and unethical to accept any risk that was avoidable at a reasonable cost. Full knowledge of product safety comes only from years of patient experience.

To test for viral contamination, general screening tests should be complemented by specific tests, the nature of which is dependent on the cell substrate. Careful consideration must be given to the entire manufacturing process for its potential to introduce a contaminating virus. Using a general in vitro virus screening assay on every lot of product manufactured from a CHO cell line, sporadic contamination by a murine parvovirus was detected. As a result, a rapid specific PCR assay was developed and instituted on production fermenters for every lot of product. This PCR assay allowed the detection of a contamination event before harvest. Contamination of production cultures has certainly been observed by others and supports the concept that appropriate control measures are necessary.

Purification processes should be designed to remove or inactivate viruses. These can be accomplished through physical separation (columns, filters) as well as inactivation (heat, solvent). A complete evaluation of any given purification process consists of two complementary parts. The first is clearance of viral agents that are known to be present in the cell substrate. The second is the demonstration of the robust nature of the purification process through the clearance of model viral agents. Such a robust recovery process was shown to remove an unexpected contaminant, minute virus of mice (MVM). The scaled-down process used for the viral clearance studies should be an accurate representation of the full scale purification process. Any validation package for this type of study should demonstrate similarity in-process performance at all scales used and should include information on yields, chromatographic profiles and purity. Still requiring further discussion is the question of a rational approach for determining a clearance/inactivation factor. Recent advances in filtration technology have shown some viral removal filters to be effective in clearing medium of large virus particles. There have been many advances in the development of filtration technology for the removal of viruses. However, as with other technologies, there are many variables that need to be considered for the use of these filtration methods in clearance. These include the nature of the product as well as characteristics of the process stream. For example, filters used in purification schemes for large molecules such as IgM will have limited use in the clearance of small viruses. The level of viral clearance appropriate to ensure safety will depend on a number of factors including the biology of the risk associated with the product and the clinical setting.

The exact manufacturing conditions have a significant impact on measures taken to control and assess viral safety. Process changes may initiate a need to re-examine the adequacy of the virus control programme and perhaps a need to repeat part or all of the viral clearance studies. Changes in their target patient population that may be more susceptible to adverse effects due to potential viral contamination may require a reassessment of the adequacy of the clearance values obtained.

There are strong economic reasons for manufacturers to consider the use of multiproduct facilities. The increased desire to use multi-use plants increases the complexity of all control procedures that include the design of facilities, equipment systems, and processes to prevent viral contamination, protein cross-contamination, and freedom from other adventitious agents. Also important are cleaning validation and process validation, analytical procedures, automation, and training. The evidence for viral contamination in biopharmaceutical manufacturing is more likely to affect only product consistency because the robust purification processes prevent the risk of exposure of the patient to live virus.

There was agreement that materials used for early clinical trials should come from a process characterized to assess the presence of infectious agents in the cell bank. Full validation studies would be required for approval and might be necessary, depending on the size and target population for pivotal trials. It was also accepted that guidance is needed for early clinical trials of less well characterized products for life threatening and/or debilitating disease.

Dr. T. Hayakawa, National Institute of Health Sciences, Kamiyoga 1-18-1, Setagaya-ku, Tokyo 158, Japan.

Brown F, Lubiniecki AS: Viral Safety and Evaluation of Viral Clearance from Biopharmaceutical Products. Dev Biol Stand. Basel, Karger, 1996, vol 88, pp 333-335

Future Directions for Ensuring Viral Safety of Biopharmaceuticals

D.W. Trent

Division of Viral Products, CBER/FDA, Rockville MD, USA

Key words: Viral safety; biopharmaceuticals.

The pharmaceutical industry is experiencing new challenges in bringing to market biologicals initially developed in the laboratory using molecular cloning and expression of genes in mammalian cell cultures. The safety and efficacy of biologicals produced and purified from mammalian cells which can support replication of adventitious viruses is an issue of primary concern. Contamination of cell cultures with adventitious viruses may occur at any of several stages during production, necessitating certification of raw materials, master cell culture seed banks, processed materials at interim steps in the purification scheme and the final product. Biosynthesis in and processing of biologicals from mammalian cell cultures requires the utmost vigilance to ensure product quality. Each biological, (i.e. hormone, cytokine, enzyme or virus) has unique chemical and physical properties which must be considered during production and purification to maintain biological activity. Therefore, issues regarding production and purification of biologicals from mammalian cell cultures can only be resolved by development of processes which will eliminate adventitious viruses while maintaining biological activity and product purity. Manufacturing principles that minimize the opportunity for viruses to enter the production pathway, methods for removal of viruses from the product and standards of safety validation must be clearly defined. The production of biologicals which are safe, potent, and efficacious requires a united effort by the pharmaceutical industry and regulatory authorities to define standards of acceptable chemical purity, risk and benefit issues, and virus safety. Sharing information about virus contamination of mammalian cell cultures during production and practical solutions to these issues is essential to progress in the use of cell culture substrates for the production of biologicals. It is highly desirable that industry and regulatory authorities develop opportunities for candid and frank discussion of the issues and support development of harmonized ICH policies for ensuring viral safety.

Additional scientific information about viral safety and clearance from biopharmaceutical products is needed and continued effort must be made to define issues of viral clearance. International standards are required for testing of cell culture substrates used for recombinant DNA-based biologicals production, efficiency of viral inactivation and removal procedures, and testing validation. Each of these areas must be considered independently and the impact of possible viral contamination on individual products or classes of products evaluated. Development of practical, sensitive and reliable tests and protocols for the detection of known and unknown mammalian viruses in biologicals is critical. Much progress in the area of virus detection and testing has been made. The development of rapid and sensitive nucleic acid-based polymerase chain reaction techniques for virus detection has increased our ability to test for specific viruses; however, many of these tests do not identify viruses which have not been well characterized and may contaminate mammalian cell culture substrates. It is therefore essential that new and improved methods for in vivo and in vitro detection and cultivation of viruses be developed to identify unknown and/or poorly characterized viruses which may be present in biologicals. Improved methods for immunochemical identification of virus antigens in mammalian cells showing no cytopathology need to be developed and supported by monoclonal and polyclonal antibody reagents and techniques. Evaluation of methods for detection of endogenous retroviruses in cell culture substrates and tissues needs to be continued and extended. These tests should resolve and define concerns about retroviruses RNA-containing viruses, recombinant DNA-containing viruses, latent DNA-containing viruses, and emerging viruses currently unidentified which may be introduced into biological products from wild animal sources.

Methods have been developed to inactivate and remove viruses from biologicals based on experience and knowledge of anticipated virus contaminants. These «model» virus systems are important for process development; however, a continued effort must be made to ensure that the methods used will effectively remove the viruses that are contaminants. Systems and processes need to be designated specifically to remove viruses with varied biological and physical properties, to ensure product safety. The use of chemicals to inactivate viruses and filtration and chromatography processes to remove viruses during the processing of certain biologicals is well documented; however, further work is needed to improve methods where biologicals are sensitive to commonly used inactivation and clearance techniques. The implementation of new methods for the production of biologicals in multiple cell substrates, using different classes of gene expression vectors, and live virus production present multiple challenges. Many biopharmaceutical manufacturers produce multiple products requiring and using dramatically different biological systems, necessitating changes in production methods in multi-purpose manufacturing plants. The complexity of manufacturing and appropriate testing requires strict safety and product quality control considerations. To avoid problems which may result because of complex manufacturing requirements, virus safety standards for production of pharmaceuticals in cell cultures needs to be harmonized industry-wide. The biological safety factors which determine the reliability of clearance need to be resolved to harmonize safety protocols in industry with those used for regulatory considerations. Resolution of these issues would answer questions about «model» virus systems used to ensure clearance and safety throughout industry. If standard methods for the calculation of virus clearance and validation can be established, efforts to produce virus-free products would not be frustrated by regulations but sustained and defended by common agreement of industry and regulatory authorities.

The risk that adventitious viruses contaminate biologicals produced in mammalian cell cultures is very small; however, lessons from the past remind us that strict standards of virus safety must be maintened. The area of biotechnology has resulted in the development of new methods for the production of biologicals and opened new avenues for gene therapy and genetic alteration of chromosomal expression. Although much has been accomplished, we stand on the threshold of scientific endeavour that will require new thought and changes in the application of ideas to solve human needs. Research on viral safety, genetic stability and product efficacy must continue to build foundations for unrealized accomplishments in biology and medicine.

Dr. D.W. Trent, Division of Viral Products CBER/FDA, 1401 Rockville Pike, HFM 445,
Rockville MD 20852, USA.

List of Speakers

ANDRE F., SmithKline Beecham Pharmaceuticals, 96 rue de l'Institut, B-1330 Rixensart, Belgium.

BAKER D., Baxter Healthcare Corporation, Hyland Division 1820 Flower Avenue, Duarte, CA 91010, USA.

BANDER N., Cornell University, 525 E. 68th Street, Box 23, New York, NY 10021, USA.

BEATRICE M.G., FDA/CBER, 8800 Rockville Pike, Bldg. N29B, Rm. 5NN10, HFM-2, Bethesda, MD 20892, USA.

BEELER J.A., FDA/CBER/OVRR, 8800 Rockville Pike, Bldg. N29A, Rm. 2B17, HFM-463, Bethesda, MD 20892, USA.

BERTHOLD W., Hoffman-la-Roche, Inc., Bldg 123/3, 340 Kingsland Street, Nutley, NJ 07110-1199, USA.

BIERLEY S.T., Pathology Associates International, 4915-D Prospectus Drive, Durham, NC 27713, USA.

BURSTYN D.G., Alkermes, Inc., 64 Sidney Street, Cambridge, MA 02139, USA.

CHEN S., Bayer Corporation, 800 Dwight Way, P.O. Box 1986, Berkeley, CA 94710-1986, USA.

DEATLEY A.M., Viral Vaccine Research, Lederle-Praxis Biologicals, 401 N. Middletown Road, Pearl River, NY 10965, USA.

DELUSTRO F., Collagen Corporation, 2500 Faber Place, Palo Alto, CA 94303, USA.

DETWILER L.A., USDA/APHIS, Veterinary Services, 320 Corporate Blvd, Robbinsville, NJ 08691, USA.

DEVINE R., FCA/CBER, 1401 Rockville Pike, Rm 200N, Rockville, MD 20852, USA.

DINOWITZ M., Genentech, Inc., 460 Point San Bruno Blvd, South San Francisco, CA 94080, USA.

ECK S., Institute of Human Gene Therapy, University of Pennsylvania, 422 Curie Blvd., Rm 409 (BRB-1), Philadelphia, PA 19104-6069, USA.

EGAN E.M., Diacrin, Inc., Charlestown Navy Yard, 13th Street, Bldg 96, Charlestown, MA 02129, USA.

FINLAYSON J., FDA/CBER/OBRR, 1401 Rockville Pike, 200 N, HFM-330, Rockville, MD 20852-1448, USA.

FINTER N.B., Tideswell, 9 Burntwood Road, Sevenoaks, Kent TN13 1PS, U.K.

FOGH J., Novo Nordisk, Niels Steensensvej 1, DK-2820 Gentofte.

FOSTER L.G., Hyclone Laboratories, Inc., 1725 South Hyclone Rd, Logan, UT 84321, USA.

GARFINKLE B.D., Biological Quality Operations, Merck & Co., Inc., P.O. Box 4, WP 36M-6, West Point, PA 19486, USA.

GARNICK R.L., Genentech, Inc., 460 Point San Bruno Blvd., South San Francisco, CA 94080, USA.

GIL P.J., Grace Biomedical, 128 Spring Street, One Ledgemont Center, Lexington, MA 02173, USA.

GRIFFITHS E., World Health Organization, CH-1211 Geneva 27, Switzerland.

HAGEMAN T., Biogen, 14 Cambridge Center, Cambridge, MA 02142, USA.

HAYAKAWA T., Div. Bio. Chem & Biologicals, National Institute of Health Science, Kamiyoga 1-18-1, Setagaya-ku, Tokyo, Japan 158.

HEHIR K., Genzyme Corporation, 1 Mountain Rd, Rockville, MD 20850, USA.

HELLMAN K.B., Center for Devices and Radiological Health, FDA, Rockville, MD 20852, USA.

HEWLETT I.K., FDA/CBER, 1401 Rockville Pike, HFM-315, 29B, Rockville, MD 20852-1448, USA

HONSTEAD J.P., FDA/CVM, Div. of Animal Feeds, 7500 Standish Place, HFV-222, MPN II, Rockville, MD 20855, USA.

HORAUD F., Institut Pasteur, 25 rue du Docteur Roux, F-75724 Paris Cedex 15, France.

HUGHES B., Glaxo-Wellcome Biotechn., Langley Court, South Eden Park Rd, Beckenham, Kent BR3 3BS, U.K.

HUGHES J.V., Quality Biotech, Inc., 1667 Davis Street, Camden, NJ 08104, USA.

KEEGAN P., FDA/CBER/OTRR, 1401 Rockville Pike, Woodmont II, Suite 2086, HFM 573, Rockville, MD 20857-1448, USA.

KHAN A.S., FDA/CBER, Div. of Viral Products, Laboratory of Retrovirus Research, 8800 Rockville Pike, Bldg. 29B, Room 4NN10, HFM-454, Bethesda, MD 20892, USA.

KOHAMA T., National Institute of Health, 4-7-1 Gakuen, Musashimuray Ama, Tokyo, 208 Japan.

KOTEWICZ M., Microbiological Associates, 9900 Blackwell Rd, Rockville, MD 20850, USA.

KOVESDI I., GenVec.

KOZAK R.W., Bayer Corporation, 800 Dwight Way, P.O. Box 1986, Berkeley, CA 94710, USA.

LEE C.K., The Salk Institute, Government Services Division, P.O. Box 250, Swiftwater, PA 18370, USA.

LEVINGS R.L., Veterinary Service Laboratories, USDA/APHIS, 1800 Dayton Road, Box 844, Ames, Iowa 50010, USA

LEWIS A., FDA/DBER/OBRR, 1401 Rockville Pike, HFM-400, Rockville, MD 20852, USA.

LOEWER J., Paul-Ehrlich-Institute, Paul-Ehrlich-Strasse 51-59, D-63225 Langen.

LUBINIECKI A.S., SmithKline Beecham Pharmaceuticals, 709 Swedeland Rd, P.O.Box 539, UE-0428, King of Prussia, PA 19406-0939, USA.

MANABE S.I., Fukuoka Women's University, Fac.of Human Environmental Science, 1-1-1 Kasumigaoka, Higashi-ku, Fukuoka, Japan 813.

MARCUS-SEKURA C., Biotechnology Assessment Services, Inc., 7413 Ottenbrook Terrace, Rockville, MD 20855, USA.

MARTIN P., Fred Hutchinson Cancer Center, 1124 Columbia Street, M718, Seattle, WA 98104, USA.

MASTRANGELO M., Thomas Jefferson University, Curtis Bldg, Rm 1024, 1015 Walnut Street, Philadelphia, PA 19107, USA.

MATSUNAGA Y., Dept of Epidemiology, National Institute of Health, 1-23-1, Toyama, Shinjuku-ku, Tokyo, Japan 162.

METZ R.A., Quality Biotech Inc., 1667 Davis Street, Camden, NJ 08104, USA.

MINOR P.D., N.I.B.S.C., Blanche Lane, South Mimms, Potters Bar, Herts. EN6 3QG, U.K.

MOREIRA A., University of Maryland, 5401 Wilkins Ave., ECS Bldg, Rm 101, Baltimore, MD 21228, USA.

MORGAN E., Microbiological Associates, Inc., 9900 Blackwell Rd, Rockville, MD 20850, USA.

MORIGUCHI H., Research & Development Division, MHW, Pharmaceutical Affairs Bureau, Kasumigaseki 1-2-2, Chiyoda-ku, Tokyo, Japan.

MURPHY B., National Institute of Allergy & Infectious Diseases, 12441 Parklawn Drive, Bldg. 7, Rm. 106TW2, Rockville, MD 20857, USA.

NOBUHARA M., JCR Pharmaceuticals Co. Ltd., Labs for Bioengineering & Research, 2-2-10 Murotani, Nishi-ku, Kobe 651-22, Japan.

NOGUCHI P., FDA/CBER, 1401 Rockville Pike, HFM-515, Rockville, MD 20852, USA.

OLIVER C.N., MedImmune, Inc., 35 West Watkins Mill Rd., Gaithersburg, MD 20878, USA.

PADILLA R., Bayer Corporation, 800 Dwight Way, P.O. Box 1986, Berkeley, CA 94710-1986, USA.

PARKINSON D.R., National Cancer Institute, 6120 Executive Plaza Blvd., EPN, Rm. 715, Bethesda, MD 20892, USA.

PARKMAN P.D., Parkman Assoc., 4101 Glenridge Street, Kensington, MD 20895, USA.

POTASH L., Dyncorp-PRI, 1 Taft Court, Suite 200, Rockville, MD 20850, USA.

PRATT B.M., Genzyme Corporation, 1 Mountain Rd, Framingham, MA 01701-9322, USA.

RAMIES D.A., Bayer Corporation, 800 Dwight Way, P.O. Box 1986, Berkeley, CA 94710-1986, USA.

RIDGWAY A., Biotechnology Section, Bureau of Biologics, HPB-Bldg 7, Tunney's Pasture, Ottawa, Ont. K1A 0I2, Canada.

ROBERTSON J.S., N.I.B.S.C., Blanche Lane, South Mimms, Potters Bar, Herts EN6 3QG, U.K.

ROBINSON M.M., Animal Production Unit, FAO/IAEA Agriculture and Biotechnology Lab., A-2444 Seibersdorf, Austria.

ROHWER R.G., Medical Research Service, Veterans Affairs Medical Center, 10 North Green Street, 3A-129, Baltimore, MD 21201, USA.

SAVAGE M., Bayer Corporation, 800 Dwight Way, P.O. Box 1986, Berkeley, CA 94710-1986, USA.

SCHLESINGER H., Tektagen, Inc., 358 Technology Drive, Malvern, PA 19355, USA.

SEAMOM K., CBER/FDA, 1401 Rockville Pike, HFM-20, Rockville, MD 20852-1448, USA.

SHADLE P.J., SmithKline Beecham Pharmaceuticals, 709 Swedeland Rd., UE 3839, King of Prussia, PA 19406, USA.

STEIN K.E., FDA/CBER, 8800 Rockville Pike, HFM-555, Bethesda, MD 20892, USA.

TOMINAGI A., Seikagaku Corporation, 3-1253 Tateno Higashiyama, To-Shi, Tokyo 207, Japan.

TRAPNELL B., Genetic Therapy, Inc.

TRENT D.W., Division of Viral Products, FDA/CBER, 1401 Rockville Pike, HFM-445, Bethesda, MD 20852, USA.

TROUVIN J.-H., French Drug Agency, 143 Bouvelard Anatole France, F-93285 St-Denis, France.

VACANTE D., Microbiological Associates, Inc., 9900 Blackwell Rd., Rockville Pike, MD 20850, USA.

VICARI G., Istituto Superiore di Sanita, Viale Regina Elena 299, I-00161 Rome, Italy.

VINCENT C.K., FDA/CBER, 5600 Fishers Lane, HFD-160, Rm. 18B08, Rockville, MD 20857, USA.

WALTER J.K., Dr. Karl Thomae GmbH, Dept Biotech Production, Birkendorferstrasse 65, D-88397 Biberach.

WEINER G., University of Iowa, C-32 General Hospital, 200 Hawkins Drive, Iowa City, IA 52242, USA.

WIEBE M.E., Genentech, Inc., 460 Point San Bruno Blvd., South San Francisco, CA 94080, USA.

WILLKOMMEN H., Paul-Ehrlich-Institut, Paul-Ehrlich-Strasse 51-59, D-63225 Langen.

WILSON C.A., FDA/CBER/OTRR, 8800 Rockville Pike, Bldg. N29A, Rm. 2B23, HPM-530. Bethesda, MD 20892, USA.

WOLFF R., FDA, 1401 Rockville Pike, HFM-594, Rockville, MD 20852, USA.

WRIGHT G., PPL Therapeutics, Roslin, Edinburgh EH25 9PP, Scotland, U.K.

YU M.W., Division of Hematology, OBRR/CBER/FDA,HFM-345, 8800 Rockville Pike, Bldg N29, Rm 305, Bethesda, MD 20892, USA.

ZIOMEK C.A., Genzyme Transgenics Corp., One Mountain Road, Framingham, MA 01701-9322, USA.

ZOOM K., FDA/CBER, 1401 Rockville Pike, Bldg. N29, Rm. 130, HFM-1, Rockville, MD 20852-1448, USA.

A full list of participants is also available at the following address: BIOSTANDARDS, P.O. Box 456, CH-1211 Geneva 4, Switzerland.

Index of Authors

Symposia Series in Immunobiological Standardization

Vol. 1: Talloires 1965	Rabies
Vol. 2: Munich 1965	Neurovirulence of viral vaccines
Vol. 3: Stockholm 1965	Biotechnical developments in bacterial vaccine production
Vol. 4: Royaumont 1965	Immunological methods of biological standardization
Vol. 5: London 1966	Laboratory animals
Vol. 6: Utrecht 1966	Adjuvants of immunity
Vol. 7: Marburg/L. 1967	Assay of combined antigens
Vol. 8: Lyon 1967	Foot-and-Mouth Disease: Variants and immunity
Vol. 9: Paris 1967	Pseudotuberculosis
Vol. 10: London 1967	Biological assay methods of vaccine and sera
22nd Symp., London 1968	Standardization of immunofluorescence. Ed. Blackwell
Vol. 11: London 1968	Rubella vaccines
Vol. 12: Tunis 1968	Brucellosis. Standardization and control of vaccines and reagents
Vol. 13: Utrecht 1969	Pertussis vaccine
Vol. 14: London 1969	Standardization of interferon and interferon inducers
Vol. 15: Berne 1969	Enterobacterial vaccines
Vol. 16: Versailles 1970	Anti-lymphocytic serum
29th Symp., Brighton 1969	Hazards of handling simians. Ed. Lab. Animals Ltd. (4)
Vol. 17: Frankfurt 1970	BCG vaccine
Vol. 18: Copenhagen 1972	HL-A reagents
Vol. 19: Utrecht 1972	Smallpox vaccines
Vol. 20: London 1972	Influenza
Vol. 21: Lyon 1972	Rabies II
Vol. 22: Monaco 1973	The use of vaccines

Developments in Biological Standardization

Vol. 23: Madrid 1973	Sterilization and sterility testing of biological substances
Vol. 24: San Francisco 1973	Preservatives in biological products
Vol. 25: Lyon 1973	Requirements for poultry virus vaccines
Vol. 26: Budapest 1973	Selected veterinary vaccines
Vol. 27: Budapest 1973	Purity of human plasma proteins
Vol. 28: London 1974	Immunity to infections of the respiratory system in man and animals
46th Symp., London 1974	Breeding of simians and their uses in developmental biology. Ed. Lab. Animals Ltd. (6)
Vol. 29: Geneva 1974 (WHO/IABS)	Standardization and control of allergens administered to man
Vol. 30: Milan 1974	Viral hepatitis
Vol. 31: Rabat 1975	Brucellosis (II)
Vol. 32: Paris 1975 (OIE/IABS)	Colstridial sera and vaccines for veterinary use
Vol. 33: Douglas 1975	Vaccination by the non-parenteral route
Vol. 34: Budapest 1976	Pyrogenicity, innocuity and toxicity test systems for biological products
Vol. 35: Lyon 1976	Foot-and-mouth disease vaccines

Vol. 36: Lyon Washington D.C. 1976	Freeze-drying of biological products
Vol. 37: Geneva 1976	Standardization of cell substrates for the production of virus vaccines
Vol. 38: London 1977	Biological preparations in the treatment of cancer
Vol. 39: Geneva 1977	Influenza immunization (II)
Vol. 40: Marburg 1977	Standardization of rabies vaccines (III) (WHO/IABS)
Vol. 41: Guadeloupe 1978	Vaccinations in the developing countries
Vol. 42: Paris 1978	Second General Meeting of European Society of Animal Cell Technology
Vol. 43: Brussels 1978	Immunization: Benefit versus risk factors
Vol. 44: Villars 1979	Test methods for the quality control of plasma proteins
Vol. 45: San Antonio 1979	The Standardization of animals to improve biomedical research, production and control
Vol. 46: Oxford 1979	Third General Meeting of European Society of Animal Cell Technology
Vol. 47: Bilthoven 1980	Reassessment of inactivated poliomyelitis vaccine
Vol. 48: Geneva 1980	Standardization of albumin, plasma substitutes and plasmapheresis
Vol. 49: Leetown 1981	Fish biologics: Serodiagnostics and vaccines
Vol. 50: Heidelberg 1981	Joint ESACT-IABS Meeting on the use of heteroploid and other cell substrates for the production of biologicals
Vol. 51: Lyon 1981	Immunization of adult birds with inactivated oil adjuvant vaccines
Vol. 52: Lyon 1981	Herpes virus of man and animal: Standardization of immunological procedures
Vol. 53: Dublin 1982	Enteric infections in man and animals: Standardization of immunological procedures
Vol. 54: Athens 1982	Viral hepatitis: Standardization in immunoprophylaxis of infections by hepatitis viruses
Vol. 55: Copenhagen 1982	Fifth General Meeting of European Society of Animal Cell Technology
Vol. 56: Algeria 1983	IIIrd International Symposium on Brucellosis
Vol. 57: Paris 1983	Monoclonal antibodies: Standardization of their characterization and use
Vol. 58: Budapest 1983	BCG vaccines and tuberculins
Vol. 59: Geneva 1983	Standardization and control of biologicals produced by recombinant DNA technology
Vol. 60: Gardone Riviera 1983	Production and exploitation of existing and new animal cell substrates
Vol. 61: Geneva 1984	Proceedings of the Fourth International Symposium on pertussis
Vol. 62: Stockholm 1985	Diagnostics and vaccines for parasitic diseases
Vol. 63: San Francisco 1985	Use and standardization of chemically defined antigens
Vol. 64: London 1985	Reduction of animal usage in the development and control of biological products
Vol. 65: Amsterdam 1985	Use and standardization of combined vaccines
Vol. 66: Baden 1985	Advances in animal cell technology: cell engineering, evaluation and exploitation
Vol. 67: Melbourne 1986	Standardization in blood fractionation including coagulation factors
Vol. 68: Geneva 1986	Cells – Products – Safety
Vol. 69: London 1987	Cytokines: Laboratory and clinical evaluation
Vol. 70: Arlington, 1988	Continuous cell lines as substrates for biologicals
Vol. 71: Utrecht, 1989	Monoclonal antibodies for therapy, prevention and *in vivo* diagnosis of human disease
Vol. 72: Annecy, 1989	Progress in animal retroviruses
Vol. 73: Shizuoka, 1990	Pertussis: Evaluation and research on acellular pertussis vaccines
Vol. 74: Bethesda, MD, 1990	Biological product freeze-drying and formulation
Vol. 75: London, 1990	Virological aspects of the safety of biological products
Vol. 76: Bethesda, MD, 1991	Continuous cell lines – Current issues
Vol. 77: Annecy, 1991	Standardization of the immunopharmacology of natural and synthetic immunomodulators
Vol. 78: Bethesda, MD, 1991	Poliovirus attenuation: Molecular mechanisms and practical aspects
Vol. 79: Ploufragan, 1992	The first steps towards an international harmonization of veterinary biologicals: 1993 and free circulation of vaccines within the E.E.C.
Vol. 80: Heidelberg, 1992	Transmissible spongiform encephalopathies-Impact on animal and human health.
Vol. 81: Cannes, 1992	Virological safety aspects of plasma derivatives.
Vol. 82: Albany, NY, 1993	Recombinant vectors in vaccine development.
Vol. 83: Annecy, 1993	Genetic stability and recombinant product consistency.

INTERNATIONAL ASSOCIATION OF BIOLOGICAL STANDARDIZATION

WHAT IS THE I.A.B.S. ?

The International Association of Biological Standardization (IABS) was founded in Lyons in 1955 by a group of independent experts who identified an urgent need for an improvement in the quality and comparability of the data being exchanged between scientists working in the research, development, production, standardization and regulation of biological products. To this end, IABS aims to bring together state controllers, manufacturers and research workers from academic institutions and public health organisations.

The IABS is a branch of the International Union of Microbiological Societies and has about 500 members in 50 countries. Every four years, its members elect the Council by postal ballot and its representatives, in their turn, elect the officers.

The duty of the Council is to organise meetings on subjects of contemporary interest in biological products being used in human and veterinary medicine and to promote the editorial activity of the IABS by publishing: i) the proceedings of the international meetings in the symposium series *Developments in Biological Standardization* published by S. Karger, ii) *Biologicals,* a major international journal published quarterly by Academic Press (London), and iii) the *Newsletter* which reports on current IABS activities.

The Council also supports the objectives of the major international public health programmes and organisations, and particularly the World Health Organization.

In 1994, the Council decided to establish an IABS Task Force on Vaccines to provide a focus for our activities in the field of vaccination and immunisation. The role of the Task Force is to foster communication between vaccine manufacturers, scientific research workers and regulatory authorities concerning the quality control, standardization and efficacy of vaccines and vaccine combinations and the exchange of information on the development of new improved vaccines. A major objective of the Task Force will be the organisation of international meetings and the promotion of high quality publications on vaccines in our journal *Biologicals* and in the *Developments in Biological Standardization*.

The Task Force has a membership of 30 leading international experts on vaccination and immunisation working together to identify the major topics in the field of vaccines that need international debate and cooperation.

INTERNATIONAL ASSOCIATION OF BIOLOGICAL STANDARDIZATION

WHY NOT BECOME A MEMBER OF THE IABS?

As a member of the IABS, it is possible to contribute directly to the activity of our Association and to support the research, development, standardization and regulation of biological products. By your membership you would:

- receive our quarterly *Newsletter* which gives information on the activities of the IABS and highlights international news of current interest;
- subscribe to our journal *Biologicals*, published quarterly by Academic Press (London), at a reduced rate;
- take advantage of the discount on booksellers' prices for the volumes in the symposium series *Developments in Biological Standardization*, published by Karger;
- participate fully in our international meetings, with a reduction in the registration fees.

If you are not already a member of our Association and wish to join us, please complete this form and send it to our permanent office:

BIOSTANDARDS
P.O.B. 456
78, av. de la Roseraie
CH-1211 Geneva 4
Switzerland
Fax: (41) 22 702 93 55

or hand it to the Secretariat, if attending a Meeting.

- -

Surname...First name(s)..

Scientific titles..Professional position..

Complete mailing address...

....

Field of expertise...

Tel.. Fax..

<u>Annual fee</u>: Sfr. 170.-, which includes membership, subscription to *Biologicals* and the quarterly *Newsletter* of the IABS;
or Sfr. 90.-, without the subscription to *Biologicals*.

<u>Method of payment</u>:

❍ I have given instructions to my bank to make a credit transfer to the Swiss Bank Corporation, P.O.B. CH-1211 Geneva 11, Switzerland, a/c Biostandards n° 131,652.0.
❍ I enclose my cheque for Sfr................................ made payable to Biostandards
❍ I authorise you to charge Sfr.................. to my card: VISA / MASTERCARD / AM. EXPRESS

N°... Expiry date........................

Signature...

ACHEVÉ D'IMPRIMER
SUR LES PRESSES DE L'IMPRIMERIE
MÉDECINE ET HYGIÈNE
GENÈVE
NOVEMBRE 1996